Advances in Geographic Information Science

The ***Advances in Geographic Information Science*** series aims to: present current and emerging innovations in GIScience; describe new and robust GIScience methods for use in transdisciplinary problem solving and decision making contexts, illustrate GIScience case studies for use in teaching and training situations, analyze GIScience tools at the forefront of scientific research, and examine the future of GIScience in an expanding knowledge-based economy. The scope of the series is broad and will encompass work from all subject disciplines that develop and use an explicitly spatial perspective for analysis and problem-solving.

Anindita Nath · Arkoprovo Biswas ·
Bappaditya Koley · Tanupriya Choudhury ·
Jung-Sup Um
Editors

Groundwater Depletion and Sustainability

A Methodology Utilizing Artificial Intelligence and Earth Observation Systems

Springer

Editors
Anindita Nath
Department of Geography
Bankim Sardar College
Tangrakhali, South 24 Parganas, West Bengal, India

Bappaditya Koley
Department of Geography
Bankim Sardar College
Tangrakhali, South 24 Parganas, West Bengal, India

Jung-Sup Um
Department of Geography
Kyungpook National University
Daegu, Korea (Republic of)

Arkoprovo Biswas
Geological Studies Unit
Physics and Earth Science Division
Indian Statistical Institute
Kolkata, West Bengal, India

Tanupriya Choudhury
School of Computer Sciences
University of Petroleum and Energy Studies
Dehradun, Uttarakhand, India

ISSN 1867-2434 ISSN 1867-2442 (electronic)
Advances in Geographic Information Science
ISBN 978-3-032-09920-4 ISBN 978-3-032-09921-1 (eBook)
https://doi.org/10.1007/978-3-032-09921-1

This Springer imprint is published by the registered company Springer Nature Switzerland AG
The registered company address is: Gewerbestrasse 11, 6330 Cham, Switzerland

Foreword

Groundwater sustains life, livelihoods, and ecosystems, serving as a critical freshwater reserve for billions of people worldwide. However, this invisible resource is under severe strain in many regions. Unsustainable extraction, inefficient management, and the compounded effects of climate variability have resulted in alarming rates of groundwater depletion. The increasing complexity of these challenges requires equally sophisticated, integrative responses that bridge disciplinary boundaries and leverage emerging technological capabilities.

In this urgent context, *Groundwater Depletion and Sustainability: A Methodology Utilizing Artificial Intelligence and Earth Observation Systems* edited by *Dr. Anindita Nath, Dr. Arkoprovo Biswas, Dr. Bappaditya Koley, Dr. Tanupriya Choudhury, and Dr. Jung-Sup Um*, makes a timely and compelling contribution. As part of the *Advances in Geographic Information Science* series by Springer-Nature, this volume provides a rigorous, forward-looking examination of how artificial intelligence (AI), remote sensing, and Earth observation systems can be systematically applied to analyze, monitor, and ultimately guide the sustainable use of groundwater resources.

The editors have formed a multidisciplinary group of scholars and practitioners who explore integrating geospatial data, machine learning algorithms, and modeling frameworks to tackle groundwater-related challenges. These chapters illustrate how advanced computational methods can enhance groundwater mapping, improve predictive capabilities, support early-warning systems, and guide evidence-based policy at both local and regional levels. Notably, the book also addresses the socio-environmental aspects of groundwater management, emphasizing the significance of equitable access, participatory governance, and long-term sustainability.

What sets this volume apart is its technological sophistication and its commitment to real-world application. By grounding complex methodologies in case studies from various hydrogeological settings, the book ensures that its insights are not merely theoretical but actionable. It provides a crucial link between research innovation and applied decision-making regarding one of our time's most pressing environmental issues.

This book will be an invaluable resource for researchers, students, and professionals in hydrology, geography, environmental science, data science, and public policy. It also reinforces the vital role of interdisciplinary collaboration in tackling global water crises, where AI and Earth observation are both tools and catalysts for more adaptive and sustainable water futures.

I commend the editors for their vision and scholarly rigor in curating this important volume. It will undoubtedly stimulate new lines of inquiry, encourage cross-sector dialogue, and contribute significantly to the science and practice of groundwater sustainability.

Varanasi, India

Dr. Prabhat C. Chandra
Ex-Regional Director, Central Ground Water Board, MoJS, Government of India
Ex-Consultant, The World Bank
Ex-Advisor, WAPCOS, MoJS, Government of India
Ex-Expert Consultant, National Geophysical Research Institute (CSRI-NGRI), Hyderabad

Preface

Groundwater is the foundation of global water security, providing the main source of drinking water for nearly half of the world's population and supporting over 40% of irrigated agriculture. In dry and semi-dry areas, it is often the only dependable water source. However, this essential resource faces increasing threats from overuse, irregular recharge patterns, urban expansion, and climate change. Groundwater depletion is no longer just a local or regional issue; it has become a worldwide challenge with broad socioeconomic and environmental effects. Addressing this complex problem requires moving beyond traditional hydrological studies and adopting interdisciplinary, technology-based solutions. In this regard, combining Artificial Intelligence (AI), Machine Learning (ML), and Earth Observation (EO) systems offers a unique opportunity to monitor, model, and manage groundwater resources more accurately, at a larger scale, and with deeper insight.

This edited volume, *Groundwater Depletion and Sustainability: A Methodology Utilizing Artificial Intelligence and Earth Observation Systems*, is a contribution to the *Advances in Geographic Information Science* series by Springer-Nature. It brings together current research and practical approaches that show the synergy between AI and EO in tackling groundwater sustainability. The book features a diverse set of chapters written by researchers and practitioners from academia, research institutions, and applied science fields. Their work highlights how emerging computational methods can transform groundwater resource assessment, making it more predictive, adaptive, and responsive to real-world changes.

The volume is organized around several thematic areas:

Predictive Analytics and Machine Learning: Several chapters explore the use of ML algorithms, such as artificial neural networks, decision trees, support vector machines, and deep learning models, for groundwater level forecasting, recharge estimation, aquifer mapping, and water quality prediction. These models enable efficient analysis of nonlinear, multi-variable systems and often outperform traditional hydrological methods.

Remote Sensing and Earth Observation Integration: A primary focus of the volume is on utilizing EO data from satellites and aerial platforms, including

GRACE, Landsat, and Sentinel missions. Authors examine how such data can be used to monitor land-use changes, evapotranspiration patterns, and spatiotemporal groundwater dynamics, providing valuable insights at local, regional, and national levels.

Geospatial Technologies and GIS Applications: Many chapters illustrate how Geographic Information Systems (GIS) serve as integrative platforms that combine AI and EO data with field observations, facilitating spatial analysis, risk assessment, and interactive groundwater management tools.

Decision-Support Systems and Policy-Relevant Models: A key strength of this volume is its focus on real-world relevance. The models and case studies shown are not just academic; they align with policy objectives. They can serve as decision-support tools for planners, water resource managers, and government agencies.

Case Studies from Various Regions: The volume features case studies from across India and other areas experiencing severe groundwater stress. These examples demonstrate how AI and EO technologies can be adapted to different hydrogeological environments, from alluvial plains and hard rock aquifers to peri-urban and agricultural zones.

This volume is intended for a broad and interdisciplinary audience. For academic readers and researchers, it offers methodological advances and clearer concepts in integrating AI and EO with hydrogeology. It identifies future research directions and serves as a valuable teaching resource in earth science, environmental engineering, geoinformatics, and sustainability science. For policymakers, water managers, and planning authorities, the book provides scalable and practical frameworks for adopting technology-based solutions. It stresses the importance of data-driven governance, predictive planning, and the role of science-policy integration in securing the long-term sustainability of groundwater.

We sincerely thank the contributors, authors, reviewers, and institutional partners who have helped shape this volume with their research and insights. Their combined expertise covers environmental science, data analytics, remote sensing, hydrology, and engineering, highlighting the interdisciplinary nature of the book.

We also thank Springer Nature for supporting this project through the *Advances in Geographic Information Science* series. This platform allows the sharing of impactful research with the global academic and professional communities. As editors, we hope this volume functions as both a scholarly resource and a practical guide, encouraging further innovation in sustainable groundwater management through the robust integration of artificial intelligence, remote sensing, and geospatial science.

South 24 Parganas, India — Anindita Nath
Kolkata, India — Arkoprovo Biswas
South 24 Parganas, India — Bappaditya Koley
Dehradun, India — Tanupriya Choudhury
Daegu, Korea (Republic of) — Jung-Sup Um

Contents

Artificial Intelligence and Machine Learning Innovations in Groundwater Resource Management

Advanced Application of Artificial Intelligence (AI) and Earth Observation Systems on Groundwater Management

Rahul Kumar, Debasis Mitra, and Tanupriya Choudhary

Abstract The sustainability of water supplies may be guaranteed by groundwater management, particularly considering rising demand and climatic unpredictability. This chapter discusses the sophisticated use of Earth observation systems and artificial intelligence in groundwater management. In groundwater monitoring, prediction, and decision-making, using AI algorithms in conjunction with satellite and remote sensing data improves process accuracy and efficiency. It starts with a synopsis of the fundamental ideas behind AI and Earth Observation and then goes into great depth about how they might work to evaluate groundwater. The authors will provide case studies to illustrate successful implementations, highlighting the advantages and difficulties faced. The development of AI-based decision support tools and predictive models, which have the potential to completely transform groundwater management techniques, will also be covered. This article also discusses ethical issues and prospects, offering a comprehensive view of the situation and new developments. It will give scholars, decision-makers, and practitioners the information and resources they need to use AI and Earth Observation to improve sustainable groundwater management.

Keywords Earth observation systems · Artificial intelligence (AI) · Groundwater monitoring · Groundwater management

R. Kumar (✉) · D. Mitra
Department of Microbiology, Graphic Era (Deemed to be University), Dehradun, 248002, Uttarakhand, India
e-mail: rkpma.dpmi@gmail.com

T. Choudhary
School of Computer Sciences, University of Petroleum and Energy Studies, Dehradun, 248007, Uttarakhand, India

A. Nath et al. (eds.), *Groundwater Depletion and Sustainability*, Advances in Geographic Information Science, https://doi.org/10.1007/978-3-032-09921-1_1

1 Introduction

Groundwater management has been transformed by the combination of artificial intelligence-based technologies and earth observation, which has provided unprecedented insight into the resource (Zhan et al. 2024). By analyzing vast volumes of data collected from satellites, ground sensors, and other sources, these technologies enable improved monitoring and forecasting of groundwater consumption and management in a sustainable manner (Singh and Sharma 2023). Recent developments in this field have greatly improved our understanding of and capacity to safeguard aquifers, forecast water supply, and lessen the impact of climate change on groundwater supplies (Scanlon et al. 2023). Machine learning algorithms now analyze multispectral satellite data to identify minute variations in soil moisture and vegetation that represent subsurface water dynamics (Du et al. 2024). The use of AI-based models that integrate data from ground measurements with orbiting gravity-measuring satellites to provide a comprehensive picture of aquifer depletion and recharge rates is another breakthrough (Rixen et al. 2024). Assessing and monitoring saltwater intrusion into freshwater aquifers along coasts is also made easier with deep learning applied to satellite radar data. Predicting and preventing groundwater pollution based on land-use, industrial activity, and geophysical data trends is now feasible thanks to sophisticated AI-driven predictive early warning systems (Perumal et al. 2024). Neural networks are being used to optimize groundwater extraction schedules while balancing sustainable aquifer management with fluctuating demand (Zare and Koch 2021). These developments show how AI and Earth observation may work together powerfully to address complicated groundwater problems and create more resilient water resources management plans (Dube et al. 2023).

2 An Overview of the Difficulties of Managing Groundwater

In the twenty-first century, managing groundwater is fraught with difficulties due to rising demand, climate change, and the effects of human activities on aquifer health (Cooley et al. 2014). To use this resource responsibly, such issues call for innovative solutions. Overuse of aquifers causes ground subsidence and dropping water levels in many arid and semi-arid locations (Gaaloul and Eslamian 2022). For instance, the Central Valley of California claims that excessive water use for agriculture caused significant soil subsidence, which can jeopardize infrastructure and change surface-water flow patterns (Ali 2020). Groundwater pollution is the main problem associated with deteriorating water quality. For example, Bangladesh's millions face extremely significant health concerns due to natural arsenic in their aquifers (Islam and Mostafa 2021). Management efforts are further complicated by climate change. All these activities are made more difficult by the fluctuation, change, and growing frequency of severe occurrences (Thornton et al. 2014). Sea level rise increases saltwater intrusion and jeopardizes the freshwater supply to coastal towns in the Mediterranean coastal aquifers. Furthermore, most aquifers, like the Guarani Aquifer System in South America, are transboundary, which presents a geopolitical difficulty for cross-border management coordination. To address the complexity of groundwater management concerns, these instances remind the authorities of the urgent need for all-encompassing, flexible management methods that grow from scientific knowledge through technology innovation to regulatory frameworks.

3 Earth Observation and Artificial Intelligence's Role in Resolving These Issues

By providing robust, competent, and precise tools for data analysis, monitoring, and decision-making in groundwater resource assessment, Earth and AI observation technologies aid in addressing the difficulties associated with groundwater management (Mashala et al. 2023). The new technologies significantly improve timely and accurate resource evaluations, improving the data that managers and policymakers use to inform their choices (Maciejewski 2017). For example, machine learning computation is now a crucial component of satellite imagery interpretation and ground sensor data used to identify land subsidence (Azarakhsh et al. 2022) caused by excessive groundwater extraction, as was recently done in the North China Plain. Another use is incorporating historical groundwater levels, land use data, and climate data into AI-powered models that forecast future aquifer conditions to determine proactive management approaches (Zhang et al. 2021). It has been successfully implemented in a U.S. High Plains Aquifer system. Gravity-measuring satellites from Earth Observations produce large-scale data on groundwater depletion, such as the GRACE project (Asrar et al. 2001). AI systems may then evaluate this data to track changes in aquifer storage across wide areas. In northern India, this method has been very helpful in tracking the fast depletion of groundwater (Ibrahim et al. 2024). For instance, combining AI and remote sensing is increasingly used to identify and track ecosystems that depend on groundwater (Shaikh and Birajdar 2024). Recent studies in Australia's Great Artesian Basin have shown that these technologies can explain intricate relationships between surface and groundwater ecosystems (Kaushik et al. 2023). As shown in (Table 1) these examples demonstrate how AI and Earth Observation have transformed groundwater management by providing previously unheard-of knowledge and forecasting capabilities.

Table 1 The application of Earth Observation (EO) and AI to groundwater management issues

Challenge	Role of AI	Role of earth observation	Reference
Engagement of stakeholder	Dashboards and visuals powered by AI enhance stakeholder communication	Visualizations of EO data encourage community participation in groundwater management	Burek et al. 2020)
Systems for monitoring and early warning	AI improves groundwater status monitoring and alarms in real time	EO makes it possible to monitor water and land resources from orbit continuously	Kumar et al. 2021)
Planning and resource management	AI optimizes predictive analytics-based management techniques and resource allocation	By mapping land use and groundwater resources, EO aids in spatial planning	Ding et al. 2019)
Integration and analysis of data	AI combines several data sources for thorough analysis and forecasting	Large-scale data from EO may be analyzed in conjunction with measurements taken on the ground	Li et al. 2020)
Impact of climate change	Groundwater reactions to shifting climatic factors are simulated using AI models	EO keeps an eye on temperature and precipitation variations that impact recharging	Taherian et al. 2020)
Degradation of groundwater quality	AI uses pattern recognition and classification to find the origins of contamination	EO finds sources of contamination and changes in land use that impact groundwater quality	Sun et al. 2020)
Lack of water	AI optimizes water consumption by forecasting groundwater levels and recharge rates	EO provides spatial data on water availability and consumption trends	Santos et al. 2019)

4 Artificial Intelligence (AI) Fundamentals for Groundwater Management

A variety of techniques that use computer capacity to analyze complicated data to find patterns and even make predictions about groundwater systems are included in artificial intelligence in groundwater management techniques (Tao et al. 2022). The way that we think about aquifers is changing because of AI techniques like deep learning, machine learning, and artificial neural networks (Fernandes et al. 2024). Fundamentally, artificial intelligence in this field involves training algorithms on enormous datasets that include meteorological, hydrological, geological, and land-use data to simulate groundwater behaviour with ever-increasing accuracy (Malakar et al. 2021). Recent applications already demonstrate the effectiveness of such methods. Researchers in California, for instance (Safford and Bischel 2019) developed a deep learning model that accurately predicted groundwater levels using information from many sources, including satellite photos, rainfall records, and well readings (Khan et al. 2023). Another illustration is the Indian team's use of AI to improve groundwater recharge structure placement by examining rainfall patterns, soil characteristics, and terrain (Prasad et al. 2020). AI is also used in groundwater water quality management. For instance, a study in the Netherlands uses machine learning algorithms to forecast groundwater nitrate concentrations based on agricultural practices and land usage (Pandya et al. 2024). Lastly, AI models will be used in Australia to predict how climate change would affect groundwater supplies, assisting water managers in making decisions in the future. These examples show how artificial intelligence (AI) may help water resource managers make better decisions by deciphering the complexities of water resource management and providing sophisticated answers to inquiries about hydraulic operations.

4.1 The Meaning and Categories of Artificial Intelligence that Are Pertinent to Groundwater Management

Using computers to carry out tasks that normally require human intellect, such as pattern recognition, learning, and problem-solving, is known as artificial intelligence (AI) in groundwater management (Adams et al. 2022). In this way, artificial intelligence encompasses several crucial categories pertinent to groundwater research (Pourmorad et al. 2024). There have been several reports of using machine learning (ML) algorithms that learn from data without explicit programming for groundwater quality and level prediction modeling (Boo et al. 2024). The field of machine learning, deep learning, artificial neural networks, etc., is well-suited to handle the high-dimensional, complicated data that hydrogeological research frequently faces (Zhao et al. 2024). NLP makes it possible to glean valuable information from the enormous amounts of textual data in scientific papers and books. Computer vision techniques are used to analyze satellite images and other visual data to evaluate how changes in land use affect groundwater. These recent applications demonstrate how well-suited the AI kinds are for managing groundwater. For example, by combining hydrogeological and meteorological parameters, machine learning has been used to the North China Plain to forecast groundwater levels (Zhang et al. 2020). Using Deep Learning in California, multispectral satellite data was used to assess the health of ecosystems that depend on groundwater (Chiloane et al. 2022). NLP has advanced our understanding of the aquifer's properties by generating insights from thousands of well records in the High Plains Aquifer. Water resource planning in India has benefited from using computer vision algorithms for satellite imagery analysis, which have helped pinpoint locations with promising groundwater recharge potential (Singh and Sharma 2023). These illustrations demonstrate how various forms of AI may completely transform groundwater assessment, monitoring, and management methods.

5 Artificial Intelligence Applications in Water Resources and Hydrology

In hydrology and water resource management procedures, artificial intelligence has become essential for developing fresh, creative solutions to challenging water-related issues (Krishnan et al. 2022). These days, machine and deep learning methods are primarily used in almost every field of water research, from water quality monitoring to stream flow prediction (Wai et al. 2022). Researchers and water managers can evaluate large volumes of data, find hidden patterns, and draw better conclusions about hydrological systems because of the sophisticated techniques. For example, a new application in the Colorado River Basin employs artificial intelligence (AI) to improve forecasting of snowfall and streamflow, which is crucial for managing water resources in the dry southwest of the United States (Fleming et al. 2023). In a different study, Chinese researchers created a deep-learning model to forecast urban water consumption based on economic indicators, demographic shifts, and weather patterns (Fu et al. 2022). One use of AI is flood prediction and management; in the UK, the model analyzes real-time sensor data and meteorological forecasts using machine learning algorithms to provide early warnings about flooding (Byaruhanga et al. 2024). Water quality management in India is aided by the monitoring and classifying algal blooms in water bodies with AI-powered image recognition tools (Saxena et al. 2024). These illustrations show how AI advances our understanding of hydrological processes, helps with improved water resource planning, and makes it possible to manage water resources more sustainably and effectively in various geographic and climatic situations.

6 Groundwater and Earth Observation Systems: The Part They Play

As water supplies become more limited owing to pollution, climate change, and population expansion, Earth Observation (EO) technologies are essential for contemporary environmental monitoring in groundwater management (Dube et al. 2023). Better decision-making is made possible by these systems, which comprise satellites and aerial platforms and offer data on groundwater levels, quality, and trends (Shaikh and Birajdar 2024). Thanks to remote sensing technologies, EO systems may get information regarding environmental factors that affect groundwater recharge and availability, such as land surface temperature, plant health, soil moisture, and precipitation (Subramoniram et al. 2013). Effective groundwater management is made simpler by EO systems' capacity to cover large, distant areas inaccessible to conventional techniques. This is especially important in arid and semi-arid regions where much of the water used for household and agricultural purposes comes from groundwater (Zamani et al. 2022). By detecting contamination-related changes in surface water, vegetation, and land cover, EO systems may also evaluate groundwater quality. This facilitates prompt response by assisting researchers and policymakers in identifying pollution sources and trends (Rose et al. 2020). Additionally, useful tools for modeling and forecasting groundwater behavior under a variety of conditions, including climate change, are EO systems. Models that forecast groundwater levels in response to changing precipitation patterns or rising water use can be created by scientists using both historical data and present observations. To prevent overexploitation and guarantee sufficient supply in the future, these models may be used for sustainable extraction strategies. When combined with Geographic Information System (GIS), Earth Observation data improves groundwater monitoring efficiency and enables stakeholders to assess groundwater resources connected to other land and water users (Gupta and Gupta 2024). This integration makes better planning and decision-making about groundwater management methods based on facts and information possible.

6.1 When Integrating Groundwater Management, Include Artificial Intelligence and Earth Observation

Earth observation (EO) and artificial intelligence (AI) technologies are transforming groundwater management by offering previously unheard-of insights into sustainable practices and water resource monitoring (Causevic et al. 2024). They make it possible to gather detailed information on the Earth's surface, such as land use, plant cover, soil moisture, and hydrological patterns, by fusing these technologies with remote sensing technology (Govender et al. 2022). Real-time data is gathered via satellites, drones, and aerial sensors to identify groundwater networks over large geographic regions. Through the analysis of EO data on land use, industrial operations, and agricultural practices, artificial intelligence (AI) plays a critical role in managing water supplies and identifying groundwater pollution (Espinel et al. 2024). With machine learning algorithms, stakeholders can effectively manage this valuable resource by classifying different forms of land cover and determining how they relate to groundwater contamination (Mariani et al. 2023). AI can also help create real-time monitoring systems that notify authorities of pollution incidents so that ecosystems and public health may be promptly addressed. Decision-making and stakeholder communication are strengthened by the growing use of AI and EO technology (Raji et al. 2024). Policymakers, water managers, and the public may be presented with complicated data using AI-driven visualization tools, which make the data easier to access and comprehend. As AI and EO technologies advance, groundwater supplies may be tracked and managed more precisely for resilient and sustainable usage in the face of environmental concerns (Mashala et al. 2023). By optimizing resource use, offering insightful information about groundwater dynamics, and developing workable conservation and preservation plans, the combination of AI and EO technologies may be viewed as a revolutionary approach to groundwater management.

6.2 Utilizing Remote Sensing for Monitoring and Exploration of Groundwater

Through effective and economical evaluation techniques, remote sensing has transformed the science of groundwater research and monitoring (Avtar et al. 2020). Aerial photography, satellite images, and remote data collecting methods that examine surface characteristics, plant patterns, and other geological structures that can point to possible groundwater resources' existence serve as the technology's foundation (Mays 2013). Contemporary methods have enhanced the possibilities of remote sensing. For example, in 2023, scientists employed a combination of multispectral satellite images and image machine learning algorithms to pinpoint potential groundwater recharge sites in parched parts of North Africa. In the end, this greatly enhanced the management of water resources. Another example demonstrated in 2022 is how InSAR technology may be used to identify even the smallest ground surface deformations. Scientists could track groundwater depletion and aquifer changes in California's Central Valley considerably more precisely thanks to this. In 2024, a team in Australia mapped the salinity and moisture content of the soil using hyperspectral remote sensing, which is helpful for groundwater farming that applies sustainable practices. In 2023, India created very comprehensive 3D models of aquifers for further groundwater exploration and management plans by combining remote sensing data, geophysical surveys, and GIS (Fig. 1).

7 Using Remote Sensing to Identify Groundwater Potential Zones

This has developed into a vital instrument for managing water resources; it provides a comprehensive and effective method for identifying groundwater resources and determining their availability. To analyze the geology, geomorphology, land use, soil types, drainage patterns, and other factors that will provide a general indication to further explore the possibility of having free-flowing or trapped groundwater, this method combines remote sensing techniques with geographical information systems (GIS) (Krishnamurthy et al. 2000). Recent advances in the field have improved the precision and sophistication of identification. For instance, MCDA, combined with

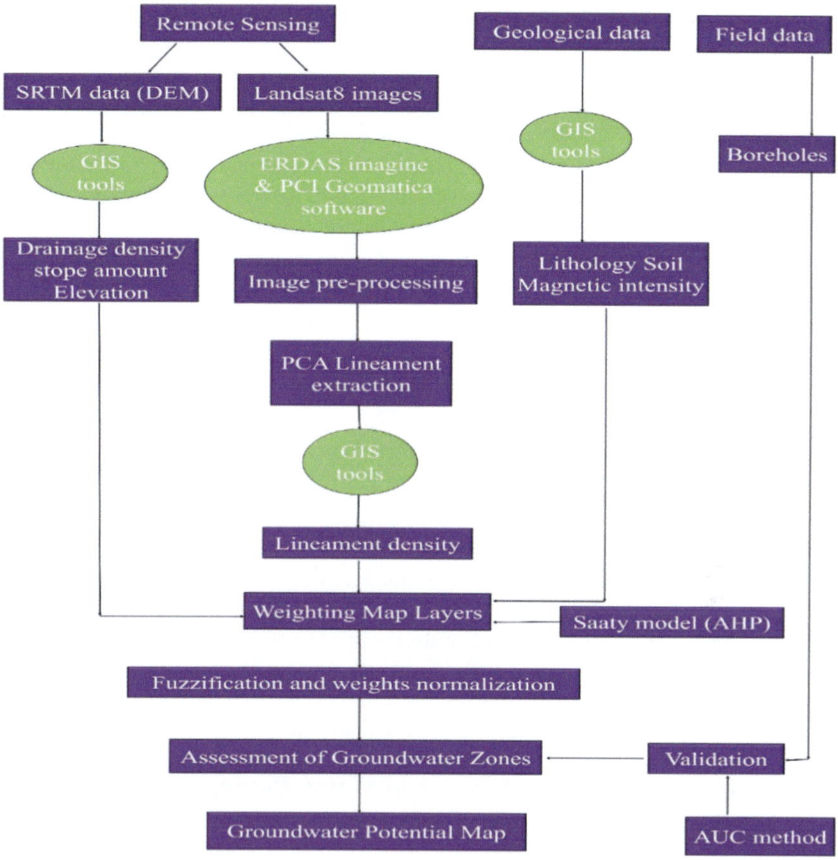

Fig. 1 Conceptual framework of applications of remote sensing using groundwater

remote sensing data, was used in a 2023 study in the semi-arid areas of western India to create a map of possible groundwater-finding zones (Kumar et al. 2024). The finding of certain unidentified aquifers was implied. Another innovative product from the Mediterranean area in 2022 identified the locations and seasonal fluctuations of coastal groundwater discharge by analyzing multispectral satellite images using machine learning techniques. By 2024, sub-Saharan African scientists had created comprehensive maps of groundwater potential by integrating high-resolution satellite data into artificial neural networks. These have significantly enhanced methods for allocating water supplies in regions affected by drought (Dikshit et al. 2022). Finally, a project in the Andes mountains was implemented in 2023 using a novel technique that combines InSAR technology with thermal infrared remote sensing to identify ecosystems that are fed by groundwater and provide crucial data on ecological preservation and high-altitude water resource management.

7.1 Groundwater: Quantity and Quality Monitoring Using Satellite Data

Because of this, it is a very effective technique for observing the quantity and quality of groundwater. It can accomplish frequent observations and extensive coverage that would be impossible with conventional ground-based techniques (Adnan et al. 2024). To explain or characterize the groundwater state, the method measures variables such as soil moisture, surface water extension, and ground subsidence by ground using a variety of data processing algorithms and satellite sensors. Recent advancements have also improved this. To help find a source of sustainable water resources in this heavily populated area, scientists, for example, tracked the depletion of the groundwater reservoir in the North China Plain in 2023 using data from GRACE-FO (Gravity Recovery and Climate Experiment Follow-On) satellites (Qu et al. 2024). Another use is from 2024, when Sentinel-1 SAR images were accepted by Australia to monitor and identify regions with ecosystems that rely on groundwater, helping to preserve them (Castellazzi et al. 2024). To enhance irrigation management and maintain soil health, California estimated water salinity levels over sizable agricultural regions in 2022 using Landsat thermal data and machine learning algorithms (Tola et al. 2024). Lastly, a 2023 study in India evaluated the effects of urbanization on groundwater recharge using spectral indices and multi-temporal satellite images, which would greatly aid in directing the creation of water conservation policies and urban planning.

7.2 Artificial Intelligence-Powered Water Resource Management Decision Support Systems

Artificial intelligence (AI) has transformed the groundwater resource management decision support system into sophisticated instruments for evaluating intricate data, identifying forecasted patterns, and creating the best possible water usage plans (Elshaikh et al. 2024). These artificial intelligence (AI) systems use a wide range of information, including sensor networks, satellite imaging, and historical records, to provide more precise insights and suggestions for long-term groundwater management (Himeur et al. 2022). Recent uses have shown their strength and adaptability. In early 2023, Californian scientists introduced a deep learning algorithm that used pumping records, land-use data, and climate data to estimate groundwater levels with exceptional precision, enabling more effective use of water resources during droughts. The next ground-breaking project, which was completed in India by 2024, used a reinforcement learning algorithm to optimize the groundwater pumping schedule throughout a sizeable agricultural region (Prabu et al. 2024). This resulted in significant water savings without lowering crop yield. In 2022, the risk assessment of saltwater intrusion in Australia's coastal aquifers was conducted using a hybrid artificial intelligence system that included fuzzy logic and neural networks. The application is for long-term groundwater maintenance management along the shore.

Policymakers could make an educated choice on water storage methods in 2023 when an AI-driven decision support system for a new initiative in the Middle East examined the viability and possible effects of managed aquifer recharge projects (Al Awadh and Mallick 2024). An AI model that offers advanced predictions for future groundwater demand and conflict regions, as well as potential proactive policy actions for fair water access, was built in 2024 by European scientists using socioeconomic integrated data with hydrogeological data.

7.3 Creation and Deployment of Systems for Decision Support

These days, decision support systems are among the most crucial instruments in industries that may provide insightful data to improve decision-making. DSS combines practical suggestions for managers and policymakers with sophisticated data analysis, predictive modeling, and intuitive user interfaces (Dasaka 2024). Further advancements have expanded DSS's capabilities and applicability in other domains. For example, a healthcare facility implemented an AI-based DSS in 2023 that examined various data sources, including patient medical records and publications about treatment outcomes, to help physicians create individualized treatment regimens for each patient. Patient care for complex procedures was improved as a result. In contrast, a DSS was put up in the financial industry in 2024, when investment managers used machine learning algorithms directly connected to market data to manage risks or improve portfolio allocations efficiently. Since environmental management as DSS was created to assist urban planners in creating climate-resilient cities by integrating climate models, infrastructure data, and socioeconomic elements, it saw significant use in 2022 (Singh and Niyogi 2024). The final but not least, is that the global corporation implemented the supply chain DSS in 2023 (Soori et al. 2024). Logistics and inventory management were optimized of using scenario modeling and predictive analytics. Significant cost reductions and increased operational effectiveness were the outcomes. Some instances, for example, demonstrate the growing significance and adaptability of decision support systems in resolving complicated problems in a variety of fields.

8 Case Studies Demonstrating the Efficacy of AI-Driven Decision Assistance

As seen by the recent dearth of case studies, AI-driven DSSs have proven to be very inventive for various businesses. These systems can provide the relevant company with essential insights and recommendations that improve the firm's decision-making process thanks to big data analytics, complex algorithms, and machine learning algorithms (Singh et al. 2022). For instance, a major retail corporation invested an AI-powered DSS in 2023 to improve supply chain and inventory management (Kondapaka 2021). An example for the health sector is this system, which reduced stock-outs by 15% by analyzing past sales data, market trends, and outside variables like weather patterns (Muranganwa 2022). According to a 2024 case study, an AI clinical decision-support system is successful throughout an extensive hospital network. Doctors to arrange therapy, recommend a diagnosis, and speed up patient outcomes follow-up by 20% thanks to the system's analysis of treatment and patient data, findings, and medical literature (Popescu et al. 2021). Implementing an AI-based DSS for fraud detection and risk assessment at one of the financial institutions in 2022 was yet another noteworthy application (Adelakun et al. 2024). It detected suspected potential dangers and trends and process large amounts of transactions in real-time. Its ability to avoid fraud rose by 30%, leading to significant cost savings (Ali et al. 2022). Finally, by combining soil sensor data, weather forecasts, and satellite images, the study provides an AI-based DSS for the agricultural sector in 2023 that has assisted farmers in improving crop management. As a result, crop yields increased by 25% and water use decreased by 20%, demonstrating unequivocally that AI-driven decision assistance supports sustainable agricultural methods (Uzunoz Altan and Nabatov 2024).

9 Conclusion

Groundwater management is being transformed by AI and EO technologies, which offer precise, effective, and predictive resource monitoring. These systems provide real-time analysis of complicated groundwater dynamics by combining AI algorithms with EO data, giving information about water availability, quality, and recharge rates. Future groundwater scenarios under various climatic and use situations may be predicted by AI drive models, which improve decision-making and encourage sustainable management techniques. When paired with machine learning and deep learning methodologies, the comprehensive data on hydrological and environmental aspects provided by EO systems might improve our comprehensive aquifer behavior. To alleviate ground water scarcity and guarantee long-term sustainability, this combine shows potential. Implementing scalable data-driven solutions in global groundwater management will require future development and multidisciplinary cooperation.

Disclose Conflict of Interest None.

References

Adams, K. H., Reager, J. T., Rosen, P., Wiese, D. N., Farr, T. G., Rao, S.,... & Rodell, M. (2022). Remote sensing of groundwater: current capabilities and future directions. *Water Resources Research*, *58*(10), e2022WR032219.

Adelakun BO, Onwubuariri ER, Adeniran GA, Ntiakoh A (2024) Enhancing fraud detection in accounting through AI: techniques and case studies. Financ Account Res J 6(6):978–999

Adnan M, Xiao B, Ali MU, Bibi S, Yu H, Xiao P, An X (2024) Human inventions and its environmental challenges, especially artificial intelligence: New challenges require new thinking. Environ ChallS 100976

Al Awadh M, Mallick J (2024) A decision-making framework for landfill site selection in Saudi Arabia using explainable artificial intelligence and multi-criteria analysis. Environ Technol Innov 33:103464

Ali A (2020) Water politics in a water-scarce landscape: examining the groundwater debate in California's Central Valley

Ali A, Abd Razak S, Othman SH, Eisa TAE, Al-Dhaqm A, Nasser M, Saif A (2022) Financial fraud detection based on machine learning: a systematic literature review. Appl Sci 12(19):9637

Asrar G, Kaye JA, Morel P (2001) NASA research strategy for earth system science: climate component. Bull Am Meteor Soc 82(7):1309–1330

Avtar R, Komolafe AA, Kouser A, Singh D, Yunus AP, Dou J, Kurniawan TA (2020) Assessing sustainable development prospects through remote sensing: a review. Remote Sens Appl: Soc Environ 20:100402

Azarakhsh Z, Azadbakht M, Matkan A (2022) Estimation, modeling, and prediction of land subsidence using Sentinel-1 time series in Tehran-Shahriar plain: a machine learning-based investigation. Remote Sens Appl: Soc Environ 25:100691

Boo KBW, El-Shafie A, Othman F, Khan MMH, Birima AH, Ahmed AN (2024) Groundwater level forecasting with machine learning models: a review. Water Res 121249

Burek P, Satoh Y, Kahil T, Tang T, Greve P, Smilovic M, Wada Y (2020) Development of the community water model (CWatM v1. 04)–a high-resolution hydrological model for global and regional assessment of integrated water resources management. Geosci Model Dev 13(7):3267–3298

Byaruhanga N, Kibirige D, Gokool S, Mkhonta G (2024) Evolution of flood prediction and forecasting models for flood early warning systems: a scoping review. Water 16(13):1763

Castellazzi P, Gao S, Pritchard J, Ponce-Reyes R, Stratford D, Crosbie R (2024) Detecting springs and groundwater-dependent vegetation in data-scarce regions of Australia combining citizen science, GRACE, and optical/radar imagery. Remote Sens Environ 313:114345

Causevic A, Causevic S, Fielding M, Barrott J (2024) Artificial intelligence for sustainability: opportunities and risks of utilizing Earth observation technologies to protect forests. Discov Conserv 1(1):2

Chiloane C, Dube T, Shoko C (2022) Impacts of groundwater and climate variability on terrestrial groundwater dependent ecosystems: a review of geospatial assessment approaches and challenges and possible future research directions. Geocarto Int 37(23):6755–6779

Cooley H, Ajami N, Ha ML, Srinivasan V, Morrison J, Donnelly K, Christian-Smith J (2014) Global water governance in the twenty-first century. World's Water: Bienn Rep Freshw Resour 1–18

Dasaka S (2024) Optimizing decision-making: balancing intuition with evidence in digital experience design

Dikshit A, Pradhan B, Santosh M (2022) Artificial neural networks in drought prediction in the 21st century–a scientometric analysis. Appl Soft Comput 114:108080

Ding J, Fang F, Lin W, Qiang X, Xu C, Mao L, Li Y (2019) N2O emissions and source partitioning using stable isotopes under furrow and drip irrigation in vegetable field of North China. Sci Total Environ 665:709–717

Du R, Chen J, Xiang Y, Xiang R, Yang X, Wang T, Chen Y (2024) Timely monitoring of soil water-salt dynamics within cropland by hybrid spectral unmixing and machine learning models. Int Soil Water Conserv Res 12(3):726–740

Dube T, Seaton D, Shoko C, Mbow C (2023) Advancements in earth observation for water resources monitoring and management in Africa: a comprehensive review. J Hydrol 623:129738

Elshaikh A, Mabrouki J, Mohamed AAO (2024) The future of water management: leveraging AI for effective decision support. In: Advancements in climate and smart environment technology. IGI Global, pp 213–225

Espinel R, Herrera-Franco G, Rivadeneira García JL, Escandón-Panchana P (2024) Artificial Intelligence in agricultural mapping: a review. Agriculture 14(7):1071

Fernandes VJ, de Louw PG, Bartholomeus RP, Ritsema CJ (2024) Machine learning for faster estimates of groundwater response to artificial aquifer recharge. J Hydrol 637:131418

Fleming SW, Zukiewicz L, Strobel ML, Hofman H, Goodbody AG (2023) SNOTEL, the soil climate analysis network, and water supply forecasting at the natural resources conservation service: past, present, and future. JAWRA J Am Water Resour Assoc 59(4):585–599

Fu G, Jin Y, Sun S, Yuan Z, Butler D (2022) The role of deep learning in urban water management: a critical review. Water Res 223:118973

Gaaloul, N., & Eslamian, S. (2022). Groundwater quality in arid environments. *Clean Water and Sanitation*, 260–272.

Govender T, Dube T, Shoko C (2022) Remote sensing of land use-land cover change and climate variability on hydrological processes in Sub-Saharan Africa: key scientific strides and challenges. Geocarto Int 37(25):10925–10949

Gupta S, Gupta SK (2024) Water resource mapping, monitoring, and modeling using geospatial approaches. Dev Environ Sci 16:575–602

Himeur Y, Rimal B, Tiwary A, Amira A (2022) Using artificial intelligence and data fusion for environmental monitoring: a review and future perspectives. Inf Fusion 86:44–75

Ibrahim A, Wayayok A, Shafri HZM, Toridi NM (2024) Remote sensing technologies for modelling groundwater storage dynamics: comprehensive review. J Hydrol X 100175

Islam MS, Mostafa MG (2021) Groundwater status and challenges in Bangladesh. Sustain Agric Rev 52:79–146

Kaushik PR, Ndehedehe CE, Kalu I, Burrows RM, Noll MR, Kennard MJ (2023) Identifying potential hotspots of groundwater-climate interaction in the Great Artesian Basin Australia. Ecol Inform 78:102354

Khan J, Lee E, Balobaid AS, Kim K (2023) A comprehensive review of conventional, machine leaning, and deep learning models for groundwater level (GWL) forecasting. Appl Sci 13(4):2743

Kondapaka KK (2021) AI-driven demand sensing and response strategies in retail supply chains: advanced models, techniques, and real-world applications. J Artif Intell Res Appl 1(1):459–487

Krishnamurthy J, Mani A, Jayaraman V, Manivel M (2000) Groundwater resources development in hard rock terrain-an approach using remote sensing and GIS techniques. Int J Appl Earth Obs Geoinf 2(3–4):204–215

Krishnan SR, Nallakaruppan MK, Chengoden R, Koppu S, Iyapparaja M, Sadhasivam J, Sethuraman S (2022) Smart water resource management using Artificial Intelligence—a review. Sustainability 14(20):13384

Kumar A, Ramsankaran RAAJ, Brocca L, Muñoz-Arriola F (2021) A simple machine learning approach to model real-time streamflow using satellite inputs: demonstration in a data scarce catchment. J Hydrol 595:126046

Kumar P, Singh P, Asthana H, Yadav B, Mukherjee S (2024) Groundwater potential zone mapping of middle Andaman using multi-criteria decision-making and support vector machine. Groundw Sustain Dev 26:101191

Li W, El-Askary H, Thomas R, Tiwari SP, Manikandan KP, Piechota T, Struppa D (2020) An assessment of the hydrological trends using synergistic approaches of remote sensing and model evaluations over global arid and semi-arid regions. Remote Sens 12(23):3973

Maciejewski M (2017) To do more, better, faster and more cheaply: Using big data in public administration. Int Rev Adm Sci 83(1_suppl):120–135
Malakar P, Sarkar S, Mukherjee A, Bhanja S, Sun AY (2021) Use of machine learning and deep learning methods in groundwater. In: Global groundwater. Elsevier, pp 545–557
Mariani C, Navrotska Y, Mancini M (2023) Unsupervised machine learning for project stakeholder classification: Benefits and limitations. Proj LeadShip Soc 4:100093
Mashala MJ, Dube T, Mudereri BT, Ayisi KK, Ramudzuli MR (2023) A systematic review on advancements in remote sensing for assessing and monitoring land use and land cover changes impacts on surface water resources in semi-arid tropical environments. Remote Sens 15(16):3926
Mays LW (2013) Groundwater resources sustainability: past, present, and future. Water Resour Manage 27:4409–4424
Muranganwa M (2022) Supply-side determinants of medicine stock-outs in South African public hospitals. Doctoral dissertation, University of Johannesburg
Pandya H, Jaiswal K, Shah M (2024) a comprehensive review of machine learning algorithms and its application in groundwater quality prediction. Arch Comput Methods Eng 1–22
Perumal M, Sekar S, Carvalho PC (2024) Global investigations of seawater intrusion (SWI) in coastal groundwaters in the last two decades (2000–2020): a bibliometric analysis. Sustainability 16(3):1266
Popescu C, Golden G, Benrimoh D, Tanguay-Sela M, Slowey D, Lundrigan E, Turecki G (2021) Evaluating the clinical feasibility of an artificial intelligence–powered, web-based clinical decision support system for the treatment of depression in adults: Longitudinal feasibility study. JMIR Form Res 5(10):e31862
Pourmorad S, Kabolizade M, Dimuccio LA (2024) Artificial intelligence advancements for accurate groundwater level modelling: an updated synthesis and review. Appl Sci 14(16):7358
Prabu RT, Sarkar M, Chaudhary D, Al Obaid S, Al-ateeq TK, Kalam MA (2024) IoT-enabled groundwater monitoring with k-NN-SVM algorithm for sustainable water management. Acta Geophys 72(4):2715–2728
Prasad P, Loveson VJ, Kotha M, Yadav R (2020) Application of machine learning techniques in groundwater potential mapping along the west coast of India. Giscience Remote Sens 57(6):735–752
Qu W, Zhang P, Chen P, Li J, Gao Y (2024) Spatiotemporal variations and sustainability characteristics of groundwater storage in North China from 2002 to 2022 revealed by GRACE/GRACE follow-on and multiple hydrologic data. Remote Sensing 16(7):1176
Raji E, Ijomah TI, Eyieyien OG (2024) Integrating technology, market strategies, and strategic management in agricultural economics for enhanced productivity. Int J Manag Entrep Res 6(7):2112–2124
Rixen M, Dowell M, Immler F, Dusart J, Kristopaitis E (2024) Earth observation strategic research and innovation Agenda
Rose DC, Mukherjee N, Simmons BI, Tew ER, Robertson RJ, Vadrot AB, Sutherland WJ (2020) Policy windows for the environment: tips for improving the uptake of scientific knowledge. Environ Sci Policy 113:47–54
Safford HR, Bischel HN (2019) Flow cytometry applications in water treatment, distribution, and reuse: a review. Water Res 151:110–133
Dos Santos JFC, Gleriani JM, Velloso SGS, de Souza GSA, do Amaral CH, Torres FTP, Dos Reis M (2019) Wildfires as a major challenge for natural regeneration in Atlantic Forest. Sci Total Environ 650:809–821
Saxena R, Srivastava V, Bharti D, Singh R, Kumar A, Sharma A (2024) Artificial intelligence for water resource planning and management. In: Innovations in machine learning and IoT for water management. IGI Global, pp 51–70
Scanlon BR, Fakhreddine S, Rateb A, de Graaf I, Famiglietti J, Gleeson T, Zheng C (2023) Global water resources and the role of groundwater in a resilient water future. Nat Rev Earth & Environ 4(2):87–101

Shaikh M, Birajdar F (2024) Advancements in remote sensing and GIS for sustainable groundwater monitoring: applications, challenges, and future directions. Int J Res Eng, Sci Manag 7(3):16–24

Singh V, Chen SS, Singhania M, Nanavati B, Gupta A (2022) How are reinforcement learning and deep learning algorithms used for big data based decision making in financial industries–a review and research agenda. Int J Inf Manag Data Insights 2(2):100094

Singh M, Niyogi D (2024) Leveraging ML approaches for scaling climate data in an atmospheric urban digital twin framework. In: Advances in machine learning and image analysis for GeoAI. Elsevier, pp 315–346

Singh D, Sharma V (2023) Transformative potential of ai and remote sensing in sustainable groundwater management. In: Proceedings of international conference on recent advancements in artificial intelligence. Singapore: Springer Nature Singapore, pp 139–156

Soori M, Jough FKG, Dastres R, Arezoo B (2024) AI-based decision support systems in industry 4.0: a review. J Econ Technol

Subramoniram SR, Joseph M, Bera AK, Sharma JR (2013) Role of earth observation (EO) technologies in dryland management towards adaptation to climate change. Ann Arid Zone 52(1)

Sun J, Jin L, He T, Wei Z, Liu X, Zhu L, Li X (2020) Antibiotic resistance genes (ARGs) in agricultural soils from the Yangtze River Delta China. Sci Total Environ 740:140001

Taherian Z, Khataee A, Orooji Y (2020) Facile synthesis of yttria-promoted nickel catalysts supported on MgO-MCM-41 for syngas production from greenhouse gases. Renew Sustain Energy Rev 134:110130

Tao H, Hameed MM, Marhoon HA, Zounemat-Kermani M, Heddam S, Kim S, Yaseen ZM (2022) Groundwater level prediction using machine learning models: a comprehensive review. Neurocomputing 489:271–308

Thornton PK, Ericksen PJ, Herrero M, Challinor AJ (2014) Climate variability and vulnerability to climate change: a review. Glob Change Biol 20(11):3313–3328

Tola D, Satgé F, Pillco Zolá R, Sainz H, Condori B, Miranda R, Espinoza-Villar R (2024) Soil salinity mapping of Plowed agriculture lands combining radar sentinel-1 and optical sentinel-2 with topographic data in machine learning models. Remote Sens 16(18):3456

Uzunoz Altan M, Nabatov E (2024) Using machine learning to enhance agricultural productivity in Turkey: insights on the importance of soil moisture, temperature and precipitation patterns. Int J Environ Sci Technol 1–18

Wai KP, Chia MY, Koo CH, Huang YF, Chong WC (2022) Applications of deep learning in water quality management: a state-of-the-art review. J Hydrol 613:128332

Zamani MG, Moridi A, Yazdi J (2022) Groundwater management in arid and semi-arid regions. Arab J Geosci 15(4):362

Zare M, Koch M (2021) Hybrid signal processing/machine learning and PSO optimization model for conjunctive management of surface–groundwater resources. Neural Comput Appl 33(13):8067–8088

Zhan C, Dai Z, Yin S, Carroll KC, Soltanian MR (2024) Conceptualizing future groundwater models through a ternary framework of multisource data, human expertise, and machine intelligence. Water Res 121679

Zhang G, Zheng W, Yin W, Lei W (2020) Improving the resolution and accuracy of groundwater level anomalies using the machine learning-based fusion model in the north China plain. Sensors 21(1):46

Zhang C, Duan Q, Yeh PJF, Pan Y, Gong H, Moradkhani H, Guo X (2021) Sub-regional groundwater storage recovery in North China Plain after the South-to-North water diversion project. J Hydrol 597:126156

Zhao T, Wang S, Ouyang C, Chen M, Liu C, Zhang J, Wang L (2024) Artificial intelligence for geoscience: progress, challenges and perspectives. The Innovation

Leveraging AI and Machine Learning for Groundwater Pollution Control and Management

Jyoti Bhattacharjee and Subhasis Roy

Abstract Groundwater pollution and management pose significant challenges in maintaining sustainable water resources, especially as global demand for clean water rises. Integrating Artificial Intelligence (AI) and Machine Learning (ML) into groundwater management offers innovative solutions to these challenges. This chapter explores the novel application of AI and ML techniques to control groundwater pollution and enhance management practices. AI and ML algorithms can analyze vast datasets from various sources, including remote sensing, geographic information systems (GIS), and historical records, to accurately predict contamination events and identify pollution sources. By leveraging predictive analytics, these technologies can forecast groundwater quality trends, enabling proactive measures to mitigate potential pollutants. Machine learning models, such as neural networks and support vector machines, are employed to develop robust groundwater contamination prediction systems. These models can learn complex patterns and relationships within the data, facilitating early detection of pollution incidents and optimizing remediation strategies. Additionally, AI-driven optimization techniques are used to design efficient monitoring networks and manage groundwater extraction sustainably. This ensures that water withdrawal does not exceed replenishment rates, preserving aquifer health. Advancements in future work will be driven by the further implementation of deep learning, explainable artificial intelligence (XAI), and other cutting-edge techniques. As discussed, these advancements portray AI and ML's application to sparsely study variables, model new or unique study areas, and implement ML techniques for comprehensive groundwater quality management. This chapter delves into case studies where AI and ML have been successfully implemented in groundwater management, highlighting their positive impact on environmental conservation and resource sustainability. Challenges such as data quality, model interpretability, and the need for interdisciplinary collaboration are discussed to provide a comprehensive understanding of the current state and future potential of AI and ML in this field.

J. Bhattacharjee · S. Roy (✉)
Department of Chemical Engineering, University of Calcutta, Kolkata, India
e-mail: subhasis1093@gmail.com

A. Nath et al. (eds.), *Groundwater Depletion and Sustainability*, Advances in Geographic Information Science, https://doi.org/10.1007/978-3-032-09921-1_2

Keyword Hybrid machine learning · Performance models · Groundwater management

1 Introduction

Groundwater is considered one of the most important resources because of its use in drinking water, agriculture, irrigation, and industrial input. Therefore, its role in supporting ecosystems and economic activities is critical, and this water source urgently needs efficient management and protection (Akhtar et al. 2021; McFarlane et al. 2020). Groundwater is constantly threatened by pollution due to urbanization, agricultural runoff, industrial discharge, or climate change. Traditional methods of monitoring and managing groundwater pollution often rely on manual sampling, chemical analysis, and rule-based modeling approaches, which can be time-consuming, expensive, and reactive rather than proactive (McFarlane et al. 2020; Bhattacharjee and Roy 2024a). The recent breakthroughs in AI and ML can be transformative regarding groundwater pollution control and management. The technologies offer tools for analyzing big data, pattern recognition, and predicting future scenarios. The optimization of management strategies will also be possible through AI and ML. Thus, Researchers and practitioners can attain accurate, real-time, and low-cost solutions to challenges involving groundwater pollution, as shown in Fig. 1.

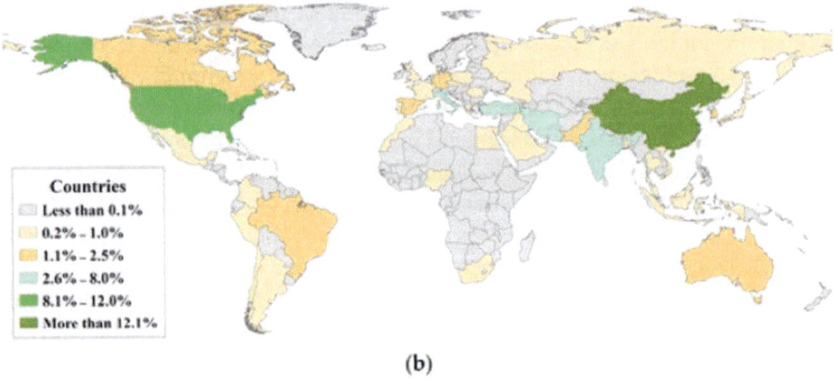

Fig. 1 Global distribution of hybrid AI models in groundwater management **a** Articles focusing on managing groundwater quality using Hybrid AI models; Number of publications for each country. **b** Articles on groundwater quantity management will be distributed using Hybrid AI models. The number of publications by country is presented. The percentage values represent the share of articles published per country from the overall number of articles published using Hybrid AI to manage water resources (Li et al. 2024) (Reproduced with permission)

1.1 Relevance of Groundwater Pollution Control

Groundwater pollution poses severe risks to human health and the environment. Nitrates, heavy metals, organic pollutants, and other contaminants could penetrate aquifers through agriculture, waste disposal, or accidental spills (Bhattacharjee and Roy 2024b). The impacts of pollution in terms of water quality degradation, ecosystem disruption, and increased treatment costs are significant. Moreover, since the residence time of these contaminants in aquifers is long, prevention and early detection are very important. Traditional methods of groundwater pollution control rely on hydrogeological models and extensive field data collection, which are limited in their complexity and inability to process dynamic data in real-time. Integrating AI and ML into groundwater management represents a paradigm shift, enabling the transition from static to dynamic, adaptive systems (Bhakta et al. 2024; Cai et al. 2015).

1.2 Role of AI and ML in Groundwater Management

AI is the simulation of human intelligence in machines, which enables them to do tasks similar to decision-making, problem-solving, or pattern recognition. ML, which falls under AI, is an algorithm that allows systems to learn from data and improve performance over time. Together, these technologies have been changing many aspects of various industries, including the management of water resources (Cai et al. 2015; Zaresefat and Derakhshani 2023).

1.3 Recent Development and Application

Technological advancements supporting the adoption of AI and ML in groundwater management include high-performance computing, IoT devices, and big data analytics. Several applications have proven the potential of these technologies through the work of researchers and practitioners (Zaresefat and Derakhshani 2023; Mukherjee and Ramachandran 2018; Tahmasebi et al. 2020).

- Source Identification of Contaminants: ML algorithms can identify pollution sources by analyzing spatial and temporal patterns in groundwater quality data. This is particularly useful in urban and industrial areas with multiple potential contamination sources.
- Estimation of Groundwater Recharge: AI models predict recharge rates based on meteorological, hydrological, and land use data, aiding sustainable aquifer management.
- Risk Assessment: AI technologies can measure aquifers' vulnerability to contamination, including land use, soil properties, and dynamics in groundwater flow.

Such risk maps determine land use planning and the application of measures to prevent pollution.

- Remote Sensing and Satellite Data: AI applications of satellite imagery and remote sensing data allow monitoring of groundwater depletion, subsidence of land, and hotspots of pollution at a large scale with the ability to create early warning systems (Zaresefat and Derakhshani 2023; Mukherjee and Ramachandran 2018).

Despite these challenges, the opportunities for AI and ML in groundwater management are immense. The ongoing advancements in computational power, data acquisition technologies, and algorithm development will continue to expand their applicability. Furthermore, the increasing emphasis on sustainable development and environmental protection provides a strong impetus for adopting these technologies. Future aspects of AI and ML in controlling groundwater pollution can be in the integrated development of systems encompassing data-driven and process-based approaches. Hybrid models will exploit the advantages of each method, and there could be enhanced prediction accuracy and improved decision-making processes. In addition, progressions in XAI would increase transparency and credibility for ML models, promoting their acceptance by stakeholders (Mukherjee and Ramachandran 2018; Tahmasebi et al. 2020).

The acceleration of AI and ML in groundwater management requires collaborative efforts from governments, research institutions, and the private sector. Investments in data infrastructure, capacity building, and regulatory frameworks will help successfully implement these technologies. The possibilities offered by AI and ML for transforming the approach to groundwater pollution control and management are unprecedented. The most significant water resource problems confronting society will be addressed using these technologies while ensuring that the environmental health associated with water resources is protected for posterity (Tahmasebi et al. 2020; Osman et al. 2022). This chapter surveys recent progress in AI and ML applied to groundwater pollution control with the potential to reshape current practices and open avenues to new solutions.

2 Groundwater Pollution: Challenges and Impacts

Groundwater is one of the most important natural resources, providing nearly half of the world's drinking water and a significant share of agricultural and industrial needs. However, it is under threat due to pollution from diverse sources. Challenges in managing groundwater pollution include the invisible nature of contaminants, delayed detection, and inadequate infrastructure for monitoring and remediation. Pollutants often move long distances through aquifers, making their removal challenging and expensive (Osman et al. 2022; Gil et al. 2021). Groundwater pollution intensifies water scarcity, jeopardizes sustainable development goals, and is a significant threat to global water security.

2.1 Sources and Types of Groundwater Contaminants

Groundwater contaminants emanate from natural and anthropogenic sources. Natural sources include arsenic, fluoride, and radionuclides leached from soil and rock formations. Human-induced sources include agricultural activities, such as the overuse of fertilizers and pesticides, and industrial effluents containing heavy metals and chemicals. Organic and inorganic pollutants are also introduced through poorly managed landfills, leaking septic tanks, and mining activities (Osman et al. 2022; Gil et al. 2021; Shaikh et al. 2024). New contaminants, including pharmaceuticals, microplastics, and endocrine-disrupting chemicals, complicate pollution dynamics. Determination and classification of these contaminants are essential for developing focused remediation measures.

2.2 Effects on Ecosystems and Human Health

The effects of groundwater contamination spill into ecosystems and public health. Contaminants have altered water quality and biogeochemical cycles in aquatic habitats, which is quite detrimental to biodiversity. To the human world, polluted groundwaters have been linked with a wide range of diseases, such as cancers, developmental abnormalities, and neurological conditions because of exposure to heavy metals, such as lead and cadmium or nitrates. For instance, arsenic-contaminated groundwater has caused widespread health crises in many parts of Asia. In addition, socio-economic healthcare costs and agricultural productivity losses due to polluted irrigation water exacerbate community vulnerabilities (Shaikh et al. 2024; Bhattacharjee and Roy 2024c).

2.3 Current Challenges in Groundwater Management

Although much progress has been made, managing groundwater pollution still poses significant challenges. There is a lack of real-time monitoring systems that can pick up contaminants early. Current remediation technologies, like pump-and-treat systems, are energy-intensive and often ineffective for diffuse pollution. The legal frameworks remain poorly defined, and the regulations are inconsistently enforced. Developing countries lack infrastructure, expertise, and funding for addressing groundwater issues. More importantly, the siloed approach to water management, which neglects the nexus of surface and groundwater systems, hinders integrated solutions (Bhattacharjee and Roy 2024c, 2024d).

The Nubian Sandstone Aquifer System is a shared aquifer in four African countries with inadequate transboundary cooperation. Additionally, low public awareness makes things worse. In many regions in Asia, people in rural areas are stuck using polluted water resources due to the unavailability of proper alternatives. Economic factors make advanced technologies like AI and machine learning unused for optimizing groundwater pollution management, whose deployment costs are too high (Bhattacharjee and Roy 2024d; Taghikhah et al. 2022; Ghannam and Techtmann 2021).

3 AI and Machine Learning: An Overview

3.1 Fundamental Concepts of AI and Machine Learning

Artificial Intelligence refers to the simulation of human intelligence by machines, which enables them to perform tasks such as reasoning, learning, and problem-solving. Machine Learning is a subdomain of AI that aims to develop algorithms that allow computers to learn and improve based on data without explicit programming. Supervised, unsupervised, and reinforcement learning are the core types of ML approaches suited to different problem domains. In groundwater management, AI models can predict pollution sources, simulate aquifer systems, or optimize resource allocation. For example, supervised learning models trained on water quality datasets can identify contamination trends. On the other hand, unsupervised learning algorithms like k-means clustering might group regions based on similar groundwater contamination profiles, aiding in targeted interventions (Ghannam and Techtmann 2021; Ghobadi and Kang 2023; Krishnan et al. 2022).

3.2 Key Algorithms Used in Environmental Studies

- Decision Trees and Random Forests: These techniques are particularly good at predicting groundwater contamination, especially in identifying sources of key pollutants. For instance, they have been applied to classify water quality levels in industrial zones.
- Neural Networks: Intense learning models have shown their ability in complex aquifer simulations. A CNN can analyze spatial data for mapping contamination.
- SVM: This can be pretty useful in land-use-based classifications of impacts on groundwater quality. SVMs have also been used to classify urban and rural pollutant profiles (Krishnan et al. 2022; Arabameri et al. 2020; Sánchez-Murillo et al. 2022), as shown in Fig. 2.

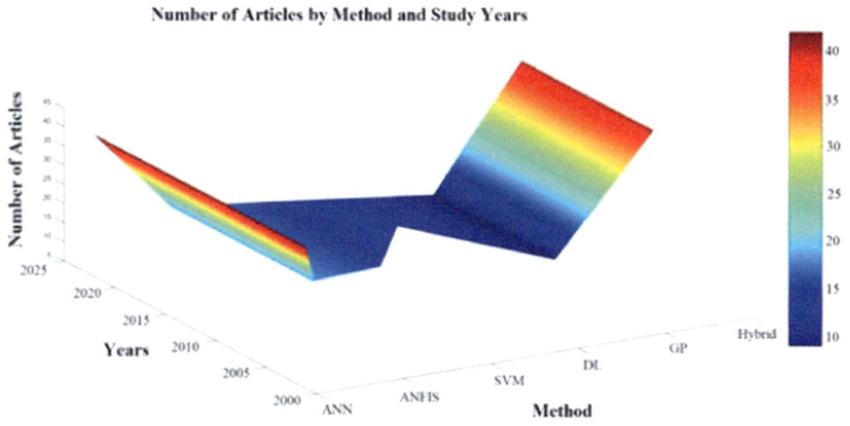

Fig. 2 Trend in the number of scientific articles published from 2001 to 2023 on using various AI-based methods in groundwater level (GWL) modeling, including Artificial Neural Networks (ANNs), Adaptive Neuro-Fuzzy Inference Systems (ANFISs), Support Vector Machines (SVMs), Deep Learning (DL), Genetic Programming (GP), and Hybrid Algorithms (Xu et al. 2022) (Reproduced with permission)

3.3 Big Data and IoT in Groundwater Management: Significance

Integration of Big Data and IoT in groundwater monitoring has genuinely transformed it. Big Data offers the capability to gather complex data at a scale never seen before. In contrast, continuous monitoring of real-time parameters such as pH, turbidity, and heavy metals is achieved through IoT sensors. For instance, water quality information from borewell sensors enabled with IoT is transmitted immediately to cloud platforms. This information is processed through ML models, and probable contamination is forecasted for early intervention. Big Data techniques like Hadoop and Spark are further used in processing datasets from various sources, such as weather stations, industrial effluents, and agricultural runoff (Amanambu et al. 2020).

Case Study: The Ganges River Basin initiative used IoT sensors and Big Data analytics to monitor and improve groundwater quality in real-time, significantly reducing pollution levels.

Nitrate contamination is an essential issue in groundwater, with the source often being agricultural runoff. ML algorithms can model how soil, water, and fertilizer application rates interact to predict nitrate leaching. A practical application in the Midwest USA is when ML models integrate meteorological, agricultural, and hydrological data and suggest optimal fertilizer application schedules to reduce groundwater contamination (Abbaszadeh Shahri et al. 2022; Zaresefat et al. 2023).

3.4 AI and Remote Sensing for Groundwater Mapping

Satellites and AI-driven remote sensing can provide the most innovative approach to water mapping. AI models of images can estimate changes in groundwater availability by detecting shifts in vegetation and soil moisture levels. For instance, the GRACE mission utilizes satellite data with AI to monitor changes in groundwater storage across the globe for drought management purposes (Javidan and Javidan 2022).

3.5 Machine Learning for Contaminant Spread Modeling

ML models reproduce the flow of contaminants through aquifers so that there is an exact identification of contaminant sources. For instance, ML models consider factors such as soil porosity, aquifers' permeability, and water flow dynamics. As an illustration, scientists from Canada are using ML to model aquifers' contamination due to arsenic, where remediation could be implemented with much fewer resources, saving time. Big Data and AI enable groundwater management systems to deliver actionable insights to decision-makers. DSS combines diverse data sources, ML predictions, and user inputs to recommend the best courses of action. A Water India project, funded by the EU, uses AI-based DSS platforms that deliver techniques to save water, mitigate contamination, and promote community engagement according to local needs (Javidan and Javidan 2022; Moayedi et al. 2023) (Fig. 3).

Despite their potential, AI, ML, Big Data, and IoT in groundwater management face challenges, including data reliability, extensive infrastructure, and high implementation costs. However, computational power and sensor technology advancements are making these tools more accessible. Future developments could include AI models trained on global groundwater datasets, enabling cross-regional insights. Furthermore, innovations in IoT, such as low-cost biodegradable sensors, could enhance monitoring in resource-constrained areas (Moayedi et al. 2023; Rajaee et al. 2019).

4 Prediction of Groundwater Contamination is One of the Most Critical Components of Managing Groundwater

The application of ML models, especially regression analysis and classification techniques, in predicting the risk of groundwater contamination has been very prevalent. Such predictions are often made by analyzing historical water quality data, land use patterns, rainfall data, and geological features (Rajaee et al. 2019; Hanoon et al. 2021).

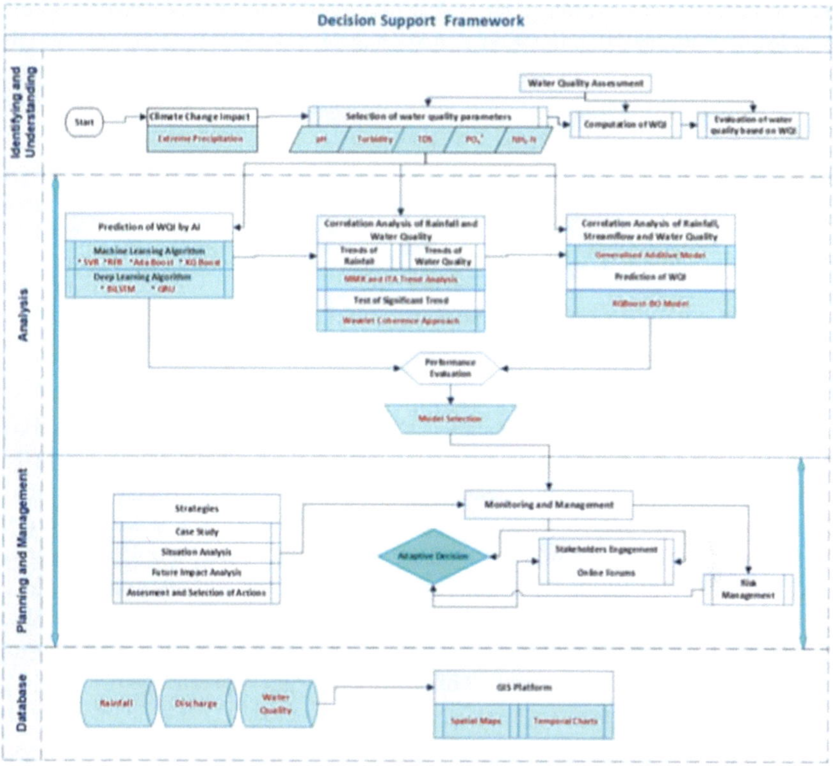

Fig. 3 Schematic diagram of the proposed AI and Machine Learning-based framework for groundwater pollution control and management, which depicts the main components such as data collection, preprocessing, model training, predictive analysis, real-time monitoring, and decision support for sustainable water resource management (Rajaee et al. 2019) (Reproduced with permission)

Application of ANNs in California Predicted contamination levels of groundwater in agricultural areas. The model analyzed the relationship between pesticide application and groundwater pollution and helped the authorities predict pollution hot spots and take preventive measures before pollution spread.

4.1 *Integration of Remote Sensing and Satellite Data*

AI/ML techniques are increasingly combined with satellite remote sensing data to monitor and manage groundwater resources. Machine learning models can analyze satellite imagery and other remote sensing data for signs of groundwater depletion or pollution across large areas. By combining data from various sources such as weather stations, sensors, and geographic information systems, AI can assess groundwater levels, identify contamination sources, and even monitor land-use changes that impact groundwater. The European Space Agency (ESA) uses AI in the "Earth Observation" satellite program to monitor groundwater levels. The system can develop real-time models by integrating AI with geospatial data, enabling authorities to track changes in groundwater across regions and predict future trends (Bhattacharjee and Roy 2023, 2024e).

4.2 Data-Driven Decision-Making for Groundwater Management

AI empowers decision-making using vast quantities of groundwater data. From collecting data to data fusion to predictive analytics, AI offers actionable insight through its tools. Using machine learning algorithms such as decision trees, random forests, and support vector machines in processing complex data sets enables decisions on managing groundwater resources. In India, AI-based tools are integrated into the Ministry of Water Resources, which evaluates data on groundwater levels taken across all the wells within India. ML algorithms help to predict fluctuations in the water table; this is useful when deciding groundwater extraction and measures for conserving water (Malekzadeh et al. 2019; Phan and Nguyen 2020).

4.3 Groundwater Contamination Source Identification

AI and ML can be effectively utilized for the identification of groundwater contamination sources that are very often complex and diffuse. Applying clustering techniques such as k-means and DBSCAN enables AI to analyze historical pollution data, identify patterns, and cluster areas where the pollution sources are concentrated. For instance, in the case of industrial waste disposal, AI models could be used to trace the contamination sources in an urban setup. By analyzing the concentration of pollutants at different monitoring points of groundwater, the system pinpointed the exact locations where there had been a leak or improper disposal, thus helping in timely remediation work (Bhattacharjee and Roy 2024f, 2024g).

4.4 Ground Water Quality Monitoring Using IoT and AI

IoT provides the means of real-time monitoring of the quality of groundwater by deploying sensors in wells, rivers, and reservoirs. These sensors gather data such as pH, temperature, turbidity, and the concentration of pollutants. AI algorithms can

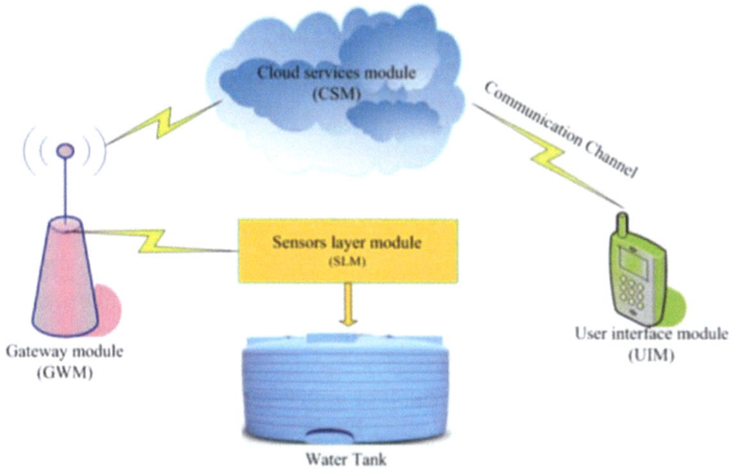

Fig. 4 High-Level IoT-Based Water Quality Monitoring System (IoT-WQMS) Model for Groundwater Pollution Control and Management. The model integrates IoT-enabled sensors for real-time water quality data collection, AI/ML algorithms for predictive analysis and anomaly detection, and cloud infrastructure for data processing and visualization. The system facilitates adaptive management and decision-making to ensure sustainable groundwater quality (Shahriar et al. 2022) (Reproduced with permission)

process sensor data to give early warning systems on groundwater contamination, predict water quality trends, and optimize monitoring processes. In the United States, an IoT-based AI system has been developed to continuously monitor groundwater in areas with high levels of agricultural runoff. The system collects real-time data on nitrate concentrations. It uses ML algorithms to detect any abnormal increases in pollutants, notifying local authorities to take action (Farzin et al. 2021; Nguyen et al. 2020; Zhang et al. 2022), as shown in Fig. 4.

4.5 AI-Based Optimization of Water Resource Management

Techniques used to optimize water resource use by AI and ML in groundwater-dependent regions have many inputs, such as temporal variation of rainfall, temporal consumption, recharge rates, and season. Using reinforcement techniques, such as deep learning, could guide the AI models on suitable extraction rates or maintain optimal usage over a more extended period, also helping mitigate the risk factors involved in over-extracting from the water reserves. In Australia, the use of reinforcement learning helped to optimize groundwater extraction to aid in irrigation. The model utilizes historical information regarding groundwater levels, weather patterns, and crop water requirements; such models help farmers make informed decisions on when and how much groundwater they should extract, which thus ensures sustainable practices (Bhattacharjee and Roy 2024g; Farzin et al. 2021; Nguyen et al. 2020; Zhang et al. 2022; Shahriar et al. 2022).

4.6 Remediation Strategies Using AI and ML

AI is highly significant in the determination of strategies to remediate groundwater. Machine learning models can forecast contaminant transport in aquifers and propose optimum locations for remediation activities, including pumping, treatment, and disposal. In addition, the models assess the performance of remediation technologies. In the case of a contaminated aquifer in New Jersey, AI was used to simulate the movement of pollutants and identify the best locations for groundwater treatment wells. By integrating data from soil characteristics, water flow patterns, and contamination levels, the system proposed a remediation plan that minimized costs and maximized cleanup efficiency (Mozaffari et al. 2022; Seidu et al. 2022; Jeihouni et al. 2019).

4.7 Predictive Maintenance for Groundwater Infrastructure

AI and ML are also used in predictive maintenance for groundwater infrastructure such as pumps, wells, and pipelines. Machine learning algorithms analyze sensor data to predict when equipment might fail or maintenance is required, reducing downtime and preventing water supply disruptions. Predictive maintenance improves the longevity and reliability of groundwater management systems. In the United Kingdom, AI-powered predictive maintenance is being used for monitoring groundwater pumping stations. By analyzing historical data from sensors installed on pumps, AI models predict wear and tear, helping to schedule maintenance before failures occur, which is critical in ensuring continuous water supply. Although AI and ML have excellent prospects in groundwater management, some significant challenges are left unaddressed (Singha et al. 2021; Alkindi et al. 2022). Poor quality and incompleteness of data, particularly in developing regions, are some such issues. Secondly, AI models need big datasets for training purposes; these are challenging to collect, especially in a spatially and temporally heterogeneous system, as in the case of groundwater. Last but not least, resource-poor settings cannot afford the high-cost implementation of AI-based systems. However, as data collection methods improve and AI technologies become more accessible, the future of AI in groundwater management seems to be bright. Collaboration between governments, research institutions, and technology companies will lead to more robust and scalable AI solutions (Bhattacharjee and Roy 2024; Bhattacharjee et al. 2024a).

AI and Machine Learning could revolutionize groundwater pollution management and resource optimization. From predictive analytics to real-time monitoring, these technologies enable better, more efficient, accurate, and sustainable management practices. Combining AI with the existing technology of IoT, satellite data, and GIS can facilitate early detection of groundwater pollution, identification of contamination sources, and optimization of remediation strategies. With future advancements in AI/ML, we expect even more sophisticated solutions to manage and protect this precious resource (Bhagat et al. 2021; Thomas et al. 2019).

5 AI-Based Groundwater Contamination Prediction Models

Groundwater is one of the most essential natural resources. It quenches the thirst of a vast majority of the world's population through drinking water. However, it is highly susceptible to contamination from some human activities such as industrial waste disposal, agricultural runoff, and landfills. Contamination of groundwater poses severe health and environmental risks. Predicting and preventing such contamination is essential for sustainable water management. Artificial Intelligence (AI) and Machine Learning (ML) have emerged as powerful environmental monitoring and management tools. AI-based prediction models can identify patterns, make forecasts, and simulate scenarios that would be difficult or time-consuming using traditional methods. In groundwater contamination prediction, AI models process large datasets, including historical contamination records, geological data, weather patterns, and land use information, to forecast contamination events and their impacts (Li et al. 2020; Dehghani and Torabi Poudeh 2022; Ye et al. 2020).

Types of AI Models Used for Contamination Prediction

5.1 Supervised Learning Models

Linear Regression is used to predict the levels of contaminants based on historical data. These models can estimate concentrations of specific pollutants over time.

– Decision Trees: They are divided into smaller manageable portions of data, leading to decisions based on the history of contamination. They can predict the likelihood of contamination at specific locations.

Random Forests and Support Vector Machines (SVM): These are ensemble methods that perform well when dealing with large data, especially in cases involving various features such as soil type, precipitation, and human activities (Beiranvand and Rajaee 2022).

5.2 Unsupervised Learning Models

– Clustering Algorithms (K-means, DBSCAN): These models can determine which areas have similar contamination levels or behaviors, help find hotspots, and understand the spatial relationships of contamination.

– Principal Component Analysis (PCA): PCA reduces the dimensionality of large datasets, thereby facilitating the analysis of multiple factors, such as industrial activities, geological formations, and groundwater quality.

5.3 Deep Learning Models

Artificial Neural Networks (ANNs): ANNs are especially adept at detecting complex patterns within large data sets. Groundwater contamination events can be forecasted based on several input factors, such as industrial activities, climate data, and reports of groundwater monitoring by ANNs (Rajaee et al. 2020; Khosravi et al. 2021; Torkashvand et al. 2021).

RNNs (Recurrent Neural Networks) are useful for time series, as in the example of groundwater contaminant level changes with time. They model temporal patterns of contamination dynamics by offering live prediction and forecasting.

Applications and Case Studies of AI Models for Ground Water Contamination Forecast

Case Study: Ground Water Contamination in Agricultural Regions

Nitrate contamination in agricultural areas due to excessive fertilizer is predicted using AI models. Supervised learning methodologies, such as Random Forest and SVM, have been followed for predicting nitrate levels under different agricultural practices, crops, irrigation techniques, and rainfall characteristics (Bhattacharjee and Roy 2024h, 2025a).

5.4 Case Study: Estimation of Heavy Metal Pollution in Industrial Areas

Deep learning models, specifically ANNs, have been applied to model the dispersion of heavy metals, for instance, arsenic, cadmium, and lead in groundwater. Here, they consider like industrial discharge, characteristics of the geology, and patterns of groundwater flows in their predictions of spread contamination.

5.5 Case Study: Contamination Forecasting in an Urban Groundwater

Predicting organic contaminants in AI models in areas of high industrial activity, particularly petroleum hydrocarbons and chlorinated solvents in urban areas, is possible using machine learning algorithms such as SVM and decision trees to identify vulnerable regions and provide an estimate of the level of contamination over time.

Although prediction models based on AI demonstrate good promise, problems such as data availability and quality, as well as the complexity of groundwater systems, pose significant obstacles. The accuracy of the model depends on the quality and quantity of data used for training. The complexity of geological formations and human intervention in certain cases also make the prediction less reliable (De et al. 2023; Bhattacharjee and Roy 2024i).

ML Applications in Groundwater Remediation Projects

Groundwater remediation refers to the process where pollution or contaminants in the groundwater are removed to restore the quality. The process is vital since such contaminated aquifers might have to be used in the future for potable water, irrigation, and many other activities. Traditional remediation methods often limit pump-and-treat as well as in-situ bioremediation to efficiencies, cost, and long duration. Machine Learning might then offer substantial improvements for that. Machine Learning can optimize groundwater remediation by improving the efficiency of the processes, predicting the outcomes of various remediation strategies, and helping in the decision-making process (Bhattacharjee and Roy 2024i; Bhattacharjee et al. 2024b).

5.6 Optimizing Remediation Strategies

Reinforcement Learning: Machine learning is particularly useful in optimizing remediation strategies because it can learn from previous interventions and make the necessary adjustments in the parameters to increase efficiency. In this regard, reinforcement learning algorithms can determine how best to deliver remediation agents like oxygen or bacteria for bioremediation or where the pumping system should be placed for efficiency.

Genetic Algorithms (GAs): GAs seek optimal solutions by simulating the natural process of evolution. In the context of remediation, GAs could optimize well placement and treatment systems, given different environmental and technical constraints (Bhattacharjee et al. 2024b; Lin et al. 2022).

5.7 *Predictive Models for Remediation Effectiveness*

Support Vector Machines (SVMs) and Random Forests: These prediction models calculate how well a given form of remediation might perform when applied to a specific contamination site. Based on these models, the anticipated effectiveness of using specific methods like bioremediation or chemical treatment on-site may be predicted by forecasting possible historical outcomes.

– Artificial Neural Networks (ANNs): ANNs may be applied to simulate the groundwater dynamics and predict contaminant transport over time. The models help analyze long-term remediation effectiveness by determining the probable recovery period of the aquifers (Mallick et al. 2022; Jamei et al. 2022).

Applications and Case Studies of ML in Groundwater Remediation

Case Study: Bioremediation of Oil Contaminated Groundwater

ML algorithms, especially neural networks, have optimized bioremediation processes, where microbes are introduced to degrade petroleum hydrocarbons in contaminated groundwater. The effectiveness of bioremediation efforts can be predicted by monitoring environmental factors such as temperature, pH, and oxygen levels and optimizing the microbial community used for degradation.

Case Study: Chemical Remediation in Heavy Metal Contaminated Sites

In heavy metal contamination, ML models can predict the best chemical treatments, such as injection of chelating agents to extract metals from groundwater. Support Vector Machines have been applied to historical data to predict the most effective chemical combinations and concentrations for site-specific conditions.

Case Study: Phytoremediation of Agricultural Runoff

Machine learning models have been used to optimize the phytoremediation process, which uses plants to absorb pollutants. In analyzing soil composition, species of plants, and contamination levels, machine learning algorithms assist in the identification of the most effective plant species in removing specific contaminants, such as nitrates or phosphates, from the groundwater (Moazamnia et al. 2020; Bhattacharjee et al. 2024c, 2024d; Samuel 1959) (Fig. 5).

Challenges and Limitations of ML in Groundwater Remediation

While ML provides powerful tools for improving groundwater remediation, several challenges need to be addressed:

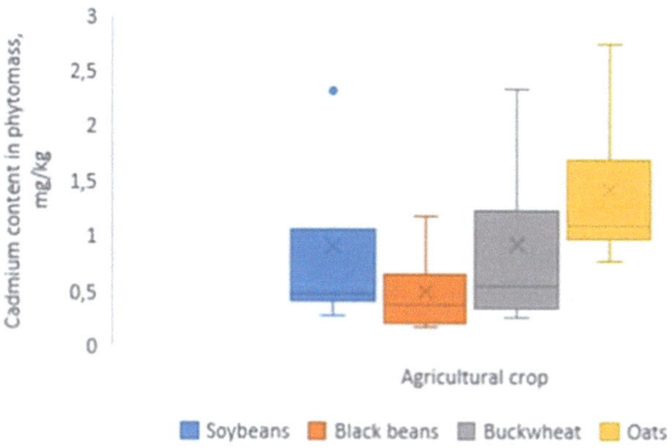

Fig. 5 Cadmium content in the phytomass of test crops based on the total indicator of soil contamination values from the vegetation experiment (Samuel 1959) (Reproduced with permission)

– Data Gaps and Uncertainty: Reliable and accurate data is crucial for effective ML algorithms. In many regions, the lack of continuous monitoring data or detailed historical records can hinder the application of ML techniques.

– Complexity of Remediation Processes: Groundwater remediation involves numerous variables, including physical, chemical, and biological factors. Modeling such complex systems requires highly sophisticated algorithms that handle uncertainty and non-linearity (He et al. 2022; Dogo et al. 2019).

– Cost and Accessibility: Implementing ML-based systems may require significant investment in software, hardware, and expert knowledge. This may be a limiting factor for some remediation projects, especially in developing regions. The future of ML in groundwater contamination prediction and remediation appears bright. Integration of AI with real-time monitoring devices, such as IoT sensors, can ensure continuous data feed to improve predictions and decision-making. Additionally, hybrid models, which combine various ML algorithms and traditional hydrological models, promises more potent and reliable solutions for effective groundwater management. AI and ML technologies are drastically changing how we forecast, monitor, and treat groundwater contamination. These enable more effective and efficient management by providing powerful tools for predicting, optimizing, and deciding. However, these approaches require overcoming challenges in data quality, model complexity, and integration of diverse technologies toward successful implementation. Continued research and development will further transform how AI and ML can play a role in revolutionizing groundwater pollution control and management strategies worldwide (Samuel 1959; He et al. 2022; Dogo et al. 2019; Bhattacharjee and Roy 2025b; Halder et al. 2025).

6 Conclusions

In conclusion, integrating AI and Machine Learning (ML) into groundwater pollution control and management represents a transformative shift in environmental science and engineering. By leveraging the advanced capabilities of these technologies, it is now possible to more efficiently monitor, predict, and mitigate the impacts of groundwater pollution. AI-driven models and ML algorithms analyze vast and complex data collections that can be obtained from sources such as sensor networks, satellite images, and data obtained by groundwater sampling. They are analyzed to monitor the contaminant and have an early warning system on site. It enables the identification of patterns and trends not found otherwise by the tremendous capacity of data processing using AI.

One of the key advantages of AI and ML in groundwater management is predictive capabilities. Machine learning algorithms, especially supervised learning models, can be trained on historical data, seasonal trends, and environmental factors to predict future contamination events. The predictive capability of these algorithms would strongly support proactive decision-making by authorities before contamination spreads or escalates, thereby minimizing public health risks and environmental damage. Moreover, reinforcement learning models can continuously improve their decision-making processes, optimizing the deployment of resources for water purification, monitoring, and mitigation activities based on real-time data. AI and ML also facilitate the development of decision support systems that integrate multiple data streams to guide policy and management decisions. These systems can simulate various scenarios, assessing the potential effectiveness of different pollution control strategies and providing policymakers with actionable insights. This enhances the efficiency of groundwater management with less reliance on expensive trial-and-error methods traditionally used in environmental management. In addition, the adaptability of AI models enables them to learn and change as new data becomes available to ensure that management strategies remain relevant in the presence of changing environmental conditions as well as emerging pollution threats. The other important feature of AI and ML in groundwater management is their potential to influence sustainable practices. Since they can correctly predict the pollution levels and susceptible areas, they allow for better resource allocation and focus on the clean-up of polluted sites while eliminating the wasteful use of chemicals or energy. This aligns with larger strategic goals of sustainable stewardship of the environment, balancing the exigencies of effective pollution control with environmental conservation. In addition to these, AI-based techniques in optimization can enhance efficiency and optimize water treatment processes, ensuring the purification of water resources using means that are cost-effective and environmentally responsive to their application. In summary, AI and ML hold massive promise for revolutionizing the control and management of groundwater pollution. Since they can process and analyze large datasets, predict future risks of contamination, optimize resource allocation, and drive sustainable solutions, these technologies present an unparalleled opportunity to improve the health and sustainability of groundwater resources. If the challenges associated with adopting AI and ML are overcome, they will significantly contribute to safeguarding one of the most vital resources for future generations.

Acknowledgements Author, S. Roy would like to acknowledge 'Scheme for Transformational and Advanced Research in Sciences [STARS]' [MoE-STARS/STARS-2/2023-0175] by the Ministry of Education, Govt. of India for promoting translational India-centric research in sciences implemented and managed by the Indian Institute of Science [IISc], Bangalore, for their support.

Conflict of Interests The authors declare no conflict of interest.

References

Abbaszadeh Shahri A, Shan C, Larsson S (2022) A novel approach to uncertainty quantification in groundwater table modeling by automated predictive deep learning. Nat Resour Res 31:1351–1373

Akhtar N, Syakir Ishak MI, Bhawani SA, Umar K (2021) Various natural and anthropogenic factors responsible for water quality degradation: a review. Water 13:2660

Alkindi KM, Mukherjee K, Pandey M, Arora A, Janizadeh S, Pham QB, Anh DT, Ahmadi K (2022) Prediction of groundwater nitrate concentration in a semiarid region using hybrid Bayesian artificial intelligence approaches. Environ Sci Pollut Res 29:20421–20436

Amanambu AC, Obarein OA, Mossa J, Li L, Ayeni SS, Balogun O, Oyebamiji A, Ochege FU (2020) Groundwater system and climate change: present status and future considerations. J Hydrol 589:125163

Arabameri A, Lee S, Tiefenbacher JP, Ngo PTT (2020) Novel ensemble of MCDM-artificial intelligence techniques for groundwater-potential mapping in arid and semi-arid regions (Iran). Remote Sens 12:490

Beiranvand B, Rajaee T (2022) Application of artificial intelligence-based single and hybrid models in predicting seepage and pore water pressure of dams: a state-of-the-art review. Adv Eng Softw 173:103268

Bhagat SK, Tiyasha T, Awadh SM, Tung TM, Jawad AH, Yaseen ZM (2021) Prediction of sediment heavy metal at the Australian bays using newly developed hybrid artificial intelligence models. Environ Pollut 268:115663

Bhakta D, Gayen D, Bhattacharjee J, Chatterjee R, Dastidar P, Roy S (2024) Repurposing industrial and agricultural wastes as sustainable alternatives in photovoltaic cells and dye degradation. In Chemistry Select, Wiley 9:38

Bhattacharjee J, Roy S (2023) Green remediation of microplastics using bionanomaterials. In: Remediation of plastic and microplastic waste, pp 240–260

Bhattacharjee J, Roy S (2024) Synergistic insights: electro-organic photocatalysis and nanostructures. In: Chemical papers, vol. 78, Issue 15. Springer Science and Business Media LLC, pp 8077–8105. https://doi.org/10.1007/s11696-024-03657-2

Bhattacharjee J, Roy S (2024) Review on green resources and AI for biogenic solar power. Energy Storage Convers 2(1)

Bhattacharjee J, Roy S (2024) Density functional waltz: Molecular theory of spin dynamics. J Phys & Opt Sci 1–4. https://doi.org/10.47363/jpsos/2024(6)242

Bhattacharjee J, Roy S (2024) Synthesis of thermoelectric nanocomposites by incorporating reduced graphene oxide. In: Archives of advanced engineering science. Bon View Publishing PTE. https://doi.org/10.47852/bonviewaaes42023514

Bhattacharjee J, Roy S (2024) Nature-inspired catalysts: a new era for water-splitting technology. In: Material science research India, vol 21, Issue 2. Oriental Scientific Publishing Company, pp 68–71. https://doi.org/10.13005/msri/210201

Bhattacharjee J, Roy S (2024) Exploration of molybdenum oxide compounds—a review. Catal Res 4(3):1–19. https://doi.org/10.21926/cr.2403011

Bhattacharjee J, Roy S (2024) Fabrication of nano-carbon-based sustainable perovskite solar cells. In: ECS meeting abstracts, vols MA2024–01, Issue 7. The Electrochemical Society, pp 790–790. https://doi.org/10.1149/ma2024-017790mtgabs

Bhattacharjee J, Roy S (2025) Soft computing techniques in wind conversion systems: a review. Soft Comput Renew Energy Technol 121–142

Bhattacharjee J, Roy S (2024) Bio-butadiene-based rubbers in damping and sound absorption applications. In: Properties and applications, of butadiene rubbers, pp-227, by Nova Science Publishers, Inc. ISBN: 979-8-89113-969–5

Bhattacharjee J, Bhakta D, Roy S (2024) Characteristics, applications, and limitations of nanocomposites in biosensing. Adv Chem Mater Eng 161–191. https://doi.org/10.4018/979-8-3693-1094-6.ch006

Bhattacharjee J, Bhakta D, Roy S (2024) Characteristics, applications, and limitations of nanocomposites in biosensing. In: Advances in chemical and materials engineering, pp 161–191. https://doi.org/10.4018/979-8-3693-1094-6.ch006

Bhattacharjee J, Bhaskar S, Roy S (2024) Green nanomaterials for antimicrobial and anticancer applications. Nano-Eng Funct Interfaces Multi-Discipl Appl 425-444. https://doi.org/10.1016/B978-0-443-21691-6.00018-4

Bhattacharjee J, Bera S, Mondal S, Roy S (2024) Heavy metals and other chemicals removal by microalgae. Adv Sustain Appl Microalgae 183–201. https://doi.org/10.1016/B978-0-443-22127-9.00008-1

Bhattacharjee J, Roy S (2024b) A review on photocatalysis and nanocatalysts for advanced organic synthesis. Hybrid Advances 6:100268. https://doi.org/10.1016/j.hybadv.2024.100268

Bhattacharjee J, Roy S (2024i) Utilizing a variable material approach to combat climate change. Mater Sci Res India 20(3):141–145

Bhattacharjee J, Roy S (2025b) A comprehensive review on integrated photo rechargeable batteries-supercapacitors, and their techno-economic feasibility. J Photochem Photobiol 25:100257. https://doi.org/10.1016/j.jpap.2024.100257

Cai X, Zhang X, Noël PH, Shafiee-Jood M (2015) Impacts of climate change on agricultural water management: a review. Wiley Interdiscip Rev Water 2:439–455

De A, Bhattacharjee J, Chowdhury SR, Roy S (2023) A comprehensive review of third-generation photovoltaic technologies. J Chem Eng Res Updat 10:1–17. https://doi.org/10.15377/2409-983X.2023.10.1

Dehghani R, Torabi Poudeh H (2022) Application of novel hybrid artificial intelligence algorithms to groundwater simulation. Int J Environ Sci Technol 19:4351–4368

Dogo EM, Nwulu NI, Twala B, Aigbavboa C (2019) A survey of machine learning methods applied to anomaly detection on drinking-water quality data. Urban Water J 16:235–248

Farzin M, Avand M, Ahmadzadeh H, Zelenakova M, Tiefenbacher JP (2021) Assessment of ensemble models for groundwater potential modeling and prediction in a karst watershed. Water 13:2540

Ghannam RB, Techtmann SM (2021) Machine learning applications in microbial ecology, human microbiome studies, and environmental monitoring. Comput Struct Biotechnol J 19:1092–1107

Ghobadi F, Kang D (2023) Application of machine learning in water resources management: a systematic literature review. Water 15:620

Gil Y, Garijo D, Khider D, Knoblock CA, Ratnakar V, Osorio M, Vargas H, Pham M, Pujara J, Shbita B et al (2021) Artificial intelligence for modeling complex systems: taming the complexity of expert models to improve decision making. ACM Trans Interact Intell Syst 11:1–49

Halder S, Bhattacharjee J, Roy S, Das BC, Biswas AB (2025) Surface modification of lignin by Cetyl Trimethyl ammonium bromide. In: Jawaid M, Ahmad A, Meraj A (eds) Handbook of Lignin. Springer, Singapore. https://doi.org/10.1007/978-981-97-2664-6_38-1

Hanoon MS, Ahmed AN, Fai CM, Birima AH, Razzaq A, Sherif M, Sefelnasr A, El-Shafie A (2021) Application of artificial intelligence models for modeling water quality in groundwater: comprehensive review, evaluation and future trends. Water Air Soil Pollut 232:1–41

He Y, Zhou Y, Wen T, Zhang S, Huang F, Zou X, Ma X, Zhu Y (2022) A Review of machine learning in geochemistry and cosmochemistry: method improvements and applications. Appl Geochem 140:1–13

Jamei M, Karbasi M, Malik A, Abualigah L, Islam ARMT, Yaseen ZM (2022) Computational assessment of groundwater salinity distribution within coastal multi-aquifers of Bangladesh. Sci Rep 12:1–28

Javidan R, Javidan N (2022) A novel artificial intelligence-based approach for mapping groundwater nitrate pollution in the andimeshk-dezful plain. Iran Geocarto Int 37:10434–10458

Jeihouni E, Eslamian S, Mohammadi M, Zareian MJ (2019) Simulation of groundwater level fluctuations in response to main climate parameters using a wavelet–ANN hybrid technique for the Shabestar plain. Iran Environ Earth Sci 78:293

Khosravi K, Bordbar M, Paryani S, Saco PM, Kazakis N (2021) New hybrid-based approach for improving the accuracy of coastal aquifer vulnerability assessment maps. Sci Total Environ 767:145416

Krishnan SR, Nallakaruppan MK, Chengoden R, Koppu S, Iyapparaja M, Sadhasivam J, Sethuraman S (2022) Smart water resource management using artificial intelligence: a review. Sustainability 14:13384

Li J, Lu W, Wang H, Fan Y, Chang Z (2020) Groundwater contamination source identification based on a hybrid particle swarm optimization-extreme learning machine. J Hydrol 584:124657

Li X, Zhang B, Jakobi T, Yu Z, Ren L, Zhang Z (2024) Laser-based bionic manufacturing. Int J Extrem Manuf 6(4):042003

Lin H, Gharehbaghi A, Zhang Q, Band SS, Pai HT, Chau KW, Mosavi A (2022) Time series-based groundwater level forecasting using gated recurrent unit deep neural networks. Eng Appl Comput Fluid Mech 16:1655–1672

Malekzadeh M, Kardar S, Shabanlou S (2019) Simulation of groundwater level using MODFLOW, extreme learning machine and wavelet-extreme learning machine models. Groundw Sustain Dev 9:100279

Mallick J, Almesfer MK, Alsubih M, Talukdar S, Ahmed M, Ben Kahla N (2022) Developing a new method for future groundwater potentiality mapping under climate change in Bisha Watershed. Saudi Arabia Geocarto Int 37:14495–14527

McFarlane D, George R, Ruprecht J, Charles S, Hodgson G (2020) Runoff and groundwater responses to climate change in South West Australia. J R Soc West Aust 103:9–27

Moayedi H, Salari M, Dehrashid AA, Le BN (2023) Groundwater quality evaluation using hybrid model of the multi-layer perceptron combined with neural-evolutionary regression techniques: case study of shiraz plain. Stoch Environ Res Risk Assess 3:1–16

Moazamnia M, Hassanzadeh Y, Nadiri AA, Sadeghfam S (2020) Vulnerability indexing to saltwater intrusion from models at two levels using artificial intelligence multiple model (AIMM). J Environ Manag 255:109871

Mozaffari S, Javadi S, Moghaddam HK, Randhir TO (2022) Forecasting groundwater levels using a hybrid of support vector regression and particle swarm optimization. Water Resour Manag 36:1955–1972

Mukherjee A, Ramachandran P (2018) Prediction of GWL with the help of GRACE TWS for unevenly spaced time series data in India : analysis of comparative performances of SVR. ANN and LRM. J. Hydrol. 558:647–658

Nguyen PT, Ha DH, Nguyen HD, Van Phong T, Trinh PT, Al-Ansari N, Van Le H, Pham BT, Ho LS, Prakash I (2020) Improvement of credal decision trees using ensemble frameworks for groundwater potential modeling. Sustainability 12:2622

Osman AIA, Ahmed AN, Huang YF, Kumar P, Birima AH, Sherif M, Sefelnasr A, Ebraheemand AA, El-Shafie A (2022) Past, present and perspective methodology for groundwater modeling-based machine learning approaches. Arch Comput Methods Eng 29:3843–3859

Phan TTH, Nguyen XH (2020) Combining statistical machine learning models with ARIMA for water level forecasting: the case of the red river. Adv Water Resour 142:103656

Rajaee T, Ebrahimi H, Nourani V (2019) A review of the artificial intelligence methods in groundwater level modeling. J Hydrol 572:336–351

Rajaee T, Khani S, Ravansalar M (2020) Artificial intelligence-based single and hybrid models for prediction of water quality in rivers: a review. Chemom Intell Lab Syst 200:103978

Samuel AL (1959) Some studies in machine learning using the game of checkers. IBM J Res Dev 3:210–229

Sánchez-Murillo R, Montero-Rodríguez I, Corrales-Salazar J, Esquivel-Hernández G, Castro-Chacón L, Rojas-Jiménez LD, Vargas-Víquez J, Pérez-Quezadas J, Gazel E, Boll J (2022) Deciphering complex groundwater age distributions and recharge processes in a tropical and fractured volcanic multi-aquifer system. Hydrol Process 36:e14521

Seidu J, Ewusi A, Kuma JSY, Ziggah YY, Voigt HJ (2022) A hybrid groundwater level prediction model using signal decomposition and optimised extreme learning machine. Model Earth Syst Environ 8:3607–3624

Shahriar SM, Bhuiyan EA, Nahiduzzaman M, Ahsan M, Haider J (2022) State of charge estimation for electric vehicle battery management systems using the hybrid recurrent learning approach with explainable artificial intelligence. Energies 15:8003

Shaikh AA, Bhattacharjee J, Datta P, Roy S (2024) A comprehensive review of the oxidation states of molybdenum oxides and their diverse applications. Sustain Chem Environ 7:100125. https://doi.org/10.1016/j.scenv.(2024).100125

Singha S, Pasupuleti S, Singha SS, Singh R, Kumar S (2021) Prediction of groundwater quality using efficient machine learning technique. Chemosphere 276:130265

Taghikhah F, Voinov A, Filatova T, Polhill JG (2022) Machine-assisted agent-based modeling: opening the black box. J Comput Sci 64:101854

Tahmasebi P, Kamrava S, Bai T, Sahimi M (2020) Machine learning in geo- and environmental sciences: from small to large scale. Adv Water Resour 142:1–33

Thomas A, Eldho TI, Rastogi AK, Majumder P (2019) A comparative study in aquifer parameter estimation using MFree point collocation method with evolutionary algorithms. J Hydroinformatics 21:455–473

Torkashvand M, Neshat A, Javadi S, Pradhan B (2021) New hybrid evolutionary algorithm for optimizing index-based groundwater vulnerability assessment method. J Hydrol 598:126446

Xu K, Qiao W, Zhu G, Feng L, Wang Z, Li W (2022) Hydrogeochemical characteristics of a multi-layer groundwater system in a coal mine area: a case study. Water 14:3146

Ye Z, Yang J, Zhong N, Tu X, Jia J, Wang J (2020) Tackling environmental challenges in pollution controls using artificial intelligence: a review. Sci Total Environ 699:134279

Zaresefat M, Derakhshani R (2023) Revolutionizing groundwater management with hybrid AI models: a practical review. Water 15(9):1750. https://doi.org/10.3390/w15091750

Zaresefat M, Derakhshani R, Nikpeyman V, GhasemiNejad A, Raoof A (2023) Using artificial intelligence to identify suitable artificial groundwater recharge areas for the Iranshahr Basin. Water 15:1182

Zhang Y, Luo X, Wang Z, Mai Q, Deng R (2022) Estimating charging demand from the perspective of choice behavior: a framework combining rule-based algorithm and hybrid choice model. J Clean Prod 376:134262

Integrating Explainable AI with Advanced Learning Models for Groundwater Management in Semi-Arid Areas

Niraj Kumar, Dhiraj Kumar, Rahul Kumar, G. V. Prajapati, and Daksh Vadalia

Abstract Groundwater quality in arid and semi-arid regions faces significant risks due to factors like over-extraction, pollution, and climate change. Traditional models often struggle to provide accurate and actionable insights, hindering effective decision-making. This chapter presents an innovative technique to improve groundwater quality prediction through the integration of Explainable Artificial Intelligence (EAI) with the use of Machine Learning (ML) and Deep Learning (DL) techniques. Transparency in the decision-making process is another benefit of the integrated approach, which helps stakeholders understand the fundamental causes of variations in groundwater quality. The chapter addresses the development of ML and DL models, the use of EAI to interpret and validate models, and the particular difficulties of managing groundwater in arid and semi-arid areas. This chapter demonstrates the practical benefits of this innovative approach through the case studies, providing a robust framework for improving groundwater quality prediction and management in water-scarce regions.

Keywords Groundwater quality · Deep learning · Explainable AI · Arid regions · Semi-arid regions

N. Kumar · D. Kumar (✉) · R. Kumar
Ph.D. Scholar, Department of Soil and Water Conservation Engineering, Junagadh Agricultural University, Gujarat, India
e-mail: dhirajsonkar.nihr@gmail.com

G. V. Prajapati
Prof. & Head, Department of Irrigation and Drainage Engineering, Junagadh Agricultural University, Gujarat, India

D. Vadalia
Research Associate, Research Training and Testing Centre, Junagadh Agricultural University, Gujarat, India

A. Nath et al. (eds.), *Groundwater Depletion and Sustainability*, Advances in Geographic Information Science, https://doi.org/10.1007/978-3-032-09921-1_3

1 Introduction

Groundwater management in semi-arid regions poses significant challenges due to escalating water demand, climate variability, and anthropogenic pressures. Limited rainfall and high rates of evaporation are characteristics of semi-arid regions, which lead to the quick depletion of aquifers and declining water quality (Eskafi et al. 2021; Hssaisoune et al. 2020). Over-extraction for agricultural, industrial, and domestic purposes further exacerbates these issues, posing threats to food security and sustainable development (Akhtar et al. 2021; Bierkens and Wada 2019). Furthermore, the effects of climate change, such as changed precipitation patterns and a rise in drought frequency, complicate groundwater management strategies (Dao et al. 2024; Green 2016). This scenario necessitates innovative approaches to enhance the sustainability and resilience of groundwater resources in these vulnerable regions (Bremer et al. 2021; Gaffoor et al. 2020).

To effectively manage groundwater, one requires a thorough understanding of how environmental groundwater depth interacts with vegetation and ecological health (Huang et al. 2019). As illustrated in Fig. 1, environmental groundwater depth can be categorized into three zones: the Lethal Zone, Sub-Healthy Zone, and Healthy Zone. These zones correspond to vegetation growth conditions ranging from "Die" to "Vigorous." The Healthy Zone represents the optimal range where groundwater availability meets vegetation needs without inducing stress due to anoxia or salinization. In contrast, the Sub-Healthy Zone reflects moderate stress conditions, often due to reduced water availability or excessive salinity. The Lethal Zone signifies extreme conditions, where vegetation either lacks sufficient water or suffers from the adverse effects of salinization, leading to its demise.

Using cutting-edge technology, especially machine learning (ML) and artificial intelligence (AI), has become a viable way to deal with these issues (Shaikh 2024; Zaresefat and Derakhshani 2023).These technologies enable the analysis of vast datasets, improving the accuracy of groundwater availability and quality predictions (Liu et al. 2022; Shiri et al. 2021). Remote sensing and geospatial technologies can monitor groundwater levels and assess recharge potential (Akhter et al. 2021), while big data analytics identifies trends to inform decision-making (Gaffoor et al. 2020). Moreover, advanced learning models can optimize agricultural water use, reducing the strain on groundwater resources (Sahoo et al. 2017). Such technologies are vital for developing sustainable management practices capable of adapting to the dynamic challenges of semi-arid regions (Hssaisoune et al. 2020). To illustrate the multidisciplinary nature of groundwater studies, Fig. 2 presents a network visualization mapping key concepts such as groundwater resources, pollution, flow, and mathematical models. These are interlinked with advanced tools like AI and ML, showcasing their role in understanding and managing groundwater in semi-arid areas.

While AI and ML offer significant predictive power, their complexity often hinders widespread adoption. Stakeholders require models that are not only accurate but also interpretable and transparent (Linardatos et al. 2020). This problem

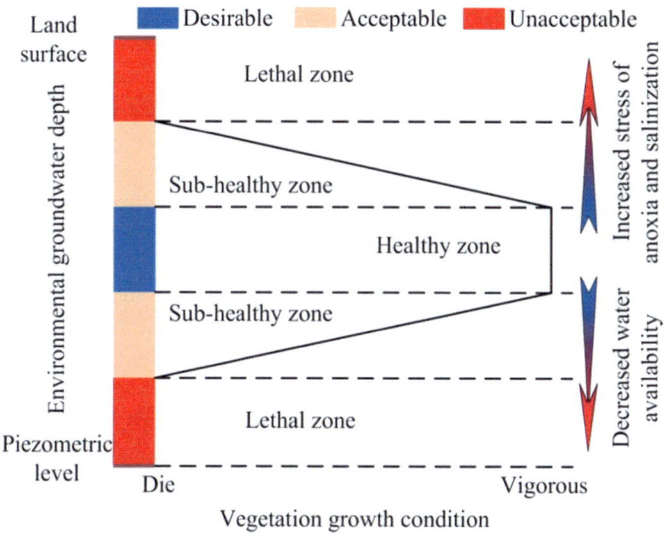

Fig. 1 A sketch map that defines the definition of environmental groundwater depth (Huang et al. 2019)

is adressed by explainable AI (EAI), which boosts trust and makes well-informed decisions easier by offering human-understandable justifications for model predictions (Başağaoğlu et al. 2022; Nallakaruppan et al. 2024). Since decisions related to groundwater management have significant socioeconomic and environmental consequences, this is especially crucial. By integrating EAI with advanced learning models, groundwater managers can leverage the predictive capabilities of AI while ensuring transparency and interpretability (Shaikh 2024). This dual emphasis on accuracy and interpretability is essential for fostering stakeholder engagement and promoting sustainable groundwater management practices in semi-arid areas.

Figure 3 demonstrates the integration of ML algorithms with EAI for classifying water quality. The process involves:

1. Splitting a dataset containing water quality parameters into training (70%) and test (30%) sets.
2. Training various ML algorithms on the training data, including Decision Trees, Random Forest, Support Vector Machines (SVM), Logistic Regression, and naive Bayes.
3. Validating model performance using test data.

enhancing interpretability with EAI tools such as SHAP (SHapley Additive exPlanations) and LIME (Local Interpretable Model-Agnostic Explanations).

This approach ensures accurate water quality classification while maintaining transparency, which is crucial for informed decision-making in groundwater management.

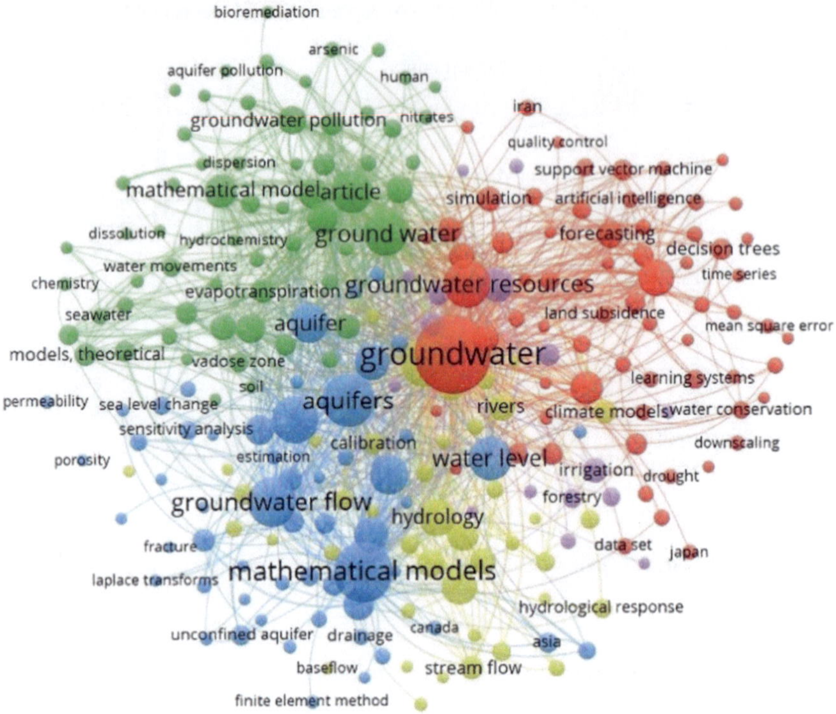

Fig. 2 Network mapping of groundwater dynamics using AI-powered insights in semi-arid regions (Afrifa et al. 2022)

1.1 Advanced Learning Models for Groundwater Management

Machine learning (ML) and deep learning (DL) have completely transformed the field of groundwater management by providing more precise forecasts of groundwater dynamics (Sahoo et al. 2017; Rohde et al. 2021). These models process large datasets comprising historical groundwater levels, rainfall patterns, land use, and water quality parameters to identify trends and predict future conditions (Liu et al. 2022; Shiri et al. 2021). Among the commonly used algorithms are:

1. **Random Forest (RF)**: An effective ensemble learning technique that boosts prediction accuracy by combining several decision trees (Raheja et al. 2021).
2. **Support Vector Machine (SVM)**: An algorithm effective for classification tasks, capable of separating groundwater quality levels into distinct categories (Derdour et al. 2023).
3. **Convolutional Neural Networks (CNNs)**: Are particularly useful for spatial data analysis, such as identifying groundwater recharge zones from satellite imagery (Camacho 2023).

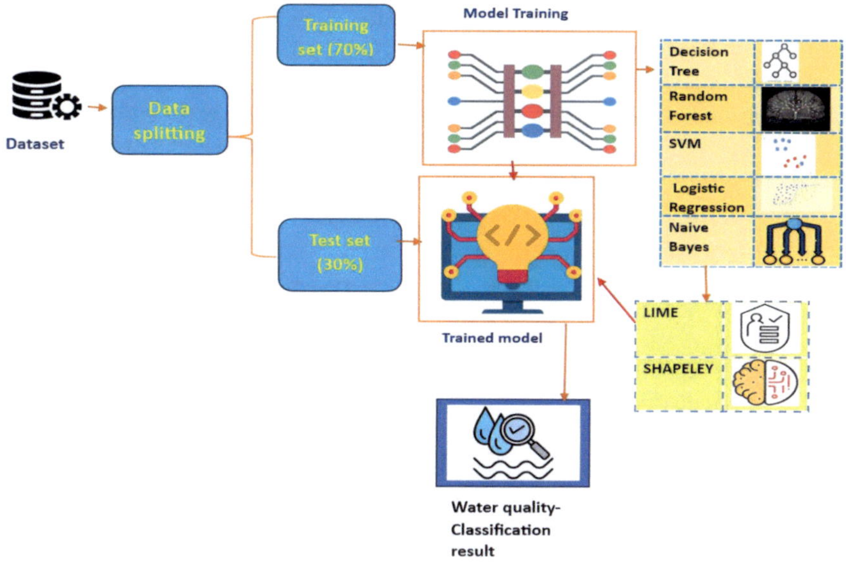

Fig. 3 Workflow integrating ML algorithms with EAI for water quality classification (Nallakaruppan et al. 2024)

4. **Recurrent Neural Networks (RNNs)**: Ideal for forecasting time series, such as trends in groundwater levels (Nur and Ruswandi 2022).

The application of these models in semi-arid regions has demonstrated improved accuracy in predicting groundwater availability and quality (Zaresefat and Derakhshani 2023). However, their "black-box" nature often leaves stakeholders skeptical, emphasizing the need for Explainable AI (EAI) tools to enhance model transparency (Başağaoğlu et al. 2022).

1.2 Role of Explainable AI in Groundwater Management

Explainable AI (EAI) fills the gap between the requirement for interpretability and the accuracy of sophisticated learning models (Linardatos et al. 2020; Singh 2024). EAI techniques, such as LIME and SHAP, allow stakeholders to understand how models arrive at particular projections, building trust in their outcomes (Nallakaruppan et al. 2024). These techniques draw attention to significant characteristics that affect groundwater forecasts, including:

- Rainfall variability.
- Land use changes.
- Over-extraction rates.
- Salinity levels.

EAI tools are particularly valuable in decision-making scenarios with high socio-economic and environmental stakes (Shaikh 2024). For instance, in a case study involving groundwater management in a semi-arid region, SHAP analyses revealed that salinity and over-extraction were the primary contributors to groundwater degradation. This insight enabled targeted interventions, such as regulating extraction rates and promoting salt-tolerant crops, to mitigate the problem.

2 Groundwater Issues in Semi-Arid Regions

Groundwater resources in semi-arid regions are critical for sustaining life, agriculture, and industrial activities. However, these regions face severe challenges in managing groundwater due to limited natural recharge, over-extraction, pollution and the escalating climate change impacts (Hssaisoune et al. 2020; Dao et al. 2024). The combination of these factors has led to unsustainable practices, exposing the quantity and quality of groundwater. This section delves into the key issues threatening groundwater sustainability in semi-arid regions.

2.1 Over-Extraction and Depletion of Resources

Over-extraction of groundwater is one of the most pressing challenges in semi-arid regions, where groundwater often serves as the lifeline for domestic consumption, agricultural irrigation, and industrial use (Bierkens and Wada 2019). These regions' lack of adequate surface water resources drives heavy reliance on groundwater, often resulting in extraction rates far exceeding natural recharge capacities (Wada et al. 2010).

The consequences of over-extraction are manifold:

- **Declining Water Tables**: Excessive groundwater pumping causes water tables to drop significantly, making extracting water increasingly difficult and costly. This trend is particularly evident in agricultural hotspots where high water demand coincides with low recharge rates (Green 2016).
- **Drying of Wells**: Prolonged depletion leads to the drying up of shallow wells, which are often the primary water source for smallholder farmers and rural communities.
- **Aquifer Compaction and Land Subsidence**: Over time, continuous extraction can cause irreversible damage to aquifers, including compaction, which leads to land subsidence and further reduces groundwater storage capacity (Pittock et al. 2016).
- **Impact on Ecosystems**: The reduction in groundwater availability affects dependent ecosystems, including wetlands and riparian zones, which rely on consistent groundwater flows for survival (Bremer et al. 2021).

As per the International Atomic Energy Agency (IAEA), a substantial amount of groundwater in semi-arid areas is fossil water, which has accumulated over thousands of years and cannot be restored in the current climate. The unsustainable extraction of such resources poses long-term risks to the socio-economic stability of these regions.

2.2 Pollution from Anthropogenic and Natural Sources

In addition to depletion, groundwater in semi-arid regions is increasingly compromised by pollution from both human and natural sources. The unique geochemical environment of these regions, combined with anthropogenic pressures, creates a challenging scenario for maintaining groundwater quality (Akhtar et al. 2021; Liu et al. 2022).

Anthropogenic Pollution

Anthropogenic pollution is primarily a result of industrial activities, agriculture, urbanization, and transportation, leading to the release of harmful substances like carbon dioxide, methane, pesticides, and heavy metals into the environment (Raheja et al. 2021; Derdour et al. 2023). This form of pollution is exacerbated by human population growth and economic development, which increase the demand for resources and energy. Figure 4 focuses on anthropogenic activities, showcasing how urbanization, agriculture, and industrial discharges contribute to groundwater contamination. These visuals underline the need for holistic pollution assessment and mitigation strategies.

- **Agricultural Runoff**: The excessive dependence on chemical pesticides and fertilizers in agriculture is a major cause of groundwater contamination. Elevated nitrate levels, a common consequence, pose severe health risks, including methemoglobinemia or "blue baby syndrome" (Akhtar et al. 2021).
- **Industrial and Urban Discharges**: Industrial activities discharge hazardous chemicals, including heavy metals, directly into soil and water systems. Urban wastewater, often untreated or poorly managed, further contaminates aquifers (Pashaeifar et al. 2021).
- **Improper Waste Disposal**: Unregulated dumping of solid and liquid wastes allows leachates to seep into groundwater, introducing toxins and pathogens (Jenifer and Jha 2022).

Natural Pollution

On the other hand, natural sources of pollution, such as wildfires, volcanic eruptions, and the decomposition of organic matter, also release pollutants into the atmosphere

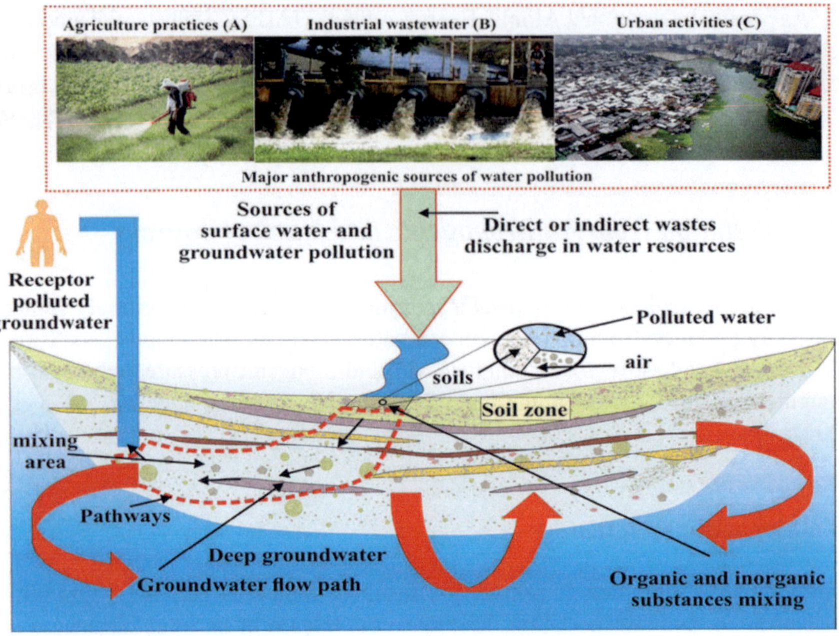

Fig. 4 Anthropogenic impacts on groundwater quality (Akhtar et al. 2021)

and ecosystems (Dao et al. 2024; Liu et al. 2022). While natural pollution occurs over long periods and is part of the Earth's natural cycles, human activities have significantly intensified its impact. Figure 5 depicts natural drivers and pathways impacting groundwater quality, highlighting processes such as soil and geological interactions, surface water intrusion, and climatic influences.

- **Salinization**: High evaporation rates in semi-arid regions exacerbate salinization, increasing groundwater salinity and rendering it unsuitable for drinking and irrigation.
- **Mineral Dissolution**: Geochemical interactions between groundwater and aquifer rocks often lead to elevated levels of natural contaminants, such as arsenic, fluoride, and iron.

The combination of anthropogenic and natural pollution poses a complex challenge to environmental sustainability and requires comprehensive strategies to mitigate and manage their effects. The intersection of these pollution sources necessitates comprehensive risk assessment frameworks. Explainable AI (EAI) can be particularly effective in identifying, predicting, and addressing contamination hotspots by providing interpretable insights into the factors driving groundwater pollution (Başağaoğlu et al. 2022; Nallakaruppan et al. 2024).

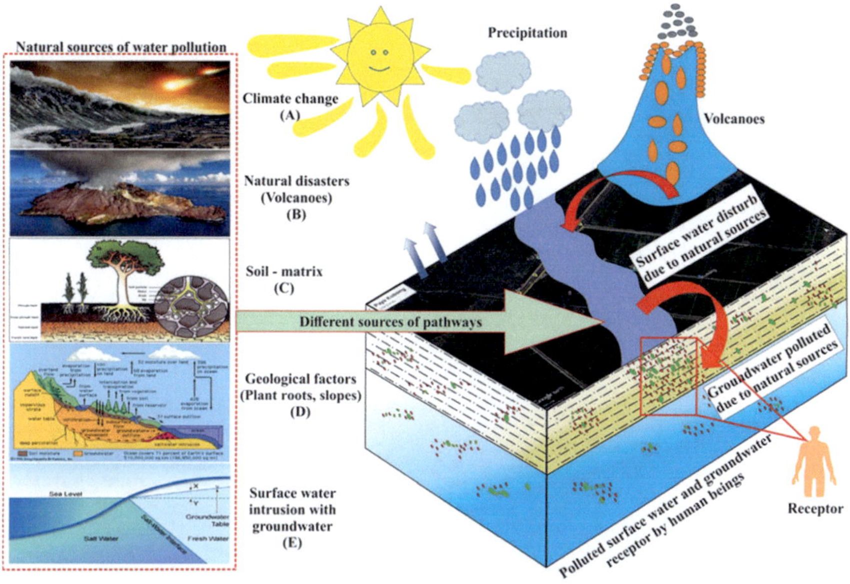

Fig. 5 Natural drivers and pathways impacting groundwater quality (Akhtar et al. 2021)

2.3 *Climate Change Impacts on Groundwater Recharge*

The difficulties in managing groundwater in semi-arid areas are further worsened by climate change. When temperatures rise, precipitation patterns change, and extreme weather events become more frequent, they all interfere with groundwater recharge processes, which reduces the amount of water available (Dao et al. 2024; Huo et al. 2010). Figure 6a highlights human interventions, such as managed aquifer recharge and biofuel production, as potential strategies to mitigate climate impacts. Figure 6b explores how high-emission (RCP 8.5) and moderate-emission (RCP 4.5) pathways influence groundwater recharge rates, offering a comparative analysis of future scenarios. These figures emphasize the urgency of adaptive strategies to ensure groundwater sustainability under changing climatic conditions.

Key Impacts

1. **Reduced Recharge Rates**: Irregular rainfall and prolonged droughts reduce the amount of water percolating into aquifers. High evaporation rates further diminish recharge potential, compounding water scarcity (Wang et al. 2021).
2. **Increased Demand**: Rising temperatures and prolonged dry spells increase agricultural and domestic water demand, intensifying pressure on already stressed aquifers (Triana et al. 2020).
3. **Saltwater Intrusion**: In coastal semi-arid areas, sea-level rise exacerbates saltwater intrusion into freshwater aquifers, degrading water quality and making it unsuitable for human use (Dao et al. 2024).

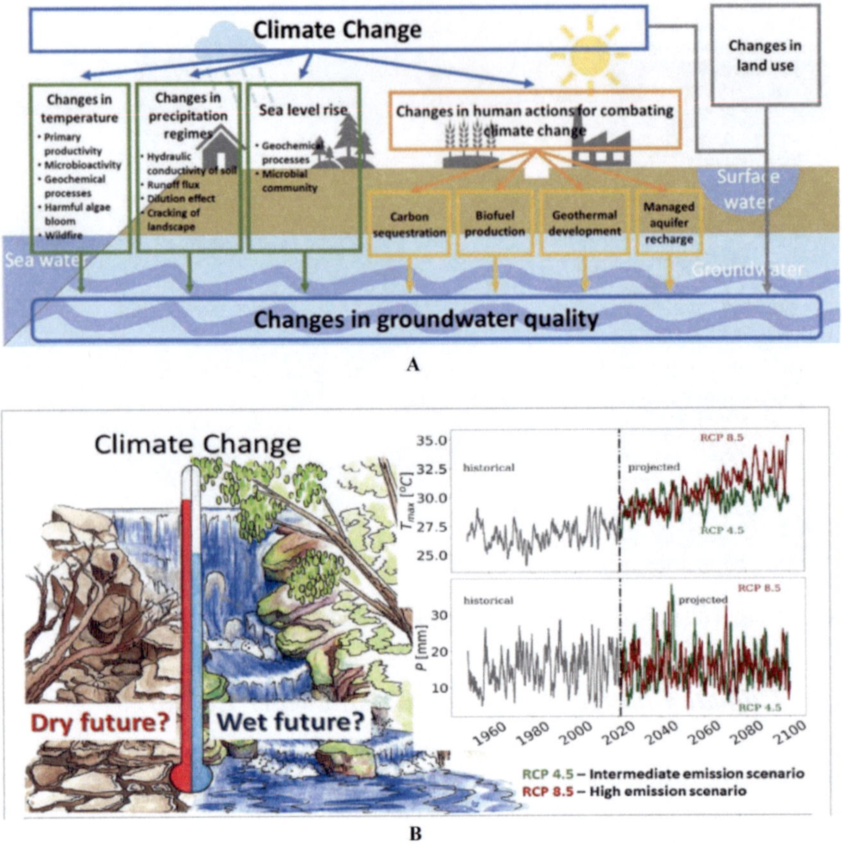

Fig. 6 **a b** Climate change impacts on groundwater quality and quantity (Dao et al. 2024)

4. **Geochemical Alterations**: Changes in groundwater temperature and recharge patterns can alter hydrogeochemical equilibria, mobilizing of harmful contaminants such as arsenic and manganese (Liu et al. 2022; Pashaeifar et al. 2021).

2.4 *Integrated Solutions for Semi-Arid Groundwater Management*

The over-extraction, pollution, and climate change interplay demands integrated, science-based solutions. Advanced technologies, like ML, DL and EAI, offer transformative potential for addressing these challenges (Shaikh 2024; Zaresefat and Derakhshani 2023).

- **AI-Powered Risk Assessment**: AI can analyze large datasets to identify pollution hotspots, predict depletion trends, and assess the vulnerability of aquifers to climatic changes (Camacho 2023).
- **Sustainable Water Use**: Smart irrigation systems powered by ML algorithms can optimize water use in agriculture, reducing over-extraction (Sahoo et al. 2017).
- **Stakeholder Engagement via EAI**: EAI's interpretability ensures that groundwater management strategies are transparent, fostering trust and collaboration among stakeholders (Linardatos et al. 2020; Nallakaruppan et al. 2024).
- **Scenario Modeling and Forecasting**: AI modeled climate and urbanization scenarios can guide policymakers in developing adaptive, forward-looking strategies for groundwater sustainability (Bremer et al. 2020; Rolnick et al. 2022).

3 Explainable Artificial Intelligence (EAI): A Primer

The term Explainable Artificial Intelligence (EAI) describes a group of techniques and tools designed to help human users comprehend and perceive the decision-making processes of AI systems (Linardatos et al. 2020; Singh 2024). EAI addresses the "black box" problem inherent in many machines learning (ML) models, which often produce highly accurate predictions but lack interpretability. This lack of transparency poses challenges in critical domains such as healthcare, environmental management, and policy-making, where trust, accountability, and understanding are paramount. EAI enhances stakeholder trust and facilitates informed decision-making by bridging the gap between accuracy and interpretability (Başağaoğlu et al. 2022).

3.1 Fundamentals of Explainable AI

The essence of Explainable AI lies in its ability to provide insights into how AI models process inputs to generate outputs (Nallakaruppan et al. 2024). Figure 7 illustrates the workflow of Explainable AI, demonstrating how transparency is incorporated into machine learning models. The process begins with data preparation and extends to applying of EAI techniques, ensuring complex models are understandable and actionable for stakeholders. The framework showcases essential components for integrating transparency into advanced machine-learning models. The process starts with data preprocessing, involving cleaning, balancing, and normalizing raw datasets to ensure reliable inputs for model training. Feature selection identifies the most relevant attributes to improve model efficiency. A variety of machine learning techniques are used for model training, including Support Vector Machines, Random Forests, and Decision Trees (Raheja et al. 2021; Derdour et al. 2023). Explainable AI techniques such as LIME (Local Interpretable Model-agnostic Explanations) are used in order to address interpretability in black-box models. These methods extract insights from complex models, breaking down predictions into understandable elements (Singh

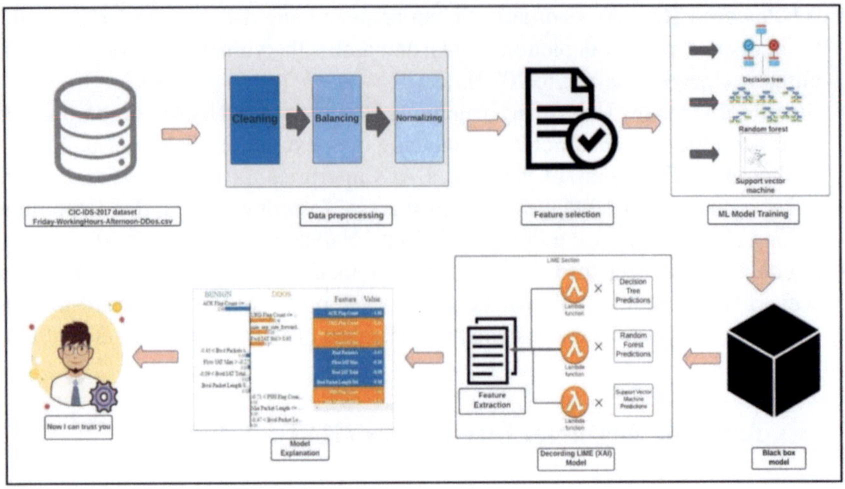

Fig. 7 Explainable AI workflow for enhancing transparency in machine learning models (Patil et al. 2022)

2024). This step enhances stakeholder trust by explaining how input features influence decisions. EAI bridges the gap between accuracy and interpretability, providing a robust framework for actionable insights in groundwater quality prediction and management (Shaikh 2024). This alignment of explainability with advanced learning models is crucial for addressing groundwater challenges in semi-arid regions. EAI focuses on demystifying complex algorithms while ensuring that predictive accuracy remains uncompromised. This is achieved through:

1. **Transparent Data Processing**: Initial steps involve data preprocessing, such as cleaning, balancing, and normalizing datasets, to provide robust inputs for model training. Feature selection is used to identify key attributes that contribute to predictive accuracy and efficiency.
2. **Model Training**: The preprocessed datasets are used to train a variety of machine learning algorithms, including Random Forests, Decision Trees, and Support Vector Machines (SVM). These models are often chosen for their performance but can be opaque regarding interpretability (Liu et al. 2022).
3. **Model Explanation Techniques**: EAI techniques such as LIME (Local Interpretable Model-agnostic Explanations) are used to overcome the interpretability issue with black-box models. LIME provides a condensed, understandable description of the decision-making process by altering input data and tracking changes in model predictions (Nallakaruppan et al. 2024).

Key Steps in LIME for Explaining Black-Box Models:

1. **Initialization**: A TabularExplainer is created with the training data or relevant statistics, along with metadata such as feature details and class names.

2. **Instance Explanation**: The explain_instance method generates explanations for specific predictions, detailing the influence of input features. These explanations are divided into:
   - **Prediction Probabilities**: Displayed on the left, showing the likelihood of each class.
   - **Feature Importance**: A middle section visualizing the top 13 features influencing the prediction. Features for one class are highlighted in orange, while those for the opposite class are in blue.
   - **Feature Values**: A list of the actual values for the top features, aiding interpretability.

LIME's ability to break down predictions into comprehensible components ensures transparency, making it a powerful tool for AI-driven groundwater management systems (Başağaoğlu et al. 2022; Singh 2024).

3.2 *Techniques for Model Interpretability*

Prominent techniques for enhancing model interpretability include:

1. **SHapley Additive exPlanations (SHAP)**:

SHAP assigns importance values to each feature in a prediction using cooperative game theory principles. It offers:

- **Global Interpretability**: Understanding overall model behavior and feature interactions.
- **Local Interpretability**: Explaining individual predictions, makes identifying key drivers of specific outcomes (Nallakaruppan et al. 2024).
- **LIME (Local Interpretable Model-agnostic Explanations)**:
- LIME approximates complex models with interpretable ones around a specific prediction. By examining how feature perturbations affect outcomes, LIME highlights the contribution of each feature to a prediction (Singh 2024).

These techniques have been widely applied across healthcare and environmental management domains. For instance, SHAP and LIME can help identify critical factors influencing water availability, enabling proactive management in groundwater quality prediction (Shaikh 2024).

3.3 *Relevance of EAI in Environmental Decision-Making*

EAI is pivotal in addressing environmental challenges, particularly in semi-arid regions where resources are scarce and decision-making is complex (Dao et al. 2024). Figure 8 presents a framework for integrating EAI into hydroclimatic prediction models, emphasizing data collection, model training, and transparent interpretation using SHAP and LIME. This alignment of explainability with advanced analytics ensures actionable insights for addressing groundwater challenges in semi-arid regions. The image illustrates a comprehensive framework for integrating Explainable AI (EAI) into groundwater management models, emphasizing transparency and accuracy. It begins with collecting and curating hydroclimatic datasets to train AI models. Hyperparameter optimization using grid search and multi-fold cross-validation ensures the selection of optimal parameters, enhancing model reliability. The AI model is trained, validated, and assessed using statistical measures. Once validated, the model is employed to project hydroclimatic scenarios, including future groundwater levels under changing climate conditions. Finally, the framework transitions into EAI using SHAP (SHapley Additive exPlanations) analysis, where predictors are ranked based on their influence on the target variable, and inflection points are identified to understand thresholds for critical predictions. This step transforms the AI model into an interpretable tool, providing actionable insights while maintaining predictive accuracy, aligning with the chapter's goal of improving groundwater management in semi-arid regions. The transparency provided by EAI fosters trust among stakeholders, making it an invaluable tool for:

1. **Groundwater Management**: EAI enables decision-makers to understand the drivers of groundwater depletion and contamination, facilitating the development of sustainable water management practices (Shaikh 2024).
2. **Climate Adaptation**: By elucidating the factors influencing hydroclimatic predictions, EAI supports the formulation of policies that mitigate the impacts of climate change (Dao et al. 2024).
3. **Stakeholder Collaboration**: Techniques like SHAP and LIME allow non-technical stakeholders to visualize and comprehend model predictions, fostering inclusivity and consensus in decision-making (Linardatos et al. 2020; Singh 2024).

4 Developing Predictive Models for Groundwater Quality

The development of predictive models for groundwater quality is an essential component of sustainable water resource management (Adimalla 2020). These models integrate various factors, including physicochemical, meteorologic, hydro-climatic, hydrogeologic, hydrologic, topographic, land use, and site-specific attributes, to predict groundwater quality effectively (Eskafi et al. 2021; Jenifer et al. 2021). Advanced models powered by artificial intelligence (AI) and machine learning (ML) are being used more frequently to improve the accuracy of groundwater quality assessments as a result of the expanding effects of industrialization, agriculture, and climate change on water supplies (Liu et al. 2022; Shiri et al. 2021). Table 1 highlights a diverse range of predictors, capturing both natural and anthropogenic influences on groundwater quality. By leveraging these factors, predictive models

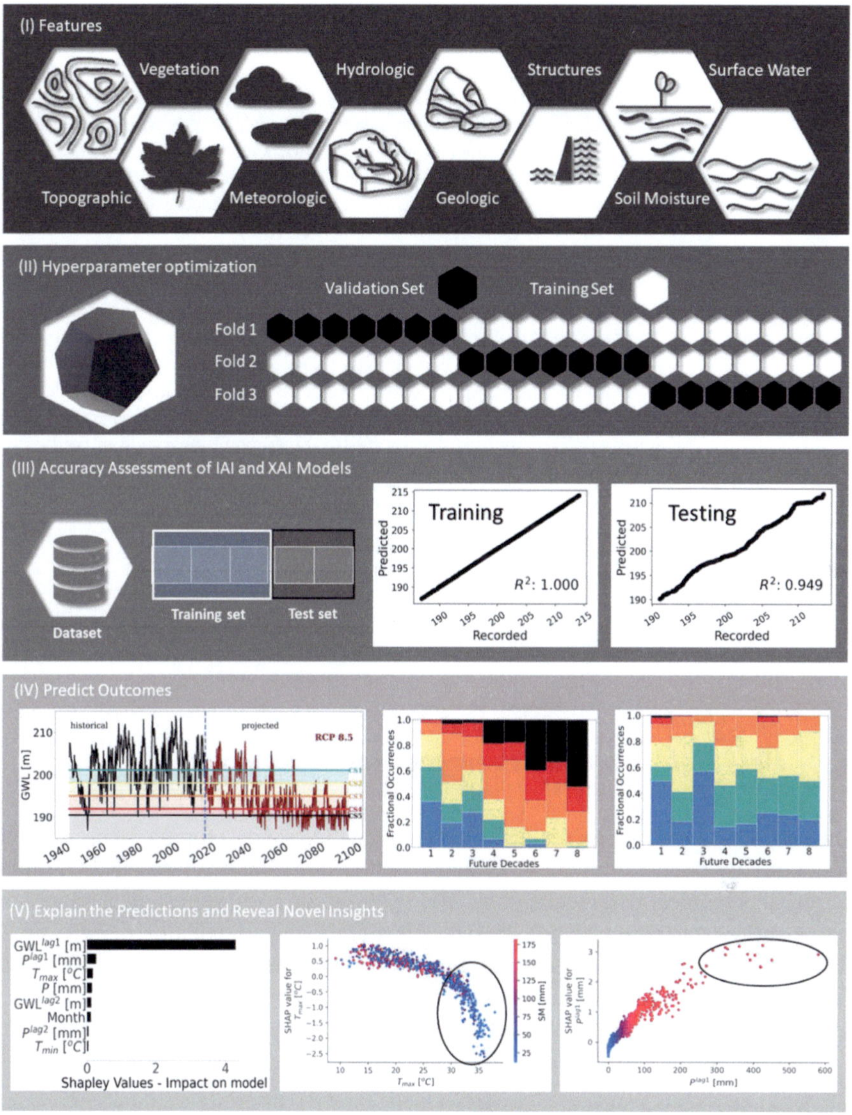

Fig. 8 Framework for explainable AI in hydroclimatic predictions and groundwater management (Başağaoğlu et al. 2022)

ensure a comprehensive assessment of groundwater contamination risks and overall quality (Derdour et al. 2023).

Table 1 Factors and predictors used in AI-based water quality predictions

Factors	Predictors
Physiochemical	pH, Total dissolved solids (TDS), Total hardness (TH), Cations, Anions, Total phosphorus (CP), Nitrate concentration (CNO3), Nitrite concentration (CNO2), Pesticide concentration (CP), Biochemical oxygen demand (BOD), Chemical oxygen demand (COD), Electrical conductivity (EC), Nitrite-Nitrogen (NO_2 ? N), Phospate ($PO_4^{3?}$), Surface water temperature (T_{SW}), Turbidity (NTU), Dissolved oxygen (DO)
Meteorologic	Precipitation (P), Temperature (T)
Hydro-climate	Evaporation (ET)
Hydrogeologic	Aquifer type, Aquifer transmissivity (T), Horizontal hydraulic conductivity (Kh), Vertical hydraulic conductivity (KV), Aquifer thickness (At), Aquitard thickness (Agt), Depth to water table(DWT), Groundwater level (GWL), Depth to screen well (DSW), Pumping capacity (Qp, max), Well density (WD), Pumping density (QP,d), Operation time length of wells (OW), Aquifer recharge (AR), Soil Type (ST)
Hydrologic	Streamflow (Qs), Stream length (STL)
Topographic	Altitude/elevation (AL), Land surface slope (SL)
Land use	Crop type, forest, urban residential land, pasture land
Site and problem specific	Presence of streams, distance from sea (DisS), Population density (PD), Distance to saline

Source (Başağaoğlu et al. 2022).

4.1 Data Collection and Preprocessing Techniques

Data collection and preprocessing form the foundation for developing robust groundwater quality prediction models. Effective data acquisition involves the integration of physicochemical parameters (e.g., pH, TDS, nitrates), meteorological data (e.g., precipitation, temperature), and site-specific factors (e.g., population density, proximity to saline sources) (Huang et al. 2019; Fakharian and Narany 2016). Data from different sources, including remote sensing, field measurements, and laboratory analyses, are aggregated to create a comprehensive dataset (Akhter et al. 2021).

Preprocessing Techniques:

1. Normalization: This process scales data to ensure uniformity, enabling the model to compare diverse parameters effectively. For example, physicochemical parameters like pH and electrical conductivity (EC) are normalized to a standard range.
2. Handling Missing Values: Missing data is a common challenge in groundwater studies. Techniques such as imputation (mean, median, or regression-based) are applied to fill gaps, preserving the dataset's integrity (Farzin et al. 2021).
3. Feature Extraction: This step identifies the most influential predictors, such as nitrate concentration and aquifer type, to reduce computational complexity and improve model accuracy (Nguyen et al. 2020).

Integrating Geographic Information Systems (GIS) with ML enables spatial visualization of groundwater quality, enhancing the understanding of contamination patterns (Jenifer and Jha 2022). Furthermore, entropy-based methods, such as the Entropy Water Quality Index (EWQI), reduce bias in parameter weighting, providing an accurate assessment of groundwater quality (Vadiati et al. 2019). The EWQI approach involves calculating each parameter's entropy weights and quality ratings, ensuring an unbiased evaluation.

4.2 Model Development for Predicting Contaminants

Predictive models for groundwater quality utilize state-of-the-art ML and deep learning (DL) algorithms to forecast contamination levels and groundwater quality indices (GWQIs) (Kulisz et al. 2021). Some commonly employed models include:

- **Artificial Neural Networks (ANNs)**: ANNs have demonstrated high accuracy in predicting groundwater contaminants. For instance, ANNs can model complex relationships between variables like nitrate concentration and aquifer properties (Shiri et al. 2021).
- **Random Forest (RF)**: This ensemble learning method is effective for groundwater quality prediction because it can handle large datasets and identify nonlinear relationships (Raheja et al. 2021).
- **Long Short-Term Memory (LSTM) Networks**: LSTMs are particularly useful for capturing temporal dependencies in groundwater quality data, enabling accurate predictions over time (Liu et al. 2022).

The integration of Explainable AI (EAI) techniques enhances model interpretability. By identifying the key predictors influencing groundwater quality, EAI aids stakeholders in understanding model outputs, making it easier to implement actionable solutions (Nallakaruppan et al. 2024; Singh 2024). For example, EAI can reveal how changes in precipitation or aquifer recharge rates affect nitrate levels in groundwater (Shaikh 2024).

4.3 Validation and Performance Metrics for Model Accuracy

Model validation ensures that predictions are reliable and applicable to real-world scenarios. Various performance metrics are employed to evaluate model accuracy:

1. **Root Mean Square Error (RMSE)**: Calculates the mean error between predicted and observed values.
2. **R-Squared (R^2)**: Assesses the proportion of variance the model explains.
3. **Receiver Operating Characteristic (ROC) Curve**: Evaluate the model's classification accuracy (Zhang et al. 2022).

Advanced models like Deep Neural Networks (DNN) and Extreme Gradient Boosting (XGBoost) have shown exceptional performance in groundwater quality prediction, yielding low RMSE values and high R^2 scores, indicating robust predictive capabilities (Liu et al. 2022; Shiri et al. 2021). Applying multi-criteria decision analysis (MCDA) further enhances groundwater quality assessments' reliability. MCDA provides a cohesive framework for evaluating model effectiveness across different datasets by integrating multiple performance metrics (Jenifer et al. 2021).

5 Application of Explainable AI in Groundwater Management

Integrating Explainable Artificial Intelligence (EAI) into groundwater management offers a revolutionary chance to improve model interpretability, model transparency, as well as understanding of complex groundwater data patterns (Başağaoğlu et al. 2022; Singh 2024). This section explores how EAI can be applied to improve stakeholder engagement, provide real-world case studies, and address the intricate patterns found in groundwater datasets.

5.1 Enhancing Model Transparency for Stakeholders

Enhancing model transparency for stakeholders, such as the public, water managers, and policymakers, is one of the main advantages of EAI in groundwater management (Linardatos et al. 2020). The complex nature of AI models often results in a "black-box" event, where people are unable to figure out the decision-making process, which might decrease user confidence and acceptance of these technologies. Stakeholders can understand the concepts behind forecasts and recommendations by applying EAI techniques, which give them insights into how AI systems make decisions (Nallakaruppan et al. 2024). This transparency is especially important for building public confidence and engagement in semi-arid areas where water shortage is a major concern. Moreover, public participation in groundwater management decisions is essential for effective resource utilization. When communities are involved in decision-making processes and can access understandable information, their awareness and engagement improve significantly (Shaikh 2024). EAI can facilitate this by providing clear explanations of AI-driven insights, thus empowering stakeholders to make informed decisions regarding groundwater management.

5.2 Case Studies: Interpretability in Real-World Scenarios

A number of case studies demonstrate how effective EAI is when used in groundwater management. For example, AI models have been employed in Texas to predict groundwater levels, demonstrating the importance of interpretability in real-world scenarios (Rohde et al. 2021). These models provided accurate predictions and included mechanisms for stakeholders to understand the underlying factors influencing groundwater fluctuations, such as climatic conditions and human activities. In Brazil, satellite-based AI techniques have been utilized to monitor large-scale groundwater resources, showcasing the flexibility of AI in handling diverse data inputs (Camacho 2023). These models' integration of EAI made it possible to spot important trends and deviations in groundwater data, which are necessary for successful management plans. The interpretability of these models enabled local authorities to respond promptly to changes in groundwater levels, thereby enhancing water resource management in semi-arid regions (Bremer et al. 2021).

5.3 Addressing Complex Patterns in Groundwater Data

Groundwater data is inherently complex, characterized by numerous variables and nonlinear relationships. EAI can significantly address these complexities by providing tools that elucidate the relationships between different predictors and groundwater levels (Shiri et al. 2021; Derdour et al. 2023). The impact of a number of variables, including irrigation techniques, precipitation, and land use, on groundwater dynamics can be recognized and explained using machine learning algorithms (Sahoo et al. 2017). Additionally, by facilitating the assessment of model performance and the detection of potential biases in data interpretation, EAI approaches can improve the resilience of prediction models (Nallakaruppan et al. 2024). This is particularly important in semi-arid regions where groundwater supplies are scarce and poor management can have serious consequences. By employing EAI, groundwater managers can better understand the implications of their decisions and the uncertainties associated with predictive models, leading to more sustainable water management practices. The application of EAI in groundwater management enhances model transparency and stakeholder engagement and provides valuable insights into complex groundwater data patterns. By integrating EAI with advanced learning models, water resource managers can improve decision-making processes and foster a more sustainable approach to groundwater management in semi-arid regions (Shaikh 2024).

6 Challenges and Ethical Implications

Integrating Explainable AI (EAI) with advanced learning models for groundwater management in semi-arid areas presents several challenges and ethical implications. These issues can be categorized into three main areas: limitations of current models and data availability, ethical considerations in AI for groundwater management, and mitigating risks related to model misinterpretation.

6.1 Limitations of Current Models and Data Availability

The limitations of existing models and the availability of high-quality data are two of the main obstacles to using AI for groundwater management (Dao et al. 2024; Liu et al. 2022). Numerous AI models currently in use primarily depend on historical data, which might not accurately predict future circumstances, particularly in semi-arid areas with high levels of climate unpredictability. Lack of high-quality, detailed data might make it difficult to create effective predictive models, which can produce inaccurate results. Furthermore, integrating AI with environmental science necessitates interdisciplinary collaboration, which can be hampered by differing terminologies and methodologies across fields (Green 2016). This lack of standardized data and models can result in inefficiencies and inaccuracies in groundwater management strategies.

6.2 Ethical Considerations in AI for Groundwater Management

The ethical implications of employing AI in groundwater management are multifaceted. As AI systems become more prevalent in decision-making processes, issues of transparency and accountability arise (Linardatos et al. 2020; Singh 2024). Users must understand the rationale behind AI-driven recommendations to trust and effectively implement these suggestions. The ethical principle of fairness is also critical; AI systems must be designed to avoid biases that could disproportionately affect vulnerable communities in semi-arid regions. Moreover, the potential for AI to exacerbate existing inequalities in resource distribution raises concerns about social justice and equity in environmental management (Shaikh 2024).

6.3 Mitigating Risks Related to Model Misinterpretation

Misinterpretation of AI models poses significant risks in groundwater management. Users may place undue trust in AI predictions without fully understanding the underlying assumptions and limitations of the models. This can lead to poor decision-making, mainly if the models are not adequately explained or if users lack the necessary expertise to interpret the results (Başağaoğlu et al. 2022). To mitigate these risks, it is essential to implement EAI techniques that provide clear, interpretable insights into model predictions (Nallakaruppan et al. 2024). By fostering a better understanding of AI outputs, stakeholders can make more informed decisions that align with sustainable groundwater management practices.

While integrating EAI with advanced learning models offers promising avenues for improving groundwater management in semi-arid areas, Addressing the challenges associated with and ethical impacts of new technologies is essential. Ensuring data availability, promoting ethical AI practices, and enhancing model interpretability are crucial to achieving effective and equitable groundwater management solutions.

7 Future Directions and Recommendations

7.1 Innovations in Explainable AI for Environmental Applications

The integration of Explainable Artificial Intelligence (EAI) within environmental applications is poised to revolutionize groundwater management, particularly in semi-arid regions (Shaikh 2024; Singh 2024). EAI enhances the transparency and interpretability of AI models, which is crucial for stakeholders involved in environmental decision-making (Linardatos et al. 2020; Başağaoğlu et al. 2022). The transformative potential of AI in environmental science is underscored by its real-time data analysis and predictive capabilities, which can significantly improve resource management strategies (Rolnick et al. 2022). For instance, AI technologies can optimize water usage by predicting groundwater levels and identifying areas at risk of depletion, thereby facilitating proactive management (Rohde et al. 2021). Moreover, the application of EAI can help in understanding the implications of AI-driven decisions, ensuring that stakeholders can trust and effectively utilize these technologies in sustainable practices (Nallakaruppan et al. 2024). Furthermore, the development of AI-driven solutions that are not only efficient but also environmentally sustainable is essential. Innovations in AI, such as machine learning algorithms tailored for environmental monitoring, can provide actionable insights into groundwater quality and availability (Liu et al. 2022; Shiri et al. 2021). These advancements can be complemented by interdisciplinary approaches that fuse AI with environmental science, leading to innovative solutions for pressing challenges like climate change and biodiversity loss (Dao et al. 2024).

7.2 Policy Recommendations for Sustainable Groundwater Management

To ensure sustainable groundwater management, it is imperative to establish comprehensive policies that incorporate the dynamic and spatial characteristics of groundwater resources (Bremer et al. 2021). Effective management requires clear property rights and enforceable regulations that consider the physical properties of groundwater (Sears et al. 2018). EAI tool integration should be a top priority for policymakers in order to improve decision-making procedures and give stakeholders the ability to understand how different management approaches might affect groundwater sustainability (Linardatos et al. 2020; Singh 2024). Also, the development of adaptive management strategies that adapt to changing environmental conditions depends on the promotion of collaboration between researchers, government agencies, and local communities (Bremer et al. 2020). Policies should also emphasize the importance of data sharing and transparency, allowing for better-informed decisions that align with sustainable development goals (Gaffoor et al. 2020). By leveraging AI technologies, policymakers can create frameworks that address current groundwater challenges and anticipate future demands and environmental changes (Camacho 2023).

7.3 Opportunities for Global Scalability

The scalability of AI-driven systems for groundwater management has great prospects for worldwide application, particularly in locations facing similar environmental concerns (Triana et al. 2020). As semi-arid areas often share common characteristics, the methodologies developed for one region can be adapted and implemented in others, enhancing global resilience to water scarcity (Dao et al. 2024). The use of AI in optimizing agricultural practices, such as precision farming, can lead to significant improvements in water use efficiency, which is critical in water-scarce regions (Sahoo et al. 2017; Zhang et al. 2022). Additionally, the global nature of climate change necessitates collaborative efforts to share knowledge and technologies across borders (Rolnick et al. 2022). AI technologies can facilitate this by providing platforms for data exchange and collaborative modeling efforts that address shared water resource challenges. The potential for AI to enhance decision-making in water management can be harnessed through international partnerships focusing on sustainable practices and resource optimization (Shaikh 2024; Zaresefat and Derakhshani 2023).

Integrating EAI with advanced learning models offers a promising pathway for improving groundwater management in semi-arid areas. By fostering innovation, implementing robust policies, and exploring global scalability, stakeholders can work towards sustainable solutions that address current and future groundwater management challenges.

8 Conclusions

Integrating Explainable Artificial Intelligence (EAI) with sophisticated machine learning (ML) and deep learning (DL) models offers a game-changing prospect for groundwater management, especially in semi-arid areas. This conclusion encapsulates the key findings from recent studies, envisions a future where EAI is seamlessly integrated into groundwater management practices and outlines a pathway toward sustainable and transparent water resource management.

8.1 Recap of Key Findings

Recent advancements in AI have demonstrated their potential to enhance groundwater management by improving predictive accuracy and decision-making processes. For instance, studies have shown that AI models can effectively simulate groundwater quality and predict contamination levels, facilitating better resource management (Zaresefat and Derakhshani 2023; Shiri et al. 2021). The use of ensemble frameworks and advanced modeling techniques has further refined groundwater potential mapping, which is crucial for assessing and managing these vital resources. Moreover, incorporating EAI into these models ensures that stakeholders can understand and trust AI-driven decisions, fostering greater acceptance and collaboration in groundwater management initiatives.

8.2 Vision for Integrating EAI with ML/DL in Groundwater Management

The vision for integrating EAI with ML and DL in groundwater management revolves around creating systems that predict outcomes and provide transparent explanations for their predictions. This integration is important in order to increase user trust and support informed decision-making among stakeholders, such as local communities and policymakers. Effective and flexible management methods can be achieved by groundwater managers through the use of EAI techniques like Shapley Additive Explanations (SHAP) and Local Interpretable Model-agnostic Explanations (LIME), which provide insights into the elements impacting groundwater dynamics. Furthermore, participatory engagement processes can enhance public understanding and involvement in groundwater management, making the science behind AI applications more accessible.

8.3 *Pathway to Sustainable and Transparent Water Resource Management*

A multi-faceted approach is necessary to achieve sustainable and transparent water resource management. This includes continuously monitoring groundwater resources, utilizing AI for real-time data analysis, and implementing community engagement strategies to raise awareness and foster collaboration. Establishing robust AI models capable of handling the complexities of groundwater systems is crucial, as these models can provide actionable insights that inform policy and management decisions. Additionally, integrating social transparency into AI systems can enhance the interpretability of AI-driven choices, ensuring that stakeholders are informed and empowered to participate in the management process. Ultimately, the pathway to sustainable groundwater management lies in leveraging advanced AI technologies while prioritizing transparency and community involvement.

The integration of EAI with ML and DL models holds significant promise for advancing groundwater management practices in semi-arid regions. By fostering transparency, enhancing predictive capabilities, and engaging communities, we can pave the way for more sustainable and effective management of this critical resource.

References

Adimalla N (2020) Groundwater quality for drinking and irrigation purposes and potential health risks assessment: A case study from semi-arid region of South India. Environ Geochem Health 42:2007–2025. https://doi.org/10.1007/s10653-019-00393-3

Afrifa S, Zhang T, Appiahene P, Varadarajan V (2022) Mathematical and machine learning models for groundwater level changes: a systematic review and bibliographic analysis. Futur Internet 14(9):259

Akhtar N, Syakir Ishak MI, Bhawani SA, Umar K (2021) Various natural and anthropogenic factors responsible for water quality degradation: a review. Water 13(19):2660

Akhter G, Ge Y, Iqbal N, Shang Y, Hasan M (2021) Appraisal of remote sensing technology for groundwater resource management perspective in Indus basin. Sustainability 13(17):9686. https://doi.org/10.3390/su13179686

Başağaoğlu H, Chakraborty D, Lago CD, Gutierrez L, Şahinli MA, Giacomoni M, Furl C, Mirchi A, Moriasi D, Şengör SS (2022) A review on interpretable and explainable artificial intelligence in hydroclimatic applications. Water 14(8):1230

Bierkens M, Wada Y (2019) Non-renewable groundwater use and groundwater depletion: a review. Environ Res Lett 14(6):063002

Bremer LL, Hamel P, Ponette-González AG, Pompeu PV, Saad SI, Brauman KA (2020) Who are we measuring and modeling for? Supporting multilevel decision-making in watershed management. Water Resour Res 56(1):e2019WR026011

Bremer L, Elshall A, Wada C, Brewington L, Delevaux J, El-Kadi A, Voss CI, Burnett K (2021) Effects of land-cover and watershed protection futures on sustainable groundwater management in a heavily utilized aquifer in Hawai'i (USA). Hydrogeol J 29(5):1749–1765

Camacho C (2023) Large-scale groundwater monitoring in Brazil assisted with satellite-based artificial intelligence techniques. Water Resour Res 59(9)

Dao PU, Heuzard AG, Le TXH, Zhao J, Yin R, Shang C, Fan C (2024) The impacts of climate change on groundwater quality: a review. Sci Total Environ 912:169241
Derdour A, Abdo H, Almohamad H, Alodah A, Dughairi A, Ghoneim S, Ali E, Hussein E (2023) Prediction of groundwater quality index using classification techniques in arid environments. Sustainability 15(12):9687
Eskafi M, Rahimi H, Faridhosseini A, Tang X (2021) Physical–chemical evaluation of groundwater quality in semi-arid areas: case study-Sabzevar plain, Iran. Sustain Water Resour Manag 7(6). https://doi.org/10.1007/s40899-021-00576-y
Fakharian K, Narany T (2016) Multidisciplinary approach to evaluate groundwater salinity in Saveh plain, Iran. Environ Earth Sci 75(7)
Farzin M, Avand M, Ahmadzadeh H, Zeleňáková M, Tiefenbacher J (2021) Assessment of ensemble models for groundwater potential modeling and prediction in a karst watershed. Water 13(18):2540
Gaffoor Z, Pietersen K, Jovanovic N, Bagula A (2020) Big data analytics and its role to support groundwater management in the southern African development community. Water 12(10):2796
Green TR (2016) Linking climate change and groundwater. In: Jakeman AJ, et al. Integrated groundwater management. Springer, Cham
Hssaisoune M, Bouchaou L, Sifeddine A, Bouimetarhan I, Chehbouni A (2020) Moroccan groundwater resources and evolution with global climate changes. Geosciences 10(2):81
Huang F, Zhang Y, Zhang D, Chen X (2019) Environmental groundwater depth for groundwater-dependent terrestrial ecosystems in arid/semiarid regions: a review. Int J Environ Res Public Health 16(5):763
Huo Z, Feng S, Kang S, Mao X, Wang F (2010) Numerically modelling groundwater in an arid area with ANN-generated dynamic boundary conditions. Hydrol Process 25(5):705–713. https://doi.org/10.1002/hyp.7858
Jenifer M, Jha M (2022) A novel GIS-based modeling approach for evaluating aquifer susceptibility to anthropogenic contamination. Sustainability 14(8):4538
Jenifer M, Jha M, Khatun A (2021) Assessing multi-criteria decision analysis models for predicting groundwater quality in a river basin of south India. Sustainability 13(12):6719. https://doi.org/10.3390/su13126719
Kulisz M, Kujawska J, Przysucha B, Cel W (2021) Forecasting water quality index in groundwater using artificial neural network. Energies 14(18):5875
Linardatos P, Papastefanopoulos V, Kotsiantis S (2020) Explainable AI: a review of machine learning interpretability methods. Entropy 23(1):18
Liu C, Xu M, Liu Y, Li X, Pang Z, Miao S (2022) Predicting groundwater indicator concentration based on long short-term memory neural network: a case study. Int J Environ Res Public Health 19(23):15612. https://doi.org/10.3390/ijerph192315612
Nallakaruppan MK, Gangadevi E, Shri ML, Balusamy B, Bhattacharya S, Selvarajan S (2024) Reliable water quality prediction and parametric analysis using explainable AI models. Sci Rep 14(1):7520
Nguyen P, Ha D, Nguyen H, Phong T, Trinh P, Al-Ansari N, Le HV, Pham BT, Ho LS, Prakash I (2020) Improvement of credal decision trees using ensemble frameworks for groundwater potential modeling. Sustainability 12(7):2622
Nur A, Ruswandi D (2022) Groundwater level forecasting in the Jakarta groundwater basin. Int J Geophys 2022:1–12
Pashaeifar M, Dehghanzadeh R, Ramazani M, Rafieyan O, Nejaei A (2021) Spatial and temporal assessment of groundwater quality and hydrogeochemical processes in Urmia lake basin, Iran. Water Sci Technol Water Supply 21(8):4328–4342. https://doi.org/10.2166/ws.2021.180
Patil S, Varadarajan V, Mazhar SM, Sahibzada A, Ahmed N, Sinha O, Kumar S, Shaw K, Kotecha K (2022) Explainable artificial intelligence for intrusion detection system. Electronics 11(19):3079
Pittock J, Hussey K, Stone A (2016) Groundwater management under global change: sustaining biodiversity, energy and food supplies. In: Jakeman AJ, Barreteau O, Hunt RJ, Rinaudo JD, Ross

A (eds) Integrated groundwater management. Springer, Cham. https://doi.org/10.1007/978-3-319-23576-9_4
Raheja H, Goel A, Pal M (2021) Prediction of groundwater quality indices using machine learning algorithms. Water Pract Technol 17(1):336–351
Rohde MM, Biswas T, Housman IW, Campbell LS, Klausmeyer KR, Howard JK (2021) A machine learning approach to predict groundwater levels in California reveals ecosystems at risk. Front Earth Sci 9:784499
Rolnick D, Donti PL, Kaack LH, Kochanski K, Lacoste A, Sankaran K, Ross AS, Milojevic-Dupont N, Jaques N, Waldman-Brown A, Luccioni AS, Bengio Y (2022) Tackling climate change with machine learning. ACM Comput Surv (CSUR) 55(2):1–96
Sahoo S, Russo T, Elliott J, Foster I (2017) Machine learning algorithms for modeling groundwater level changes in agricultural regions of the U.S. Water Resour Res 53(5):3878–3895
Sears L, Lim D, Lawell C (2018) The economics of agricultural groundwater management institutions: the case of California. Water Econ Policy 4(3):1850003
Shaikh M (2024) Artificial intelligence in groundwater management: innovations, challenges, and future prospects. Int J Sci Res Arch 11(1):502–512
Shiri N, Shiri J, Yaseen Z, Kim S, Chung I, Nourani V, Zounemat-Kermani M (2021) Development of artificial intelligence models for well groundwater quality simulation: different modeling scenarios. Plos One 16(5):e0251510. https://doi.org/10.1371/journal.pone.0251510
Singh D (2024) Enhancing transparency and interpretability in deep learning models: a comprehensive study on explainable AI techniques. Interantional J Sci Res Eng Manag 08(02):1–13
Triana J, Chu-Agor M, Guzmán J, Moriasi D, Steiner J (2020) Evaluating the risks of groundwater extraction in an agricultural landscape under different climate projections. Water 12(2):400
Vadiati M, Nalley D, Adamowski J, Nakhaei M, Asghari-Moghaddam A (2019) A comparative study of fuzzy logic-based models for groundwater quality evaluation based on irrigation indices. J Water Land Dev 43(1):158–170
Wada Y, Beek L, Kempen C, Reckman J, Vasak S, Bierkens M (2010) Global depletion of groundwater resources. Geophys Res Lett 37(20). https://doi.org/10.1029/2010gl044571
Wang W, Zhang Z, Yin L, Duan L, Jinting H (2021) Topical collection: groundwater recharge and discharge in arid and semi-arid areas of china. Hydrogeol J 29(2):521–524. https://doi.org/10.1007/s10040-021-02308-0
Zaresefat M, Derakhshani R (2023) Revolutionizing groundwater management with hybrid ai models: a practical review. Water 15(9):1750
Zhang J, Jing H, Dong K, Jin Z, Ma J (2022) The effect of drip irrigation under mulch on groundwater infiltration and recharge in a semi-arid agricultural region in china. Water Sci Technol Water Supply 22(4):4043–4054. https://doi.org/10.2166/ws.2022.033

Artificial Intelligence and Machine Learning Applications in Groundwater Pollution Control and Resource Management

A. Saranya, Vedprakash Maralapalle, Aijaz Ahmad Zende, Shital Marlapalle, and Rajiv Iyer

Abstract Managing and protecting this crucial resource has become even more difficult due to industrialization and agricultural and urban hazards such as direct groundwater pollution. For many environmental factors, conditions complicating heterogeneity and instability hinder the application of conventional ΓM techniques to pollute groundwater. AI and ML-integrated solutions provide unprecedented solutions to these challenges with better means of predicting, detecting, and managing contaminated water sources. More specifically, this chapter discusses how advanced technologies such as AI and ML are integrated into efforts to regulate contamination in groundwater and improve its management. It equally examines different types of AI-based models applicable to risk estimation of contamination, effective pollution interventions, and enhanced organizational decisions. Furthermore, the chapter presents successful examples of AI and ML applications for monitoring groundwater quality, identifying pollution sources, and developing sustainable management strategies. In this chapter, the authors demonstrate how AI and ML technologies can change the approach to groundwater pollution prevention and the management of water resources.

Keywords Groundwater pollution · AI and machine learning · Environmental management · Predictive modeling · Contamination risk assessment · Pollution control strategies · Groundwater quality monitoring

A. Saranya · R. Iyer
Department of CSE, Amity University Mumbai, Mumbai, India
e-mail: saranyaa@mum.amity.edu

V. Maralapalle (✉)
L. S. Raheja School of Architecture, Bandra, Mumbai, India
e-mail: civilved@gmail.com

A. A. Zende
Department of Civil Engineering, BLDEA's V P Dr P G Halakatti College of Engineering and Technology, Vijayapur, India

S. Marlapalle
Vivekanand Education Society's College of Architecture, Mumbai, India

A. Nath et al. (eds.), *Groundwater Depletion and Sustainability*, Advances in Geographic Information Science, https://doi.org/10.1007/978-3-032-09921-1_4

1 Introduction

AI and ML advance groundwater management by offering unique and groundbreaking approaches relevant to the environment. These emergent technologies have abilities in observation, modeling, and managing aquifer pollution and overextraction that have not been seen before. Current problems that have arisen following the management of modern groundwater include pollution, impacts from climate change, and shortage of resources (Li et al. 2021). These new forms of addressing challenges replace conventional first-channel strategies by offering full-scale data linkages, continual surveillance, and projections. For the improved and efficient formulation of enhanced groundwater flow models, machine learning algorithms incorporate multiple data types, such as satellite images,

The simulation of the probability of such outcomes regarding groundwater levels and diffusion and spread of contaminants (Samantaray et al. 2021). Using IoT-related gadgets and AI models to monitor the quality in real-time shows a need to identify areas that are likely to be affected by pollution. Improving the arrangements for allocation of the resources and models for extraction of the resources Sophisticated risk evaluation and control tools, Unity Engine Risk assessment and management is another area of improvement for those firms that seek an edge over the others. As such, advanced tools for risk evaluation and control are now available. Thus, the participation of artificial intelligence in decision support of groundwater management introduces a shift, the shift that makes the stakeholders develop more preventive and suitable approaches. Through machine learning, researchers and policymakers are better placed to produce more accurate estimates, apply appropriate therapies, and build cooperative forums that interchange scientific analysis and applicable administration.

1.1 Importance of Groundwater Resources

The importance of Groundwater cannot be overemphasized because it is the world's most crucial water resource that supports human, agricultural, and livestock drinking needs and wildlife consumption. It contains about 99% of all accessible freshwater used in various sectors and domains worldwide. Groundwater serves multiple crucial functions: provides potable water to somewhere in the region of 50% of humanity and supplies approximately 40 percent of the irrigation needs of crops, promotes the health of an ecosystem, and the stability of the environment helps to regulate river and wetland water during dry periods despite its importance, groundwater faces significant threats to over-exploitation by the increasing population and climate change rising industrialization, and expanded agriculture specifically, how climate change affects natural recharge mechanisms (Xiang et al. 2021). The sustainable management of resources is a favorable idea and, therefore, calls for innovation. Artificial intelligence and machine learning offer transformative solutions by increasing the possibility of accurate assessment of the character of waters in the aquifer contamination risks prediction the identification of sophisticated methods for predicting resource demands they recommended the development of real-time monitoring methods of aquifer stock health. Combining AI technologies with earth observation systems is a key approach to mitigating the problem of groundwater depletion so that this vital but often overlooked resource continues to be safeguarded for future generations.

1.2 Challenges in Groundwater Pollution Management

MCP involves complex problem-solving, and it requires credible technological solutions to address the diverse issues that characterize groundwater pollution. Some major issues cause significant and critical challenges to the stability and quality of groundwater.

a. Industrial Discharges: Inefficient hazardous waste dumped by textile industries, chemical industries, and pharmaceutical companies pollutes water sources.
b. Agricultural Practices: Using fertilizers and pesticides leads to water pollution by nitrates and phosphates Fed.
c. Urbanization: Poor waste disposal results in untreated wastewater and solid wastes seeping into water-bearing formations.
d. Land Use Changes: Through many examples, the author can clearly illustrate why deforestation is a problem within their analysis and how it destroys the natural ability of soil to filter pollutants.

Groundwater pollution presents complex management challenges, lack of knowledge about groundwater systems, unpredictability in tracking and monitoring contamination, slow up and down and sideways shifting of water in aquifers, and challenges in diagnosing or eradicating subsurface pollution (Huang et al. 2021). Depletion of water sources contaminates pollutants when they are restricted in predefined areas as extraction goes further; heavy metal pollution raises natural mineral interactions and how it has to do with geogenic pollution and many diseases, especially prevalent throughout the third world. Addressing these challenges requires enhancing corporate governance systems and governance structures for biologically friendly methods of farming. They noted that current monitoring technologies have become more complex, and advanced public awareness campaigns have integrated management plans.

1.3 The Role of AI and ML in Environmental Management

AI and ML are revolutionizing environmental management by offering intelligent patterns of measuring, analyzing, anticipating, and solving numerous ecological issues, especially in groundwater management. AI technologies enable advanced environmental interventions through clear water usage dissection and real-time quality monitoring, influencing Confounding Factors Predictive Contamination Risk Assessment resource decision-making processes. Machine learning techniques offer unprecedented capabilities in groundwater management, analyzing disparate data, satellite images, geological maps, and sensor systems to make the right predictions of the groundwater flow. They include the following: Vulnerable areas with high chances of pollution, identifying the dispersion of contaminants (Singha et al. 2021). AI-driven approaches revolutionize environmental monitoring by considering multiple channels feeding into the evaluation for data consolidation, creating accurate real-time tracking software development of a conceptual framework for forecasting ecosystem health, and facilitating preventable intervention plans. The incorporation of AI with earth observation systems is considered a major revolution in the management of the environment. The use of more extensive and sophisticated algorithms and, approaching the data processing, the scientists potent to design the most delicate and elaborate methods for groundwater protection and less pollution.

2 Overview of AI and Machine Learning

Artificial Intelligence (AI) and Machine Learning (ML) are frontier technologies that introduce intelligence to systems that learn from the environment and make decisions on suitable actions to take in managing groundwater resources (Shah et al. 2021). Artificial intelligence is an extensive discipline that seeks to bring forth intelligent machines capable of undertaking human-like tasks. Artificial intelligence has several subfields of which machine learning forms a core segment that allows the computer to learn from data without coding. Some use structural calculation methods to analyze enormous amounts of data and make correct presumptions.

Machine learning approaches offer multiple strategies for environmental management:

a. **Supervised learning**: The choice of training algorithms with the help of the labeled datasets.
b. **Unsupervised learning**: Relating to the discovered features, there is a need to identify the underlying data structures.
c. **Reinforcement learning**: Preliminary Action Plan to transform UOP performance through feedback mechanisms.

In Fig. 1, the real-time monitoring of water quality risk prediction of contamination is enhanced by multidimensional data integration, a system of ecosystem health monitoring. AI and ML also consider complex environmental information that results in the proactive conservation of groundwater (Barzegar et al. 2018). These are among the most essential up-and-coming technologies for coping with contemporary ecological issues, providing fresh ways to solve problems involved in managing natural resources.

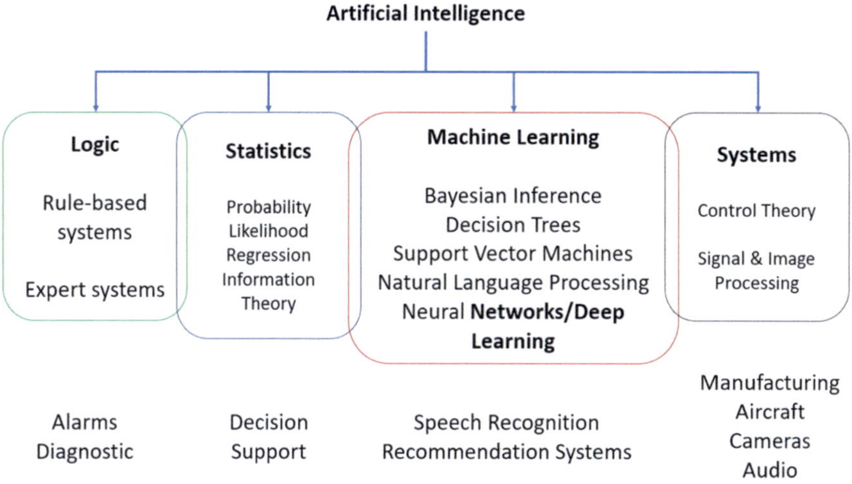

Fig. 1 Overview of AI and machine learning

2.1 *Fundamentals of AI and ML*

Machine Intelligence (MI) and Machine Learning (ML) are complex computational tools that help in designing intelligent systems and getting accustomed to their environment. They can also solve problems of various extents of groundwater management. AI fundamentally consists of key technological elements and problem-solving algorithms. As such, data computation is an interpretative mechanism in data processing new ways of learning its ability to use adaptive decision-making. He has provided an account of its ability to use adaptive decision-making capabilities as a strength.

Machine learning encompasses three primary approaches:

a. **Supervised Learning**: In predictive modeling, utilizes the labeled/ supervised
b. **Unsupervised Learning**: Learned how to discover new left–right data patterns that are not labeled.
c. **Reinforcement Learning**: Italicizes strategies using various feedback formation loops.

AI and ML technologies operate through sophisticated processes and neural networks that replicate the structure of other biological learning networks, some of the most complex operations involved in data processing and cleaning techniques information processing of the inferred algorithmic pattern recognition continuous model refinement (Chakraborty et al. 2020). In environmental contexts, these technologies enable contamination risk assessment with high accuracy water quality assessment in real-time using behavioral data analysis Resource Management enhanced Resource

Management Techniques. AI and ML have the tremendous potential of giving sophisticated computational tools to the understanding and control of complex groundwater systems, which are needed to meet some of the lingering ecological problems.

2.2 AI and ML Techniques in Environmental Science

Big data Artificial Intelligence and machine learning have emerged as strategic tools in environmental science as they allow the identification of the most suitable solutions to ecological problems.

Machine learning provides multiple advanced methodologies for environmental analysis:

a. **Neural Networks**: Borrowing from the natural ways of learning systems.
b. **Support Vector Machines (SVM)**: Pattern relationship complexity.
c. **Random Forest Algorithms**: Enabling practical construction of accurate predictive models.
d. **Deep Learning Architectures**: Analyzing complex environmental data sets

These techniques enable innovative environmental management strategies, environmental Time series analysis, probability distribution of system transitions, and prediction of the evolution of the system. Image enhancement techniques for observing the pattern of the environment in terms of challenge, the structuring of optimal resource allocation strategies is a major category.

AI methods in environmental science have transformed dramatically:

a. **1980s**: Rule-based algorithm present in expert systems.
b. **1990s–2000s**: Traditional statistical and modern artificial neural network models.
c. **Current Era**: The application of multiple layers of artificial neural networks as well as methods of ID technique integration

The groundwater management feasibility is the deciding factor in the Groundwater Management Potential. These include the accuracy of contamination risk assessment and water quality monitoring in real time. Coupled with this, a view was created for predictive modeling of the aquifer dynamics. There are challenging aspects of data integration from various sources of Complexity (Muharemi et al. 2019). Here, data integration from various datasets is complex. These sorts of complex computational methods allow researchers to design harder, predictive methods for addressing and managing many environmental issues.

2.3 Integration with Groundwater Management

The governmental and non-governmental organizations worldwide are implementing AI & Machine Learning for groundwater management as they offer technically advanced solutions to several environmental issues. AI technologies enable comprehensive groundwater management through data integration methodology that does not segregate components of data that could be aggregated. Predictive modeling per the concept of advanced analytics real-time monitoring, and analysis preemption is a way of identifying the contaminants in the water.

In Fig. 2, machine learning techniques offer unprecedented advantages in groundwater management, integrating data captured by the sensors, other remote real-time data, and real-time data that the system had recorded, determining complicated relationships linking different sets of data Hy (packaged). Developing integrated hydrogeological models to reduce uncertainty in the simulations of groundwater flow should be achieved by doing the following: AI-driven approaches provide critical environmental management capabilities in the case of groundwater what is most important is the high accuracy of forecasting its levels identifying initial exposure threats depending on the present resource function, and resource allocation strategies should be optimized enhancing the company's ability to respond quickly to changes in the environment (Yaseen 2021). The integration of AI with groundwater management involves sensors installed in home-based Internet of Things (IoT) networks, machine learning algorithms, continuous model refinement, and self-organized decision-making processes. AI technologies promise transformative potential in groundwater management by developing a joint working environment and giving stakeholders the information they need to act on. Examples of improvement through innovation include Resource sustainment, developing new and more sensitive environmental management mechanisms. The advanced computational techniques can provide more effective and preventable groundwater conservation and pollution control strategies to the researchers.

3 AI and ML in Groundwater Pollution Prediction

Artificial intelligence (AI) and Machine Learning are perfecting the science of predicting groundwater pollution from complex models and systems, thus making environmental management more effective. Machine learning offers sophisticated approaches to groundwater pollution prediction, getting to grips with Complex Hydrogeological Patterns, determining temporal-spatial contamination patterns, and estimating the likelihood of pollution dangers (Singh and Sharma 2023). It involves integrating one or many other datasets, containing environmental information.

3.1 Predictive Modeling for Contamination Risks

AI and Machine Learning have transformed the prognosis of possible groundwater pollution threats to develop complex environmental control strategies. Through

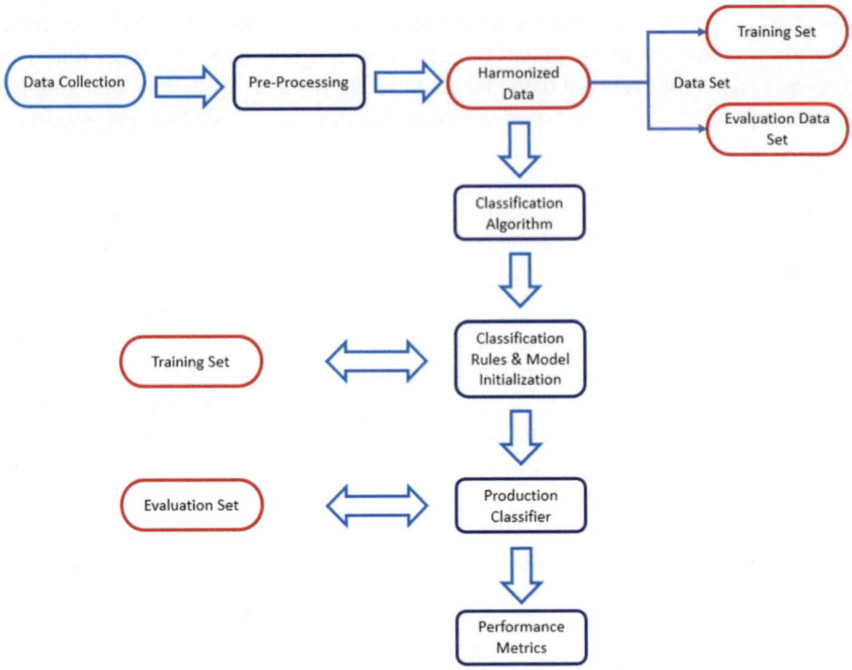

Fig. 2 Integration with groundwater management

access to high-level computational methods, the authors can design and implement more effective, granular, and responsive prevention and control of groundwater contamination, thus revolutionizing how people think about environmental conservation.

3.2 AI-Driven Detection of Pollution Sources

The ability to detect pollution sources through AI is a revolutionary way of managing groundwater to detect the origins of contamination (Iyer et al. 2024). Technological breakthroughs in computational sciences provide potential to build improved, flexible, precision-based approaches for prospection and control of groundwater pollution sources, redefining and revolutionizing environmental watch and preservation procedures.

3.3 Case Studies in Groundwater Pollution Prediction

AI and Machine Learning systems have shown excellent performance in predicting groundwater pollution through practical application in different geographical areas.

a. **IIT Kharagpur's Arsenic Detection Model**: The model successfully predicted the concentration of arsenic concentration in the groundwater in the context of the Ganges delta.
b. **Saudi Arabian Water Quality Index Prediction**: The hybrid machine learning approach for water quality analysis also incorporated.
c. **Iranian Groundwater Level Prediction**: Applied accurate neural network algorithm for monitoring underground water level.

Case study methodologies involved increased use of more sophisticated and the forecasting methods for true-time monitoring strategies, and this kind of filter requires more complex algorithm techniques (Anne et al. 2023). In Fig. 3 the research demonstrated critical capabilities for self-identifying the areas that emit pollutants Paleo Economic Quantifying human health risks generating exact contamination maps Importantly, they help support targeted remediation strategies. Case studies revealed AI's potential to forecast groundwater contamination trajectories, making it possible to start the use of early warning systems following Policy-making procedures, a sound approach to the allocation of resources (Bhunia and Chatterjee 2023). Scholars are changing how the prediction of groundwater pollution is done using complex computation methods, with an endemic view to efficient and effective protection of the environment.

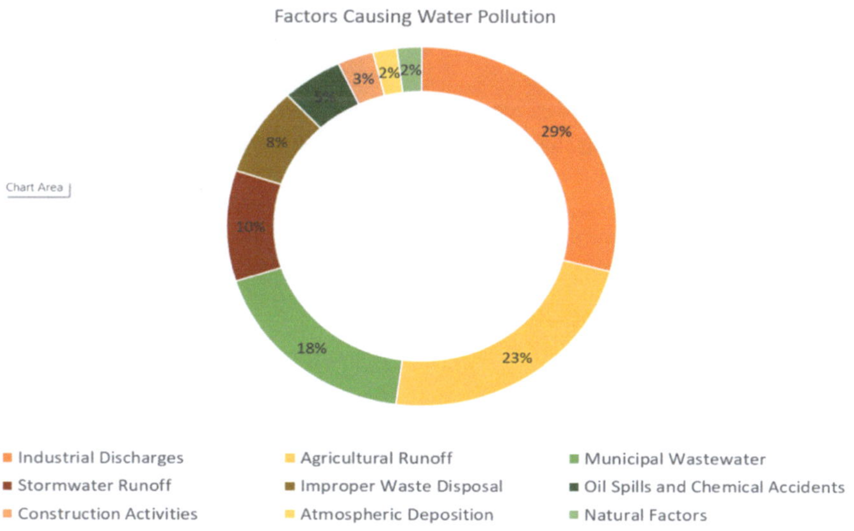

Fig. 3 Factors causing water pollution

4 Optimization of Pollution Control Strategies

AI combined with Machine Learning is enhancing the modern approaches to pollution control with a specific focus on groundwater through advanced optimization. Advanced AI techniques enable comprehensive pollution control through Precise predictive modeling, Observation and control systems, active in real-time Special approaches, dynamic resource allocation Machine learning optimization strategies including fully automated pump schedule optimization People's interchange and merger, interaction, and union, have minimization effect on the environment of those countries energy Conservation methods of the ecosystem (Talukdar et al. 2024).

AI-driven optimization provides critical environmental management capabilities. One goal of mining management is to avoid over-extraction of the groundwater resources. Vulnerability assessment of hazardous contamination areas developing adaptive management protocols enabled supporting the proactive decision-making process. Optimization techniques leverage Complex data integration, Multi-dimensional methods of environment scanning, and Iteratively learning mechanisms (Kumari and Pandey 2023). By introducing sophisticated computational methods, scholars can create enhanced interaction and accurate approaches to regulate groundwater contamination, improve the processes of environmental asset maintenance, and manage these issues efficiently. These advanced optimization algorithms represent a significant step in the company's commitment to the effective management of ecology so that the needful can be done for ecology while fulfilling the needs of the people.

4.1 AI and ML for Monitoring and Early Warning Systems

AI and Machine Learning are changing groundwater monitoring in a way that effective early warning systems for environmental management can work. Real-time monitoring technologies provide critical environmental insights through IoT multiplex sensing systems. Prolonged monitoring of the river water quality is highly effective in attack discovery. Machine learning enables sophisticated detection strategies and real-time analysis of the sensor's data, enumerating typical trends of water quality and catalyzing prompt contamination notifications. IT can predict potential risks damaging the environment. AI-driven monitoring systems offer unprecedented advantages. It focuses on monitoring small variations of the water parameters (Elshall et al. 2022). Multidimensional data processing Complex multidimensional data is one of the most important indicators for determining a given software environment's performance and efficiency rates. integrating the model into the breast cancer algorithm Generates accurate predictive models Understanding supporting intended proactive intervention measures. Advanced systems monitor critical indicators, pH levels, chemical concentrations, temperature variations, and dissolved oxygen content. Using advanced mathematics, academics can construct dynamic early indicators that revolutionize groundwater control and abatement by allowing for a decisive, pinpoint reaction to possible hazardous conditions (Bagheri et al. 2019). All these innovative monitoring technologies in the application of artificial intelligence are a significant step up in the safety measures offered to stakeholders in the environmentalism of resources.

4.2 Decision-Support Tools for Pollution Management

AI decision support systems have greatly enhanced groundwater pollution and control since they bring along modern, smart approaches to environmental management. Decision-support tools enable advanced pollution management through monitoring systems in real-time. This paper will also cover other related areas, including predictive contamination modeling (Abbaszadeh Shahri et al. 2022). AI-driven tools provide critical environmental management capabilities managing multi-parametric hydro geologic data Self Contamination Zones increasing understanding of the Effects and Contamination Zones the definition of contamination, and the importance of the vulnerable contamination zones can be elaborated as follows (+) P–Creating accurate forecasting models the main reason behind translating these measures is to support stakeholder decision-making. Advanced decision-support systems leverage machine learning algorithms IoT sensor networks multi-source data analysis dynamic environment modeling. Decision-support tools can effectively help predict the dangers of contaminating groundwater. For example, to achieve operational efficiency, organizations must improve availability and avoid wasting time, money, or effort to support regulatory compliance (Ahmed et al. 2019). These techniques enable AI tools to enhance groundwater pollution management and provide state-of-the-art approaches for accurate environmental protection. These unique technologies are a major achievement in environmental management since they would allow stakeholders to make intelligent decisions on water management for sustainability from available data.

4.3 Case Studies on Optimization of Control Measures

AI and Machine Learning have been incredible in performing case studies on rationalizing environmental control profiles of groundwater pollution across various environments. Optimization case studies revealed critical capabilities for the reduction of energy consumption during the pumping of wastewater from boreholes, stopping unreasonable exploitation of water sources, and defining Zones of Contamination vulnerability underpinning targeted formative interventions. Advanced AI techniques enabled writing decisions for real-time contamination monitoring dynamic resource allocation. Contemplative modus operandi of action-oriented intervention provisions Conservation measures of ecosystems. Case studies demonstrated AI's potential to this results in generating of the best pump schedules. The Regulation shall help to achieve an appropriate balance between human demand for water and conditions in the environment, decreasing possible undesirable impacts on the ecosystem, and Sustainable water resource management. Through the use of advanced computational methods, authors are revolutionizing the controlling of groundwater pollution hence significantly improving the knowledge relating to environmental management and conservation.

5 AI and ML in Groundwater Quality Monitoring

Modern computing methodologies are changing AI and Machine Learning in Groundwater Quality Monitoring, where precise proactive management of the environment is achievable. Machine learning provides sophisticated monitoring approaches to live water quality estimation. According to the among value, it further offers continuous contamination detection (Baduge et al. 2022)—risk assessment for different types of pollution. Key technological innovations include implementing Internet of Things (IoT) sensor networks Neural network pattern recognition flexible 'sophisticated modes of operation' Buffers, bound, and change attractiveness scores. AI-driven systems enable quick identification of water quality trends, monitoring the fluctuations in the amount of chemicals present, discovering the factors that may cause contamination, and contributing to the identification of special population student needs for focused support efforts Emerging Research Perspectives. Recent studies demonstrate AI's potential in forecasting Groundwater Quality Indices Explaining sophisticated hydrogeological information maximizes the selection of monitoring parameters for fast response capacity to these disasters, adaptable management protocols should be set with the help of high technologies, a breakthrough in the field of water quality investigation is being observed, which can significantly improve the outlook on the environment and resources. These emerging AI tools are a significant revolution in the sphere of eco-consciousness, as they help different actors, users, or decision-makers make intelligent decisions based on analyzed information.

5.1 Real-Time Monitoring Systems

Real-time monitoring systems refer to progressive management of groundwater systems with the help of artificial intelligence and machine learning for efficient environmental conservation. Real-time monitoring systems enable ongoing tests and monitoring of water quality assessment of contamination right after the event dynamic data processing Monitoring Techniques advanced systems utilize IoT sensor networks strategic sensor placement pluralistic data acquisition Google translator algorithmic applied processing by AI critical environmental indicators include pH levels chemical concentrations temperature variations dissolved oxygen content microbial pathogen presence, AI-powered real-time monitoring provides early warning mechanisms which means effective and fast response possibilities about the contamination risk assessment contaminating the environment is one of the main issues throughout the world and its assessment is a vital result of numerous studies prioritizations of resource deployment. Thus, this paper aims to discuss how real-time monitoring systems innovate groundwater management by combining complex computational methods and innovative sensor technologies (Solomatine et al. 2008). Such measures help management create preventive rather than responsive strategies for environmental conservation and optimal resource use. Groundwater monitoring's future is found in these innovative self-adaptive systems that can yield a better and more comprehensive understanding of the ecological condition.

5.2 Data Analysis and Interpretation

With the help of AI and Machine learning, it is possible to exploit data and its analysis in effective groundwater management to comprehensively understand environmental degradation. Machine learning provides sophisticated data analysis approaches interacting with extensive multi-dimensional hydrogeological data sets to determine complex environmental structures. This consists of relating many different ecological variables to each other, lowering the level of chance in the simulation of groundwater key data interpretation methodologies including any type of pattern recognition that uses neural networks tree-based AI models Data from multiple sources Statistical regression analysis. Critical Analysis Parameters climate change incidence analysis climate analysis the demographic change correlation a record on the ration of water extraction. AI-driven data analysis enables, therefore, the following sub-section will present an investigation of the factors that can be used for the prediction of the variation in the groundwater level; the four essential elements of contamination risks include enhancing the approaches to proactive management creating usable information about the environment the current study demonstrates how high computation can revolutionize the analysis of groundwater data with the ultimate goal of improving the understanding of intricate environmental frameworks toward effective resource management.

5.3 Case Studies in Groundwater Quality Assessment

AI and Machine Learning are driving radical changes in groundwater quality analysis and, consequently, in its control and prediction methods by applying new technologies. In various ways, these technologies provide innovative methods for remediation and management of contaminated groundwater. Using machine learning models gives extraordinary potential in observing, estimating, and controlling groundwater actions.

a. **Real-Time Monitoring**: AI techniques applied with IoT devices to monitor water quality facilitate constant monitoring of groundwater quality for effective response in case of pollution.
b. **Data Integration**: Machine learning approaches compile multiple datasets to provide increased levels of detail in hydrogeological models and improve understanding of groundwater systems.
c. **Optimization Algorithms**: Self-learning technology can help determine the most efficient pumping time, conserve energy, and avoid over-pumping wells.
d. **Hybrid Machine Learning Models**: The Authors have found that integrating ANN, SVM, and GPR enhances predictive accuracy.
e. **Vulnerability Assessment:** Agent-based models can advise on places that are likely to be sources of air and water pollution and natural resource exploitation.

6 AI and ML for Sustainable Groundwater Management

Sophisticated calculations are used to predict the extent of use of groundwater and thus avoid its use in the future. These models solve issues such as rainfall distribution, variations in land utilization, and extraction rates in an exhaustive capacity to predict groundwater behavior. Spreading Internet of Things (IoT) sensors and AI algorithms will make it possible to monitor the groundwater quality in real time, which can be used for quick reaction to the threats connected with contamination. This technology enables continuous data collection, such as pollution detection from potential pollution sources and constant change of management approaches. AI methods involve presenting various forms of data, and this serves the following purposes improve the reliability of the hydrogeological model optimize pumping schedules, and this serves the following purposes: improve the reliability of the hydrogeological model, optimize pumping schedules, respect recycling, Lower energy use in the pumping of the groundwater resources. Artificial intelligence methods may detect the areas that require protection from pollution, contribute to growing an effective preventive approach, and provide inclusive interfaces for the interaction between scientists, authoritative decision-makers, and citizens. These technologies, through refined data analysis, avail data that can inform sustainable utilization and management of this vital resource. It is about time water resource management leveraged AI's strengths to create better, stronger, and more sustainable methods of groundwater use.

6.1 Development of Sustainable Management Plans

Machine learning (ML) and artificial intelligence (AI) are the key forms that are helping reinvent groundwater management through new approaches in the domain of prediction, monitoring, and control. These technologies provide integrated tools for the management of problems associated with groundwater pollution and availability. ML techniques provide advanced capabilities for groundwater quality assessment, including prediction of Ground Water Table and Contamination threats. Monitoring reporting in real time also through the use of IoT devices and AI algorithms, the day-to-day application of hydrogeological information needs to be enhanced through the integration of other categories of data that will help in optimizing hydrogeological models. AI-driven models can use evaluation to examine intricate hydrogeological designs and groundwater level predictions with a high degree of accuracy to determine where the environment is most susceptible to pollution and exhaustion many of the solutions should allow for proactive intervention measures to be put into practice. Researchers are developing sophisticated ML approaches, including neural network applications in groundwater quality modeling. There are two categories of machine learning-based approaches: Gaussian process regression, adaptive neuro-fuzzy inference systems, and decision trees. New approaches mainly unveil unsupervised and deep learning algorithms to boost the data analysis. These technologies support sustainable management by pumping time schedules of the well Environment Protection Organization (EPO) reporting the early presence of bacteria in a vessel utilizing data for decision-making, this knowledge confirms that sustainable and innovative development of AI and ML will pave a new way of managing the future of groundwater.

6.2 Integration with Policy and Regulatory Frameworks

Realizing the potential of artificial intelligence (AI) and machine learning (ML) methodologies in policy and regulatory contexts for sustainable groundwater resources and reduced pollution. They offer tremendous and heretofore undisputed opportunities for predictive Modeling and Early Warning Systems. AI-driven models can revolutionize regulatory approaches by determining the amount of time with high precision for forecasting groundwater levels. The problem lies in identifying areas that could be easily contaminated to facilitate rapid response before such pollution is commonly realized to be severely problematic.

Real-time supervision and Data Synonymous ML algorithms enhance regulatory effectiveness through geophysical monitoring and constant evaluation of the quality of water within aquifers. A way of quickly identifying that the contaminant levels are deviant is the integration of data from another continuous source. The last strategic initiative of the COP is the optimization of resource management. AI technologies support sustainable policy implementation by planning on the ideal frequency of the pump. Strategies for a business include reducing exposure to environmental risks and exhausting options to a bare minimum water requirements for human beings, which must be measured with the necessity of protecting the environment. Smart applications and sophisticated methodologies, including predictive analytics models, IoT-based monitoring, and other superior algorithms, help the stakeholders of this industry to get valuable solutions. These technologies enable better and more accurate hydrogeological forecasts, provide tools for scientific collaboration between scholars and policymakers, and help develop preventive environmental policies.

6.3 Long-Term Benefits and Challenges

Machine learning and Artificial Intelligence are, fascinating tools used in analyzing and providing new solutions to problems related to groundwater resources management. These technologies bring long-term values in line with other types of long-term costs and Log Hell risks for implementation. AI-driven innovations provide unprecedented capabilities in groundwater management Predictive Modeling has the superiority of AI algorithms can predict the groundwater levels, and this information can be helpful in the decision-making process on resource deployment these models involve assessment of prior data and intricate hydrogeological behaviors that assist in avoiding over-exploitation of water and forecasting likely scarcity in the future real-time monitoring AI endless scanning helps in the identification of contamination threats as they emerge Operations can efficiently be designed to sound alarms when concentrations of contaminants go high to protect people's health, Despite promising potential, AI groundwater management faces several critical challenges Technical Expertise Requirements deploying complex AI models requires expertise, which is in keeping with the understanding of artificial intelligence data Limitations to achieve good performance from the AI, high-quality datasets are required and extensive data is usually complex to find model Interpretability since the algorithms used by intelligent systems are rather complicated, their understanding may be an issue when going mainstream. The management of groundwater in the future should be a coordinated effort of artificial intelligence platforms that integrate science, policy, and people.

7 Emerging Trends and Future Directions

AI and Machine Learning Adaptation for Controlling Ground Water Pollution and its Management. The opportunity to manage groundwater resources is being revolutionized by artificial intelligence and machine learning and with its favorable trends for pollution control and sustainable management. These approaches completely transform how we can assess, forecast, and manage the threats posed to groundwater resources. Neural networks and deep learning algorithms have become more complex and powerful, so machine learning tools are helping to create new, more accurate models of groundwater quality. Researchers are developing predictive models that can forecast groundwater levels. The methodology yields high accuracies from the above research; this study finds out that the following areas are most vulnerable to pollution and gives information on concentration levels of contaminants in the environment in real-time, contributing to the decisions about the efficiency of the overall usage of resources and the extraction systems.

New machine learning methods are applying several algorithmic methods with some of them using the approach of a combination of methodologies. For instance, researchers combine Non-parametric Gaussian Kernel learning Fuzzy adaptive neuro networks, and Decision tree algorithms. AI's potential in groundwater management relies on the shared platforms that strengthen the cooperation between the scientific community, policymakers, and residents. Advanced AI models will continue to enhance predictive accuracy, maintain environmental effects, and encourage other-directed preventive actions to implement steps to enhance the efficiency in the conservation of groundwater. As such, these technological developments point towards one of the ways to address sustainable water resource management as the environmental regime continues to shift.

7.1 Advances in AI and ML for Groundwater Management

AI and Machine Learning are reshaping the groundwater management field, which involves the use of sophisticated predictive modeling and monitoring tools. Such innovative techniques make it possible to develop integrated measures for protection against water pollution and efficient use of resources through predictive modeling. These models utilize statistics and advanced mathematics to predict the presence and volumes of water beneath the ground surface a task that has become more precise thanks to artificial intelligence by analyzing complex hydrogeological patterns, these models can quickly implications of the analysis will allow expecting fluctuations in groundwater access prevent over-extraction facilitate the practice of making decisions before the crisis hit real-time Monitoring Advanced AI technologies, combined with Internet of Things (IoT) sensors, provide critical capabilities that means learning about it early enough to be on the lookout for any contaminants that may cause harm to us on-going assessment of water quality response capacity of likely pollution threats machine learning techniques offer transformative potential in groundwater management by knowing areas that are most likely to be affected by pollution using resources to meet its goals comprehensively and efficiently using multiple datasets for analysis and data communication evidence shows that the efficiency of AI and ML algorithms can enhance the methods of groundwater management. In the Arkavathi Watershed study, machine learning techniques helped concern understand changes in the water level of aquifers where artificial recharge structures should be established to improve forecast capacity for aqueous levels, and the estimation of the extent of the recharge these innovations are a significant contribution to the sustainable water resource management because they equip the stakeholders with tools addressing increasing water problems.

7.2 Potential for AI-Driven Innovations

Groundwater management is improving of integrating artificial intelligence (AI) and machine learning (ML), which are coming up with unique solutions to environmental issues. These technologies have the potential to revolutionize approaches to foresee, control, and prevent aquifer pollution and exhaustion. From the present study, the use of state-of-the-art ML techniques in the form of neural networks, deep learning, and unsupervised algorithms are observed to model groundwater quality parameters. Interestingly, Iran and the United States of America are the pioneers of such technological interferences. Still, more emphasis is being placed on nitrate modeling and integrated approach to groundwater management.

7.3 Future Research Directions

Application of Artificial Intelligence and Machine Learning to Adaptation: Controlling Groundwater Pollutions and Management. Groundwater management in the future will depend on the effective use of artificial intelligence and machine learning that enable great changes in water pollution and resource management. Emerging research directions will focus on several key innovative approaches, and the articles all relate to the subtopics of predictive modeling and advanced techniques. Modern-day machine learning approaches are quickly improving and can produce better groundwater quality information. Future research will emphasize explaining deep learning and other AI mechanisms that could be put to practice enhancing better forecasting to complex hydrogeological structures extending modeling extent regarding little praised factors and specific areas of real-time monitoring with the capability for early detection Advanced AI systems will enable consequent sampling of the groundwater quality using IoT gadgets Contaminants of a nematic nature heavy metals, color, and nitrates, microbial contamination, resistant to certain disease germs causing auto signals for possible pollution threats fully Integrated Data Innovative AI approaches will minimize model data uncertainty in the conceptualization of groundwater flow this approaching enriches the hydrogeological models with input from other data source leads that may be more suitable for the task at hand resource utilization and protection and global resources have to be managed efficiently collaborative Platforms Future further exploration research will create AI-based systems that can integrate scientists; policymakers, and the local community for better project management to deal proactively with effectual groundwater management.

8 Conclusion

Due to technological development, specifically artificial intelligence and machine learning, groundwater has been efficiently managed through enhanced predictive models, monitoring, and data integration processes that solve the significant issues of pollution and sustainability perspectives. Novelties of Groundwater Care machine learning techniques have demonstrated remarkable capabilities in detect with high accuracy the fluctuation in the groundwater levels and the probability of contamination. For instance, Biocon cops with high-tech algorithms to allow for monitoring of water quality in the real-time application of multiple data inputs to achieve a high level of accuracy in hydrogeological modeling technological Advancements. Modern AI approaches leverage multiple techniques, including neural networks, deep learning algorithms, and non-parametric kernel Gaussian learning, possessing adaptability in the neuro-fuzzy environment. These AI-driven innovations provide transformative benefits that led to proactive contamination detection Swift intervention instruments recommending the right place to invest < input type = ” /text” id = ” resource allocation” name = ” resource allocation” placeholder = ”-” > reducing the negatives of the environment the mathematical approach that forms the basis for the calculations in groundwater flow simulations is written out in this article. Groundwater management in the future will be a collective effort of Artificial Intelligence that brings together scientists, policymakers, and the people on the ground. AI algorithms will advance even further; with better raster and vector predictions, precise vulnerability maps, and effective ways of sustaining water resource management we believe that thanks to modern machine learning approaches, it is possible to create better and more effective methods of using groundwater and preventing its polluting by water.

8.1 Summary of Key Findings

The specifications of the predictive modeling capabilities are subsequently presented machine classifiers are now capable of determining groundwater levels in reserves because they simulate intricate hydrogeological tendencies. These models take into account such factors as rainfall, changes in land use, and extraction rates to give a proper forecast of the state of the groundwater real-time surveillance and Pollutant Identification Advanced AI algorithms enable continuous monitoring of groundwater quality, allowing for multicomponent and rapid detection of contaminants that hit the permissible thresholds early disturbing intervention processes the tracking of water quality trends in operation, a clear distinction can be made between preventive action types that address potential risks data Integration and uncertainty reduction one of the most significant advantages is AI’s ability to integrate diverse data sources, which reduces the variability of parameters in the simulation of groundwater flow improve the reliability of the obtained hydrogeological models enhances predictions of the flow of groundwater and a contaminant plume. These include optimization, which is the scientific approach towards the implementation of a specific formula to improve production. At the same time, sustainability is the degree to which one is willing to endure in an undertaking to make significant progress in the achievement of their set goals and objectives AI optimization algorithms this page will help you construct ideal pumping routines, enter your details into the given boxes, and we will build the ideal pumping routine for you minimize energy consumption prevent over-extraction Preserving water sources for individuals who depend on wells for their water supply.

8.2 The Impact of AI and ML on Groundwater Pollution Control

Artificial Intelligence and Machine Learning are transforming the methods of groundwater pollution control by using improved prediction and monitoring tools. This paper demonstrates how cutting-edge advances in machine learning are revolutionizing our awareness of, and ability to forecast and control, groundwater pollution threats through real-time monitoring applying IoT sensors, AI systems embody permanent monitoring and provide information about the state of groundwater. These technologies can also allow the detection of contaminant deviations in real-time a prompt warning for either heavy metals, nitrates, or microbial pathogens, and can track environmental changes with hitherto unknown levels of accuracy advanced Analytics Techniques researchers are developing specialized ML tools for groundwater management, including we also used unsupervised learning for the detection of the spatiotemporal contamination patterns supervised learning to predict the contaminant concentrations Algorithms for solving the optimal allocation of needs and resources with the protection of ecosystems innovative Prediction Methods there is a discussion showing the potential for using recent non-parametric kernel Gaussian learning, neuro-fuzzy and neuro-wavelet models, based upon the decision tree algorithm, which identified higher levels of accuracy than parametric models when attempting to predict the water quality indices.

8.3 Recommendations for Future Applications

Machine learning and artificial intelligence are the new hopes for groundwater management because they offer new solutions that counter the complicated problems faced by the environment. Future applications should focus on several key strategic developments with the use of advanced evolving techniques and concepts in predictive modeling. AI-driven predictive models can revolutionize groundwater management, which would make a small advance in forecasting future groundwater levels more accurately, determining places prone to pollution and loss of productivity. Advantages that enhance its application include the ones described below, promoting active prevention measures. Advanced Patient Checking Devices Next-generation AI applications should prioritize using IoT gadgets for real-time contamination detection. It also involves a more advanced algorithm for continuous quality monitoring of the implemented health care solutions and quick response procedures concerning possible environmental hazards; extended data convergence future research must concentrate on presenting an overview of machine learning packages for groundwater data analysis ready for field use development of usable Python/R tools for various environments reducing uncertainty in modeled end members using multiple data sources for simulating groundwater new technology in the field promising techniques include deep learning algorithms explainable artificial intelligence neural networks classifiers, decision trees and kernel-based technology integrated classifiers.

References

Abbaszadeh Shahri A, Shan C, Larsson S (2022) A novel approach to uncertainty quantification in groundwater table modeling by automated predictive deep learning. Nat Resour Res 31(3):1351–1373. https://doi.org/10.1007/s11053-022-10062-9

Ahmed AN et al (2019) Machine learning methods for better water quality prediction. J Hydrol 578:124084. https://doi.org/10.1016/j.jhydrol.2019.124084

Anne P et al (2023) The role of space-based observations for groundwater resource monitoring over Africa. Surv Geophys 44(1):123–172. https://doi.org/10.1007/s10712-023-09747-4

Baduge SK et al (2022) Artificial intelligence and smart vision for building and construction 4.0: machine and deep learning methods and applications. Autom Constr 141:104440. https://doi.org/10.1016/j.autcon.2022.104440

Bagheri M, Akbari A, Mirbagheri SA (2019) Advanced control of membrane fouling in filtration systems using artificial intelligence and machine learning techniques: a critical review. Process Saf Environ Prot 123:229–252. https://doi.org/10.1016/j.psep.2019.01.026

Barzegar R, Asghari M, Soleimani H, Nazari S (2018) Mapping groundwater contamination risk of multiple aquifers using multi-model ensemble of machine learning algorithms. Sci Total Environ 621:697–712. https://doi.org/10.1016/j.scitotenv.2017.11.198

Bhunia GS, Chatterjee U (2023) Groundwater depletion and climate change: Role of geospatial technology for a mitigation strategy. In: climate change, community response and resilience. Elsevier, pp 291–304

Chakraborty M, Hossain S, Dutta S, Chakrabarti A (2020) Modeling regional-scale groundwater arsenic hazard in the transboundary Ganges river delta, India and Bangladesh: infusing physically based model with machine learning. Sci Total Environ 748:141107. https://doi.org/10.1016/j.scitotenv.2020.141107

Elshall AS, Ye M, Wan Y (2022) Groundwater sustainability in a digital world. In: water and climate change. Elsevier, pp 215–240

Huang R, Zhang Y, Ma H, Zhao C, Chen J (2021) Machine learning in natural and engineered water systems. Water Res 205:117666. https://doi.org/10.1016/j.watres.2021.117666

Iyer R, Maralapalle V, Patil D, Irfan M (2024) The future of smart contracts: pioneering a new era of automated transactions and trust in the digital economy. In: Irfan M et al (eds), AI-Driven decentralized finance and the future of finance. IGI Global, pp 229–256. https://doi.org/10.4018/979-8-3693-6321-8.ch010

Kumari N, Pandey S (2023) Application of artificial intelligence in environmental sustainability and climate change. In: visualization techniques for climate change with machine learning and artificial intelligence. Elsevier, pp 293–316

Li L, Fu G, Gan Y, Zhang J, Zheng L (2021) Recent advances in artificial intelligence and machine learning for nonlinear relationship analysis and process control in drinking water treatment: A review. Chem Eng J 405:126673. https://doi.org/10.1016/j.cej.2020.126673

Muharemi F, Logofătu D, Leon F (2019) Machine learning approaches for anomaly detection of water quality on a real-world data set. J Inf Telecommun 3(3):294–307. https://doi.org/10.1080/24751839.2019.1639254

Samantaray S, Sahoo A, Ghose DK (2021) Watershed management and applications of AI. CRC Press

Shah MI, Javed MF, Abunama T (2021) Proposed formulation of surface water quality and modelling using gene expression, machine learning, and regression techniques. Environ Sci Pollut Res 28:13202–13220. https://doi.org/10.1007/s11356-020-10814-x

Singh D, Sharma V (2023) Transformative potential of AI and remote sensing in sustainable groundwater management. In: proceedings of the international conference on recent advancements in artificial intelligence. Springer Nature Singapore

Singha S, Roy P, Ghosh S, Das R (2021) Prediction of groundwater quality using efficient machine learning technique. Chemosphere 276:130265. https://doi.org/10.1016/j.chemosphere.2021.130265

Solomatine D, See LM, Abrahart RJ (2008) Data-driven modelling: concepts, approaches and experiences. In: practical hydroinformatics: computational intelligence and technological developments in water applications, pp 17–30

Talukdar S, Mondal S, Singh A, Paul S (2024) Recent trends in application of geospatial technologies and AI for monitoring and management of water resources. In: water resource management in climate change scenario: innovations in geospatial techniques and models. Springer Nature Switzerland, pp 1–11

Xiang X, Zhou W, Huang Z, Zhang X (2021) Urban water resource management for sustainable environment planning using artificial intelligence techniques. Environ Impact Assess Rev 86:106515. https://doi.org/10.1016/j.eiar.2021.106515

Yaseen ZM (2021) An insight into machine learning models era in simulating soil, water bodies and adsorption heavy metals: review, challenges and solutions. Chemosphere 277:130126. https://doi.org/10.1016/j.chemosphere.2021.130126

Groundwater Management in a Changing World: Geospatial Techniques, Environmental Risks, and Adaptive Solutions

Groundwater Resilience in the Era of Climate Change: The Promise and Perils of Advanced Technologies

Gouri Sankar Bhunia

Abstract The worldwide water supply depends heavily on groundwater yet this vital resource remains defenseless against climate change intensification. The investigation combines the analysis of climate change effects on groundwater systems with modern technological assessments of remote sensing along with Geographic Information Systems (GIS) and big data analytics and cloud computing and Artificial Intelligence (AI) and Machine Learning (ML) to boost groundwater sustainability. The study utilizes VoSviewer to analyze trends in scholarly literature from 2000 to 2024 and networks between authors, co-citations and performance indicators via Field Citation Ratio and Relative Citation Ratio based on a bibliometric review of this time period. The research shows that interdisciplinary studies using technology have dramatically increased their focus on addressing groundwater stress criteria. The research pinpoints that arid and semi-arid zones represent specific regional hotspot areas because climate change impacts recharge patterns and leads to increased salinization as well as heavy aquifer extraction. The combination of artificial intelligence modeling and remote sensing groundwater mapping together with IoT-enabled sensors provides essential services for monitoring and prediction as well as adaptive management systems. The progress of sustainable groundwater management is constrained because institutions block its implementation and some regions lack equal access to technology and their data exists in isolated pieces. The research demonstrates the need to combine digital tools in groundwater policy structures alongside international cooperative efforts to establish long-lasting hydro-governance infrastructure. The presented work provides essential data-supported foundations that lead to sustainable groundwater management policies under conditions of climate change pressure.

Keywords Groundwater sustainability · Climate change · Remote sensing · GIS · Big data · Cloud computing · Artificial intelligence · Machine learning · Hydrogeology · Bibliometric analysis

---

G. S. Bhunia (✉)
Independent Researcher, Kharagpur, West Bengal, India
e-mail: rsgis_gouri@rediff.com

A. Nath et al. (eds.), *Groundwater Depletion and Sustainability*, Advances in Geographic Information Science, https://doi.org/10.1007/978-3-032-09921-1_5

1 Introduction

Climate change refers to significant and enduring shifts in global temperatures and weather patterns, resulting not just from natural cycles but, more importantly, from human activities like industrial emissions, deforestation, and large-scale land use changes. Its impacts are profound, particularly on water resources. Today, about 4 billion people—over half the world's population—experience limited access to safe drinking water for at least one month each year. A large proportion, approximately 1.5–3 billion people, depend heavily on groundwater for their daily needs (Velis et al. 2017; López-Morales et al. 2017). Groundwater also plays a critical role in agriculture, with 60–70% of global groundwater withdrawals dedicated to irrigation. Currently, groundwater supplies about half of the world's potable water, acting as a lifeline for both human communities and ecosystems (Davamani et al. 2024). However, future projections are alarming: by 2050, 57% of the global population may reside in areas facing severe water scarcity (Boretti and Rosa, 2019). Climate change exacerbates this issue by increasing precipitation variability and intensifying extreme weather events like droughts and floods. During prolonged droughts, the over-extraction of groundwater heightens the risk of aquifer depletion, especially in arid and semi-arid regions where surface water sources are minimal (Dao et al. 2024). This depletion not only stalls economic development but also deteriorates health, educational opportunities, and agricultural productivity. Thus, safeguarding groundwater resources is becoming increasingly urgent, and climate change has been recognized by the Intergovernmental Panel on Climate Change (IPCC) as a critical threat to the sustainable management of groundwater (Intergovernmental Panel on Climate Change and Edenhofer 2014).

Several hydrological models exist today that researchers utilize for assessing climate change effects on both surface and groundwater resources. A significant focus of this research has been on groundwater recharge, which depends not only on hydrological processes like precipitation and evapotranspiration but also on surface characteristics such as land structure and soil properties. Historically, most studies examining the effects of climate change on the hydrological cycle have concentrated on surface water elements—including precipitation patterns, atmospheric moisture, evapotranspiration, river flow, and snow cover. This focus is largely because these surface processes are more visible and accessible, making them easier to observe, measure, and study (Abiy and Melesse 2017; Pitz 2016; Scanlon et al. 2012). In contrast, the hidden nature of groundwater systems has resulted in fewer investigations into how climate change impacts groundwater resources until more recently (Garner et al. 2017; Haldorsen et al. 2016; Hu et al. 2019; Taylor et al. 2013a, b). Recent review studies have increasingly turned attention to the physical changes induced by climate change, examining aspects like alterations in groundwater recharge and storage (Earman and Dettinger 2011; Ramasamy and Lee 2017; Smerdon 2017), shifts in temperature and precipitation patterns (Kurylyk et al. 2014; Ramasamy and Lee 2017), and transformations within the broader hydrological circulation (Refsgaard et al. 2016). This growing body of research underscores the complex and critical relationship between climate dynamics and subsurface water resources.

The intersection of climate change, cutting-edge technologies, and groundwater sustainability is rapidly emerging as a critical area of study and innovation. As global temperatures rise and precipitation patterns shift, groundwater resources—which serve as vital sources of freshwater for agriculture, industry, and domestic use—are increasingly under threat. Groundwater systems are increasingly vulnerable due to shifts in climate patterns. Studies (Taylor et al. 2013a, b; Döll et al. 2018) reveal a strong correlation between rising global temperatures and groundwater recharge variability. Climate models predict that arid and semi-arid regions, where groundwater dependence is highest, will experience more significant recharge reductions. Climate change exacerbates groundwater depletion through heightened evapotranspiration, reduced recharge rates, and increased frequency of extreme weather events (Amanambu et al. 2020). Various high-tech solutions including remote sensing together with artificial intelligence (AI) and machine learning (ML) and Internet of Things (IoT) sensors and advanced geospatial systems are now deployed to streamline groundwater observation and modeling and management. IoT-based monitoring systems allow for real-time measurement of water levels and quality, enhancing the understanding of groundwater dynamics (Jain et al. 2021). Satellite-based tools such as GRACE (Gravity Recovery and Climate Experiment) have revolutionized global groundwater assessment. AI models now forecast groundwater availability and recharge using historical climate and hydrological data. Hybrid models combining deep learning with hydrological process models (Khandelwal et al. 2022) show improved predictive accuracy for aquifer responses to climate-induced stress. Climate-resilient approaches like Managed Aquifer Recharge (MAR) are increasingly favored. The combination of Multipurpose Aquifers Recharge system with solar-powered pumping technologies consistently demonstrates research results which enhance groundwater recharge rates in arid regions and minimize energy usage (Gleeson et al. 2020). Real-time data and predictive analytics along with actionable insights provided by these technologies assist policy decision-making and sustainable extraction method development as well as recharge plan execution.

While cutting-edge technologies offer mitigation pathways, they also come with embedded carbon emissions, rare earth usage, and infrastructural trade-offs. Recent assessments (Gupta et al. 2022) highlight the need for life cycle assessments (LCAs) to determine the climate footprint of groundwater technologies. However, the deployment of such technologies is not without environmental and socio-economic implications. While they offer immense potential for conservation and adaptive management, they can also lead to increased energy use, digital divide, and unintended overexploitation if not properly regulated or integrated into a broader climate-resilient water management framework.

2 Recent Developments/Latest Research

In recent years, climate change has increasingly threatened groundwater sustainability by altering recharge rates, increasing evapotranspiration, and intensifying droughts and floods. As groundwater supplies more than half of the world's drinking water and much of its irrigation, protecting this hidden resource has become critical. Satellite missions like GRACE (Gravity Recovery and Climate Experiment) and its successor GRACE-FO have revolutionized the way groundwater storage changes are monitored at regional and global scales. These satellites detect variations in Earth's gravity field caused by shifts in mass, including groundwater depletion, allowing scientists to directly assess aquifer health across continents (Rodell et al. 2018). International climate models, satellite measurements, land and soil properties combined with land usage information feed into advanced algorithms which forecast groundwater patterns for varying climate predictions (Shen 2018; Sun et al. 2021). Smart meters combined with blockchain can ensure real-time recording of groundwater extraction, preventing over-extraction and illegal pumping—critical for sustainable groundwater management under climatic pressures (Zhao et al. 2021). New-generation coupled climate–hydrology models (such as integrated surface–subsurface hydrological models like ParFlow, MODFLOW-6) are providing more realistic simulations of groundwater systems under climate change. These models account for dynamic interactions between groundwater, surface water, land use changes, and climate variables, offering more holistic assessments of groundwater sustainability (Maxwell and Condon 2016; Ferguson and Maxwell 2012). As groundwater recharge declines in many areas, Managed Aquifer Recharge (MAR) has emerged as a promising strategy (Table 1). Techniques such as spreading basins, infiltration ponds, and aquifer storage and recovery (ASR) are now supported by real-time monitoring sensors and predictive modeling to ensure efficiency under variable climatic conditions (Dillon et al. 2019).

3 Rationale of the Study

Climate change is increasingly altering the hydrological cycle, leading to erratic rainfall patterns, prolonged droughts, and heightened stress on freshwater resources—particularly groundwater, which serves as a critical buffer during water shortages. Groundwater resource monitoring and management and utilization benefit from state-of-the-art technology adoption such as AI along with IoT and remote sensing along with advanced desalination and precision agriculture. These technologies show two distinct functions in the realm of climate change by participating in both cause and solution areas but their combined effect requires further investigation. The implementation of better monitoring systems and efficient approaches helps decrease groundwater stress yet certain new technologies could create expanded water utilization or

Table 1 Recent research on climate change, technological innovations, and groundwater sustainability

Scope/Objectives	Geography/Region	Scale	Methods	Key outputs	Reference
AI-based modeling to assess climate and human impacts on groundwater fluctuations	Punjab, India	Regional	Artificial Intelligence, Time-Series Modeling	Predictive groundwater fluctuation models under climate and anthropogenic stressors	Singla et al. (2023)
Development of a physics-informed surrogate model for groundwater contamination under uncertain climate models	United States	National	Multi-scale Digital Twin, U-Net Enhanced Fourier Neural Operator	Rapid prediction of groundwater contamination scenarios influenced by climate variability	Wang et al. (2022)
Daily groundwater monitoring using vehicle-based Distributed Acoustic Sensing (DAS)	California, USA	Local	Vehicle-DAS, Elastic Full-Waveform Inversion	High-resolution, daily monitoring of aquifer responses to climatic events	Li et al. (2025)
Identification of groundwater-dependent vegetation using Synthetic Aperture Radar (SAR)	South Australia	Regional	SAR Imaging, XGBoost Classification	Efficient mapping of groundwater-dependent ecosystems for sustainable management	Terrett et al. (2020)
Assessment of climate change and land use impacts on groundwater resources using deep learning in data-scarce regions	Ganga Basin, India & Canada	Regional	Deep Learning, Climate Modeling	Enhanced understanding of groundwater dynamics in response to climate and land use changes	IIT Hyderabad & CSIR-NGRI (2023)
Evaluation of groundwater depletion rates under warming scenarios	India	National	Climate Modeling, Groundwater Data Analysis	Projection of tripling groundwater depletion rates by 2080 without intervention	Bhattarai et al. (2023)
Exploration of emerging technologies for sustainable groundwater management	Global	Global	Literature Review, Case Studies	Overview of AI, IoT, and other technologies in groundwater conservation	Borgohain et al. (2024)

(continued)

Table 1 (continued)

Scope/Objectives	Geography/Region	Scale	Methods	Key outputs	Reference
Analysis of farmers' perceptions of climate change impacts on groundwater-based livelihoods	Soan River Basin, Pakistan	Regional	Surveys, Statistical Analysis	Insights into local adaptation strategies and sustainable practices	Khan et al. (2023)
Implementation of AI-driven models for real-time groundwater monitoring and management	Bangalore & Hyderabad, India	Urban	AI, Machine Learning, IoT	Real-time analytics for groundwater quality and quantity management	The Groundwater Company (2024)

climb carbon emissions levels (e.g., energy-demanding water extraction operations or water processing procedures).

This research seeks to evaluate the net impact of such technologies on groundwater sustainability under the lens of climate change. The study conducts a critical analysis to establish if technology implementations match the requirements for long-term resilience improvement in regions where groundwater provides their main water supply. The research outcomes will identify successful methods and policy shortcomings as well as innovative avenues that combine climate adaptation with sustainable aquifer management.

4 Materials and Methods

The methodology employed in this study is rooted in a comprehensive bibliometric analysis aimed at evaluating global scholarly trends at the intersection of groundwater sustainability and climate change, with a particular focus on the role of cutting-edge technologies such as remote sensing, GIS, big data, cloud computing, and AI/ML. The temporal scope of the analysis spans a 25-year period from 2000 to 2024, providing a longitudinal understanding of research evolution. Literature was retrieved exclusively from English-language publications, ensuring consistency and accessibility in analysis. The bibliographic dataset was curated using a well-defined set of keywords: "Groundwater", "Sustainability", "Climate change", "Remote sensing", "GIS", "Big data", "Cloud computing", "Artificial Intelligence", and "Machine Learning". These terms were selected to encompass both the environmental and technological dimensions of the topic. The dataset was obtained from a reputable bibliographic database (such as Scopus or Web of Science), and records included bibliographic details, citations, publication year, and author affiliations.

The analytical section utilized VoSviewer as a standard software to build and depict bibliometric networks. The tool enabled researchers to detect co-author collaborations by visualizing cross-institutional relationships and to perform co-citation analyses that showed paper and author interrelations. The research methods generated knowledge about the intellectual framework as well as communitarian activities within this academic domain. Field Citation Ratio (FCR) and Relative Citation Ratio (RCR) served as normalized citation metrics to advance scholarly performance assessment as well as influence evaluation. FCR enabled researchers to understand how individual publications performed compared to their field peers during the given year and RCR measured their citation count in relation to the typical NIH-funded publications allowing assessment of interdisciplinary influence. The analysis selected ten leading scholarly journals that published the highest number of publications regarding this subject matter and assessment was made through volume measurements combined with citation impacts to determine influential research outlets. Determining the most prolific and influential authors was achieved through three criteria: publication frequency, total citations received, and the position in co-authorship networks.

Additional visualizations and tabular summaries were created to map year-wise publication trends, the correlation between publication count and citation accumulation, and the yearly shifts in FCR and RCR values. These visualizations helped trace not just growth but also quality and impact over time. This methodological approach provides a holistic, data-driven foundation to understand how climate change and groundwater sustainability research has evolved, who leads it, where it is published, and how emerging technologies have shaped the scholarly dialogue.

5 Results and Discussion

5.1 *Year-Wise Publication and Citation Trends (2000–2024)*

The bibliometric data on climate change, cutting-edge technologies, and groundwater sustainability reflects a notable evolution in research intensity, impact, and scholarly engagement from 2000 to 2024. In the early 2000s, research activity was relatively modest, with fewer than 50 publications per year and citation counts under 700. Despite the lower volume, the proportion of publications receiving at least one citation was high, often exceeding 80% (e.g., 85.71% in 2003 and 88.89% in 2007), indicating strong relevance and foundational influence of the early studies.

The period from 2008 onward marks a significant expansion in research output, beginning with a sharp rise to 196 publications in 2008 and peaking in later years such as 2022 (1,424 publications) and 2023 (1,229 publications). Notably, this growth coincides with the global intensification of climate discourse and the integration of emerging technologies in water resource management. While citation counts grew proportionately—reaching over 22,000 citations in 2024—there is a visible decline in the percentage of publications receiving at least one citation in many recent years. For instance, only 41.45% of 550 publications in 2016 were cited at least once, and just 45.83% of the 2024 publications had been cited, which may be attributed to the time lag in citation accrual or a dilution effect due to the sheer volume of publications.

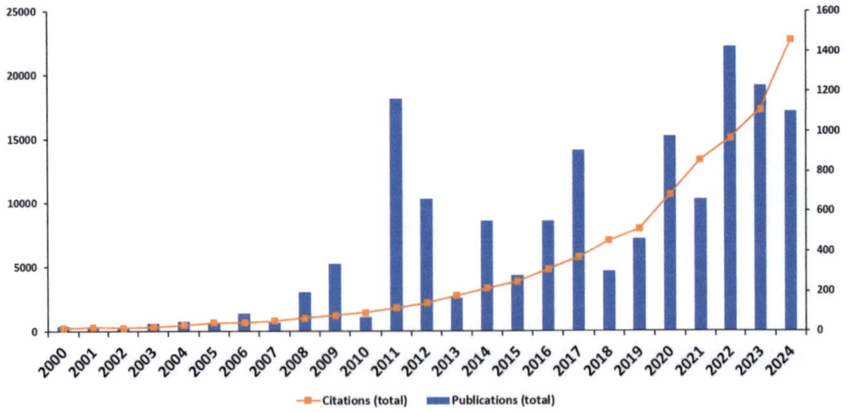

Interestingly, certain years like 2013 and 2018 stand out not only for their publication volume (166 and 300 respectively) but also for remarkably high citation efficiency, with 89.16% and 84.33% of publications being cited. This suggests that these years produced particularly impactful or widely relevant studies. Conversely, in years of high publication output such as 2011 (1,162 publications) and 2022 (1,424 publications), citation rates dropped to around 40%, highlighting a possible saturation or varied quality of contributions.

In the early 2000s (2000–2005), although the total number of publications remained under 50 per year, the percentage of cited publications was consistently high, peaking at 88.37% in 2005, indicating that early contributions were relatively impactful and well-integrated into subsequent literature. The FCR values during this period were remarkably strong, ranging from 1.6 in 2000 to over 7.8 in 2005, showing that these papers outperformed most of their field contemporaries. However, the RCR remained at zero, likely because the metric was either not available or applicable for those publications at that time, especially given its NIH origin and recent uptake. From 2006 to 2012, research activity began to grow, but citation impact showed mixed signals. In 2006 and 2008, FCR values dropped sharply to 1.21 and 0.46 respectively, and the percentage of cited publications fell below 55%, suggesting a dilution effect or possibly lower quality or relevance in certain outputs. However, there was a rebound in impact in later years, such as 2010, with 92.96% of papers cited and a high FCR of 5.65. The year 2011 stands out for its enormous publication volume (1,162 papers) but comparatively low impact—only 39.93% received at least one citation and FCR was just 0.26. This may reflect a surge in quantity over quality, or a lag in citation accrual due to the time required for scholarly uptake. A similar pattern is seen in other high-output years like 2020 and 2022, where over 1,000 publications were recorded but citation performance remained moderate.

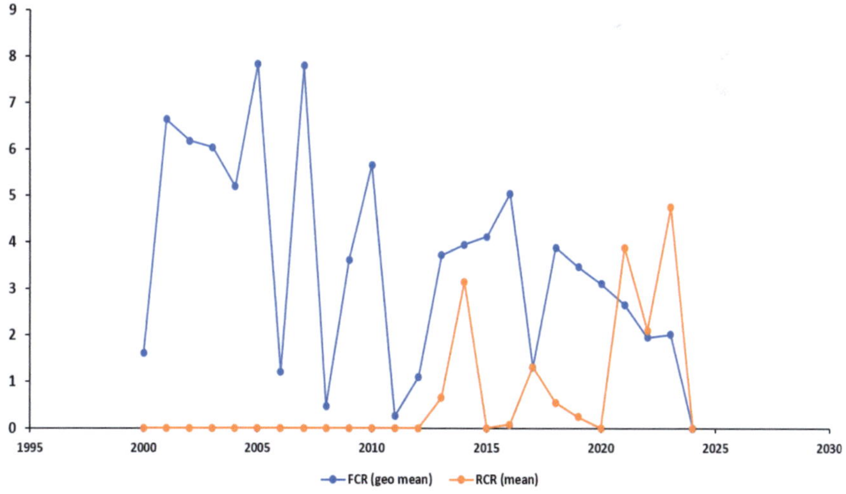

A clear turning point begins from 2013 onwards, as both RCR values become non-zero and FCR values remain consistently strong, especially in years like 2014

and 2015 with FCRs of 3.94 and 4.11, respectively. The RCR in 2014 is particularly noteworthy at 3.14, implying that the research was more than three times as influential as the average NIH-funded paper—an indication of both quality and relevance. Interestingly, 2018 emerges as a standout year: despite a moderate number of publications (300), it achieved 84.33% citation incidence, a high FCR of 3.87, and a positive RCR (0.55). This suggests that both visibility and impact of the work were optimal, reflecting a possible alignment of research with global priorities such as the SDGs and climate adaptation technologies. The impact of research based on RCR demonstrates a continuous upward trend in the time span of 2021 through 2023. In 2021, the RCR jumped to 3.876, and in 2023, it peaked at 4.75, despite a moderate FCR of 2.02. This implies that while papers might be performing average compared to their field (FCR), they are exerting much broader interdisciplinary or policy impact (reflected in the higher RCR).

6 Top Contributing Authors and Institutions

The research analysis of the ten most productive authors working with advanced technologies in groundwater sustainability together with climate change demonstrates how scientific research spreads globally while showing different research direction selection. The list is led by Iztok Podbregar from the University of Maribor, Slovenia, with 41 publications and a modest citation count of 10, averaging 0.24 citations per publication. This indicates high research output, though the influence or reach, in terms of citation, is relatively low. Following closely is Yun-Chul Hong of Seoul National University, South Korea, who has authored 36 papers with 11 citations, slightly higher in citation impact with an average of 0.31. Several other prolific contributors also hail from the University of Maribor—Jurij Krope, Darko Goričanec, and Stanislav Božičnik—each with 24 publications and very low citation averages (~0.13), suggesting a strong institutional focus on publishing within this thematic area, albeit with limited global scholarly engagement or visibility (Table 2). Similarly, Abdul Ghani Olabi from the University of Sharjah, UAE, and Mi Na Ha from Dankook University, South Korea, also contributed significantly, with 24 publications each. Notably, Mi Na Ha shows a higher citation performance (19 citations, 0.79 mean), reflecting better recognition or relevance of the work. A standout on the list is Dorian Q Fuller from University College London, UK, who, despite having fewer publications (16), garnered 153 citations, resulting in a striking mean of 9.56 citations per publication. This indicates a highly impactful body of work, likely foundational or extensively referenced in the field. Seung-Do Yu and Ho-Jang Kwon, both based in South Korea, also make the list with 16 and 15 publications respectively, maintaining a moderate citation average of 0.69 and 0.4.

Table 2 Top ten authors published research work

Sr. no	Name	Organization	Country	Publications	Citations	Mean
1	Iztok Podbregar	University of Maribor	Slovenia	41	10	0.24
2	Yun-Chul Hong	Seoul National University	South Korea	36	11	0.31
3	Jurij Krope	University of Maribor	Slovenia	24	3	0.13
4	Abdul Ghani Olabi	University of Sharjah	United Arab Emirates	24	6	0.25
5	Darko Goričanec	University of Maribor	Slovenia	24	3	0.13
6	Stanislav Božičnik	University of Maribor	Slovenia	24	3	0.13
7	Mi Na Ha	Dankook University	South Korea	24	19	0.79
8	Dorian Q Fuller	University College London	United Kingdom	16	153	9.56
9	Seung-Do Yu	National Institute of Environmental Research	South Korea	16	11	0.69
10	Ho-Jang Kwon	Dankook University	South Korea	15	6	0.4

7 Top Journals and Their Influence

The top ten journals publishing research at the intersection of groundwater, sustainability, climate change, and emerging technologies reveal a diverse spread of disciplinary focus and publication strategies. Leading the list in terms of sheer volume is HortScience, with a remarkable 1,959 publications, though the journal records only 564 citations, yielding a low citation mean of 0.29. This suggests that while the journal has hosted a large number of related articles, their individual impact or visibility is relatively limited, potentially reflecting either a narrow readership or thematic misalignment with high-citation research clusters. Second in line is Epidemiology, with 1,089 publications and 802 citations, resulting in a moderate mean citation of 0.74 (Table 3). While not specifically a journal focused on groundwater or technology, its inclusion may reflect interdisciplinary studies linking climate change and public health outcomes.

The Encyclopedia of Earth Sciences Series stands out for its high impact, with 493 publications accruing 2,836 citations, delivering a mean citation rate of 5.75. This suggests that it serves as a credible reference source, frequently cited across multiple studies. Similarly, the Lecture Notes series—appearing multiple times in the top ten—shows strong performance. For instance, Lecture Notes in Computer Science

Table 3 Top ten most productive journals published in this field of the study

Sr. no	Name	Publications	Citations	Mean
1	HortScience	1959	564	0.29
2	Epidemiology	1089	802	0.74
3	Encyclopedia of earth sciences series	493	2,836	5.75
4	Lecture notes in networks and systems	268	572	2.13
5	Lecture Notes in civil engineering	217	540	2.49
6	Lecture Notes in computer science	144	877	6.09
7	Advances in intelligent systems and computing	142	1,252	8.82
8	Encyclopedia of the UN sustainable development goals	128	435	3.4
9	Lecture notes in electrical engineering	117	385	3.29
10	Communications in computer and information science	103	299	2.9

and Advances in Intelligent Systems and Computing display high mean citation rates of 6.09 and 8.82, respectively, despite comparatively fewer publications (144 and 142). This indicates that technologically-oriented research, especially involving AI, big data, and computing in environmental systems, garners substantial academic attention and relevance. Lecture Notes in Networks and Systems and Lecture Notes in Civil Engineering, with 268 and 217 publications respectively, also contribute significantly, having mean citation rates of 2.13 and 2.49, pointing to their role in documenting engineering and systems approaches to sustainability. Meanwhile, Encyclopedia of the UN Sustainable Development Goals, despite a smaller volume (128 publications), presents a solid citation mean of 3.4, reflecting the global policy interest and academic relevance of sustainability-themed content. Rounding out the list are Lecture Notes in Electrical Engineering (117 publications, 3.29 mean citation) and Communications in Computer and Information Science (103 publications, 2.9 mean citation), both reinforcing the importance of interdisciplinary technology-oriented platforms in disseminating research on groundwater and climate change.

8 Co-Authorship Network Analysis

The co-authorship analysis conducted using VoSviewer provided significant insights into the collaborative landscape of research on groundwater sustainability and climate change in conjunction with advanced technologies. The analysis revealed a total of 1,037 co-authorships among contributing researchers, connected through 327 co-authorship links, reflecting a moderately dense collaboration network within the field. The present analysis indicates major academic interest in groundwater sustainability exists though additional collaborative opportunities between institutions and disciplinary fields exist for development.

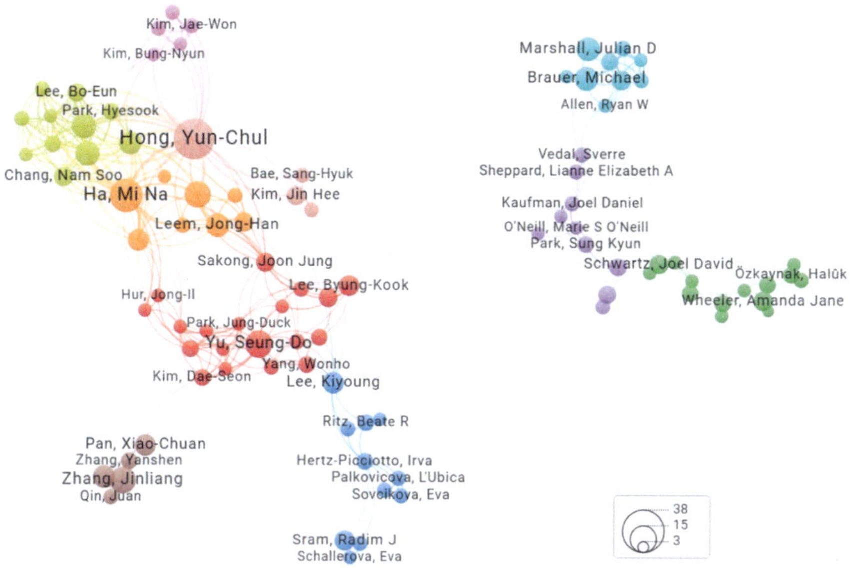

Fig. 1 Co-authorship analysis

The visual network generated by VoSviewer identified 10 distinct clusters, each representing a community of researchers frequently publishing together. These clusters often reflect thematic or geographic groupings—for instance, some clusters were characterized by strong institutional affiliations, such as national groundwater research centers or universities leading in climate-technology integration. Others were shaped by disciplinary expertise, where hydrologists, remote sensing specialists, and AI researchers formed cohesive subgroups (Fig. 1).

9 Co-Citation Analysis and Knowledge Clusters

The co-citation analysis performed using VoSviewer offers a clear depiction of the intellectual structure underlying the research on groundwater sustainability and climate change, particularly with the integration of emerging technologies. The analysis identified a total of 396 co-citations, with 53 co-citation links indicating the frequency with which pairs of documents are cited together across different studies. This relatively moderate number of co-citation links suggests that while the field draws upon a shared set of influential sources, citation behavior is somewhat dispersed, pointing to a multidimensional and evolving knowledge base.

The clustering output revealed three distinct citation clusters, each representing a thematic nucleus within the research landscape. One cluster predominantly consists of foundational works on groundwater hydrology and climate impact modeling, often

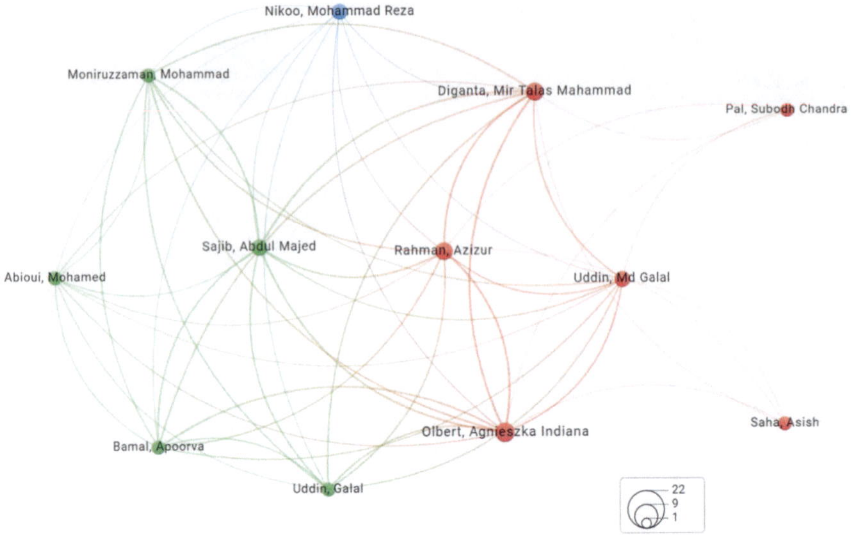

Fig. 2 Co-citation analysis using VoSViewer

citing core hydrogeological frameworks. A second cluster centers around technological integration, highlighting studies that apply remote sensing, GIS, and big data analytics to water resource monitoring and management. Climate-driven water stress and adaptation strategies receive focus through the third cluster which draws from sustainability frameworks and climate science models from vital global reports and research findings (Fig. 2).

10 Perspectives

10.1 Groundwater Systems and Vulnerability to Climate Change

The vulnerability of groundwater systems to climate change is fundamentally governed by the interplay of hydrogeological characteristics, the sensitivity and exposure of aquifers, and the spatial distribution of regional stress hotspots. Hydrogeological characteristics such as aquifer depth, storage capacity, permeability, recharge mechanisms, and the presence of confining layers play a pivotal role in determining an aquifer's resilience to external pressures. For instance, shallow unconfined aquifers, which are directly recharged by precipitation, are more immediately impacted by variability in rainfall and evapotranspiration due to rising temperatures, as compared to deep confined aquifers that recharge slowly and may be less immediately reactive but more difficult to replenish once depleted (Taylor et al. 2013a, b). The sensitivity and exposure of aquifers vary based on regional climate regimes and anthropogenic stress. A system demonstrates sensitivity by its reaction to precipitation and exposure through the combination of stress event magnitudes and frequency such as droughts and heatwaves. For example, aquifers in arid and semi-arid regions—such as those in the Middle East, western India, or parts of Sub-Saharan Africa—are particularly sensitive due to minimal recharge and high demand. In coastal areas, such as the Ganges–Brahmaputra Delta or Nile Delta, climate-induced sea-level rise and over-extraction result in saline intrusion, compromising freshwater availability (Werner et al. 2013a, b). The aquifers in high-altitude regions specifically those in the Hindu Kush-Himalaya face growing threats because of glacial reduction as well as changing precipitation patterns. Regional hotspots of groundwater stress are evident in multiple geographies where climate pressures intersect with unsustainable usage. India, for instance, is a global hotspot, with 60% of its agricultural water coming from groundwater and over 30% of blocks in a semi-critical or over-exploited state (Rodell et al. 2009). Similarly, California's Central Valley and Iran's arid plains are examples where over-pumping, combined with prolonged droughts, have led to severe groundwater depletion. Groundwater stress indices, developed using satellite-based tools like GRACE (Gravity Recovery and Climate Experiment), reveal alarming declines in groundwater storage in these regions (Richey et al. 2015). The development hotspots emphasize that it is now essential to integrate climate resilience principles into groundwater governance systems through programs that enhance water recharge and manage water consumption and apply technological monitoring systems.

11 Impact Pathways of Climate Change on Groundwater

The impact pathways of climate change on groundwater systems operate through several interlinked hydrological and geochemical processes, including recharge rate variability, changes in groundwater quality, groundwater–surface water interactions, and coastal aquifer salinization due to sea-level rise. These mechanisms collectively determine the availability, usability, and long-term sustainability of groundwater resources in different climatic and geological contexts. The recharge rate variability is one of the most direct pathways through which climate change affects groundwater. Shifts in precipitation patterns—characterized by more erratic rainfall, longer dry spells, and increased intensity of extreme weather events—disrupt the natural replenishment of aquifers. Infiltration-based recharge in semi-arid regions like the Sahel or western India is particularly vulnerable, where declining rainfall intensity and rising temperatures reduce both surface runoff and percolation (Scanlon et al. 2006). Snow-dominated recharge zones, such as those in Central Asia or the Rocky Mountains, are also at risk due to reduced snowpack and earlier snowmelt, shortening the recharge window. Moreover, the climate-induced changes in groundwater quality are becoming more pronounced. Warmer temperatures and reduced dilution during droughts can lead to higher concentrations of nitrates, heavy metals, and other pollutants in shallow aquifers (Treidel et al. 2012). In agricultural regions, higher evapotranspiration and reduced recharge may concentrate residual agrochemicals in groundwater. Furthermore, flooding events can lead to the infiltration of contaminants from landfills or industrial sites, degrading water quality. The groundwater–surface water interactions are disrupted by climate change, impacting both availability and ecosystem health. Reduced baseflow contributions to rivers during prolonged dry periods lead to the drying of streams and wetlands that depend on groundwater discharge. Conversely, declining river levels can reverse gradients, drawing pollutants into aquifers from surface water bodies (Winter et al. 1998). This interaction is critical in integrated water management, particularly in basins like the Murray-Darling in Australia or the Indo-Gangetic Plain, where groundwater and surface water are tightly interdependent. Moreover, sea-level rise leads to saltwater intrusion into coastal aquifers, one of the most pressing concerns for low-lying deltaic and island regions. As sea levels rise, the freshwater-saltwater interface in coastal aquifers moves landward and upward, especially when combined with groundwater over-extraction. Regions such as the Nile Delta, Mekong Delta, and Pacific Island nations are already witnessing salinization of potable groundwater sources, threatening agriculture, domestic supply, and ecosystem stability (Werner et al. 2013a, b).

12 Socioeconomic Implications

The socioeconomic implications of climate change-driven groundwater stress are profound, influencing sectors ranging from agriculture to urban infrastructure, and even contributing to migration and conflict dynamics. These implications unfold across several key dimensions, including agriculture and irrigation, urban water security, rural dependency, and climate-induced displacement and conflict. The agriculture and irrigation, which account for over 70% of global groundwater withdrawals, are particularly vulnerable. In arid and semi-arid regions such as northwestern India, California's Central Valley, and northern China, intensified droughts and reduced recharge are pushing groundwater tables deeper, increasing energy costs and lowering yields (Rodell et al. 2009). The decline of water availability leads to cropping pattern shifts, reduced agricultural productivity, and, in extreme cases, the abandonment of farmland. Studies also indicate that excessive groundwater pumping for irrigation is unsustainable under future climate scenarios, particularly in the absence of adaptation measures like precision agriculture or regulated abstraction (Famiglietti 2014). In urban areas, the pressure on groundwater as a backup or primary source is increasing due to erratic surface water availability and rapid population growth. Cities like Chennai, Mexico City, and Jakarta heavily rely on aquifers, which are being overdrawn, leading to land subsidence, declining water quality, and infrastructure damage (Erban et al. 2014). The interplay of heatwaves, infrastructure stress, and inadequate recharge exacerbates urban water insecurity, particularly in informal settlements and peri-urban zones. Moreover, rural communities—especially in developing countries—are often wholly dependent on shallow groundwater for both drinking and irrigation. These areas lack the financial, technical, and institutional capacity to adapt to declining groundwater levels. As climate extremes become more frequent, rural livelihoods become increasingly precarious, pushing households into poverty traps and migration (Sadoff et al. 2015). The situation is aggravated where groundwater governance is weak or absent, and where extraction is unregulated. The climate migration and resource conflicts are emerging as critical consequences of groundwater depletion. Regions with high dependence on aquifers and limited alternative sources—such as parts of Sub-Saharan Africa, Central Asia, and South Asia—are witnessing displacement of populations due to water scarcity. Groundwater stress can also fuel conflicts, especially in transboundary aquifer regions or areas where access is politically or ethnically contested. In Yemen and parts of northern Nigeria, groundwater depletion has been linked to agricultural collapse and localized conflict escalation (Gleick 2014).

13 Technological Interventions for Monitoring and Adaptation

Technological interventions have become essential tools for monitoring, managing, and adapting groundwater systems under climate change stress. These tools—ranging from remote sensing and GIS, to big data platforms, AI/ML, and smart sensor networks—enable a transition from reactive to proactive water resource management. Remote sensing and GIS technologies have revolutionized groundwater monitoring by offering spatial and temporal insights that were previously unattainable through conventional means. Satellite-based missions like GRACE (Gravity Recovery and Climate Experiment) have been instrumental in detecting regional groundwater depletion trends, particularly in hotspots such as northern India, California, and Iran (Rodell et al. 2009). GIS platforms enable integration of hydrological, geological, and climatic data to map aquifer vulnerability, monitor land-use change impacts, and guide site-specific recharge and extraction strategies (Scanlon et al. 2002). Complementing these are big data and cloud computing platforms, which facilitate real-time data storage, access, and analytics at scale. Open data initiatives, cloud-hosted repositories (e.g., Google Earth Engine), and hydrological modeling platforms are enabling collaborative and scalable analysis of groundwater-surface water interactions, recharge patterns, and aquifer health. These tools also support policy-making through dashboards and scenario modeling for groundwater abstraction, pollution tracking, and recharge potential (McDonald et al. 2011). In recent years, artificial intelligence (AI) and machine learning (ML) have gained traction for predictive modeling and decision support in groundwater management. Algorithms like support vector machines (SVM), random forests, and deep neural networks are being used to forecast groundwater levels, identify contamination risk zones, and optimize pumping strategies. Studies have shown that ML-based groundwater level predictions outperform traditional statistical models under complex hydroclimatic variability (Mohan et al. 2021). AI-driven models are also used to develop early warning systems for overexploitation and drought resilience. Moreover, smart sensors and the "Internet of Water" represent a frontier in real-time groundwater monitoring. These technologies include IoT-enabled piezometers, flow meters, and water quality probes that transmit high-frequency data to centralized platforms. Real-world deployments in the U.S., Netherlands, and India demonstrate how such networks improve transparency, enable dynamic groundwater accounting, and engage stakeholders in participatory governance (Perrone and Jasechko 2019). The Internet of Water concept advocates for standardized, interoperable data infrastructure to connect public and private groundwater data sources for informed water stewardship.

14 Limitations and Future Implications of the Study

The study offers valuable insights into the intersection of groundwater sustainability, climate change, and emerging technologies; however, it also acknowledges several critical limitations and outlines directions for future research and policy integration. One of the foremost challenges lies in the complexity of groundwater management under escalating climate pressure, as shifting precipitation patterns, increased evapotranspiration, and extreme weather events severely impact groundwater recharge and quality. In many regions, particularly arid and semi-arid zones, over-extraction continues unchecked, exacerbated by the lack of real-time monitoring and enforcement mechanisms (Taylor et al., 2013a, b; Wada et al. 2010). This makes adaptive groundwater governance not only difficult but also context-specific, requiring localized interventions and cross-sector coordination.

Another pressing concern is the persistent data gap. Although the proliferation of sensors, satellites, and remote sensing platforms has increased data availability, there remains a lack of high-resolution, long-term, and standardized groundwater datasets, especially in the Global South. Fragmented databases, incompatible data formats, and absence of open-access repositories hinder large-scale modeling and comparison (Famiglietti 2014). These technological barriers also extend to the limited integration of tools such as AI/ML, big data, and cloud-based platforms into government-led groundwater monitoring systems. Often, these tools remain confined to academic research and pilot projects, lacking scalability and institutional uptake.

To bridge these gaps, the study recommends a stronger integration of cutting-edge technologies into groundwater policy frameworks. For instance, AI/ML models can predict aquifer stress zones, while cloud-based dashboards can support real-time data-sharing and early warning systems for depletion or contamination. Governments should consider investing in technology-transfer programs, cross-institutional training, and public–private partnerships to mainstream such innovations (WWAP 2019). Additionally, promoting citizen science and mobile-based data collection can democratize monitoring and supplement formal datasets, particularly in rural settings.

Moreover, international case comparisons provide valuable lessons. For example, California's Sustainable Groundwater Management Act (SGMA) has established a legal framework incorporating satellite data, groundwater modeling, and community-based governance, while Australia's Murray-Darling Basin Authority uses remote sensing and digital platforms to manage transboundary aquifers effectively (Gleeson et al. 2020; Scanlon et al. 2016). These models show that while technology is a crucial enabler, its success hinges on institutional commitment, legal backing, and multi-level stakeholder engagement. Moving forward, the convergence of science, technology, and policy will be vital to ensure equitable, adaptive, and sustainable groundwater use under a changing climate.

15 Conclusions

The trend suggests a maturing discipline with developing scholarly interest and impact, yet leaving a gap for high-quality targeted research to remain relevant in citations amidst rapidly increasing number of publications. The data point toward the global acknowledgment of the theme of technological innovation toward groundwater sustainability in climate change scenarios while giving hints concerning the challenge posed to uphold academic mainstream and influenceability in a landscape that keeps expanding with such vigor. With data full-scale from 2000 to 2024, the rising field with a growing interest of the scientific community and a few impact years worthy of mention (2005, 2013, 2018, 2021, and 2023) opened the door midway, to applied and now highly visible research. While the number of publications is on the rise, ensuring citation efficiency and relevance remain challenges, especially in the high-output recent years. Through indicators like FCR and RCR, such direct citation numbers are dissected, reinforcing the need for targeted and impactful research aimed at more than just increasing numbers in groundwater science, policy, and sustainability in the glistening light of climate change. Indeed, such technological advances are not only helping fine-tune and upscale groundwater monitoring but also aiding in predictive modeling, real-time decision-making, and participatory governance via remote sensing, GIS, big data platform, AI/ML models, and IoT-enabled smart sensors. Despite their promise, technological integration into policy frameworks still remains uneven given the data gaps, inertia in institutions, and infrastructural tiptoeing. The study asserts cross-sectoral collaboration, open data ecosystem building, inclusive digital governance, and the embedding of these technologies into adaptive groundwater management methodology are necessary to ensure groundwater sustainability in an increasingly climate-stressed world. Ultimately, climate science, hydrogeology, and digital innovation together create a solid channel for preserving groundwater for the years to come.

Disclose Conflict of Interest No conflict of interest.

References

Abiy AZ, Melesse AM (2017) Impact of climate change on groundwater resources: a review. Environ Monit Assess 189(1):1–19. https://doi.org/10.1007/s10661-016-5706-7

Amanambu AC, Obarein OA, Mossa J, Li L, Ayeni SS, Balogun O, Oyebamiji A, Ochege FU (2020) Groundwater system and climate change: present status and future considerations. J Hydrol 589:125163

Bhattarai N, Misra V, Risser MD (2023) Warming triples groundwater depletion across India by 2080. Sci Advan 9(49):eadi1401

Boretti, A., & Rosa, L. (2019). Reassessing the projections of the World Water Development Report. npj Clean Water, 2, Article 15. https://doi.org/10.1038/s41545-019-0039-9

Borgohain A, Sharma S, Gupta S (2024) Emerging trends and technologies for conservation and sustainable approach in groundwater management. https://www.researchgate.net/publication/386877443_Emerging_Trends_and_Technologies_for_Conservation_and_Sustainable_Approach_in_Groundwater_Management

Dao PU, Heuzard AG, Le TXH, Zhao J, Yin R, Shang C, Fan C (2024) The impacts of climate change on groundwater quality: a review. Sci Total Environ 912:169241

Davamani V, John JE, Poornachandhra C, Gopalakrishnan B, Arulmani S, Parameswari E, Santhosh A, Srinivasulu A, Lal A, Naidu R (2024) A critical review of climate change impacts on groundwater resources: a focus on the current status, future possibilities, and role of simulation models. Atmosphere 15(1):122. https://doi.org/10.3390/atmos15010122

Dillon P, Page D, Vanderzalm J (2019) Evolution of managed aquifer recharge in Australia. Water 11(10):2002

Döll, P., Trautmann, T., Gerten, D., Müller Schmied, H., Ostberg, S., Saeed, F., & Schleussner, C.-F. (2018). Risks for the global freshwater system at 1.5 °C and 2 °C global warming. Environmental Research Letters, 13(4), 044038. https://doi.org/10.1088/1748-9326/aab792

Earman S, Dettinger M (2011) Potential impacts of climate change on groundwater resources–a global review. J Water Clim Chang 2(4):213–229

Erban LE, Gorelick SM, Zebker HA, Fendorf S (2014) Groundwater extraction, land subsidence, and sea-level rise in the Mekong Delta. Vietnam Environ Res Lett 9(8):084010

Famiglietti JS (2014) The global groundwater crisis. Nat Clim Chang 4:945–948

Ferguson IM, Maxwell RM (2012) Human impacts on terrestrial hydrology: climate change versus pumping and irrigation. Environ Res Lett 7(4):044022

Garner G, Hannah DM, Sadler JP, Orr HG (2017) River temperature and aquatic ecology: climate, hydrology, and ecology interactions. Prog Phys Geogr 41(1):47–75

Gleeson T et al (2020) Managed aquifer recharge: global perspectives and ecosystem considerations. Water Resour Res 56(3):e2019WR025259

Gleick PH (2014) Water, drought, climate change, and conflict in Syria. Weather, Climate, and Society 6(3):331–340

Gupta R et al (2022) Environmental implications of groundwater technologies under climate constraints. Sci Total Environ 818:151744

Haldorsen S, Heim M, Van Der Perk M (2016) Impacts of climate change on groundwater in permafrost areas: a review. Earth Sci Rev 152:14–27

Hu C, Huang G, Wang S, Leng G, Wang L (2019) Assessing groundwater drought vulnerability under climate change scenarios: a case study in a semi-arid region. J Hydrol 568:176–190

Indian Institute of Technology Hyderabad & CSIR-National Geophysical Research Institute (2023) Understanding the impact of climate change and land use change on groundwater resources using deep learning techniques. India Science, Technology and Innovation

Portal. https://www.indiascienceandtechnology.gov.in/research/understanding-impact-climate-change-and-land-use-change-groundwater-resources-future-using-deep
Intergovernmental Panel on Climate Change; Edenhofer, O., Pichs-Madruga, R., Sokona, Y., Farahani, E., Kadner, S., Seyboth, K., Adler, A., Baum, I., Brunner, S., Eickemeier, P., Kriemann, B., Savolainen, J., Schlömer, S., von Stechow, C., Zwickel, T., & Minx, J. C. (Eds.). (2014). Climate Change 2014: Mitigation of Climate Change. Contribution of Working Group III to the Fifth Assessment Report of the Intergovernmental Panel on Climate Change. Cambridge University Press. https://www.ipcc.ch/report/ar5/wg3/
Jain M et al (2021) Smart groundwater monitoring using IoT for climate resilience. Environ Monit Assess 193(5):281
Khan A, Khan IU, Ullah S, Khan I, Arshad M (2023) Farmers' perception towards the impact of climate change on groundwater-based livelihoods: evidence from the Soan River Basin. Pakistan. Water 15(7):1287
Khandelwal H et al (2022) Hybrid AI models for groundwater prediction under climate variability. J Hydrol 610:127956
Kurylyk BL, MacQuarrie KTB, McKenzie JM, Voss CI, Moule AJ (2014) Climate change impacts on groundwater and soil temperatures in cold and temperate regions: Implications, mathematical theory, and emerging simulation tools. Earth Sci Rev 138:313–334
Li Z, Biondi B, Karrenbach M (2025) Daily groundwater monitoring using vehicle-based distributed acoustic sensing (DAS) and elastic full-waveform inversion. arXiv:2501.10618
López-Morales CA, Mesa-Jurado MA (2017) Valuation of hidden water ecosystem services: the replacement cost of the aquifer system in central Mexico. Water 9:571
Maxwell RM, Condon LE (2016) Connections between groundwater flow and transpiration partitioning. Sci 353(6297):377–380
McDonald RI, Green P, Balk D, Fekete BM, Revenga C, Todd M, Montgomery M (2011) Urban growth, climate change, and freshwater availability. Proc Natl Acad Sci 108(15):6312–6317
Mohan C, Western AW, Wei Y, Saft M (2021) Predicting groundwater levels using machine learning: a review. Hydrol Earth Syst Sci 25(3):1069–1095
Perrone D, Jasechko S (2019) Deeper well drilling an unsustainable stopgap to groundwater depletion. Nature Sustainability 2(8):773–782
Pitz CF (2016) Climate Change Effects on Groundwater Recharge. In: Melesse AM et al (eds) Hydrology and best practices for managing water resources in arid and semi-arid lands. Springer, pp 273–288
Ramasamy A, Lee TM (2017) Climate change impact on groundwater resources: a review of modeling approaches. Environ Earth Sci 76(4):1–14
Refsgaard JC, Arnbjerg-Nielsen K, Drews M, Halsnaes K, Jeppesen E, Madsen H, Christensen JH (2016) Projections of future flood risk in Europe: climate scenarios and models. Clim Change 138(1):149–162
Richey AS et al (2015) Quantifying renewable groundwater stress with GRACE. Water Resour Res 51(7):5217–5238
Rodell M, Velicogna I, Famiglietti JS (2009) Satellite-based estimates of groundwater depletion in India. Nature 460(7258):999–1002
Rodell M, Famiglietti JS, Wiese DN, Reager JT, Beaudoing HK, Landerer FW, Lo MH (2018) Emerging trends in global freshwater availability. Nature 557(7707):651–659
Sadoff CW, Hall JW, Grey D, Aerts JC, Ait-Kadi M, Brown C, Whittington D (2015) Securing water, sustaining growth. Report of the GWP/OECD task force on water security and sustainable growth
Scanlon BR, Healy RW, Cook PG (2006) Choosing appropriate techniques for quantifying groundwater recharge. Hydrogeol J 10(1):18–39
Scanlon, B. R., Healy, R. W., & Cook, P. G. (2002). Choosing appropriate techniques for quantifying groundwater recharge. Hydrogeology Journal, 10(1), 18–39. https://doi.org/10.1007/s10040-001-0176-2

Scanlon BR, Faunt CC, Longuevergne L, Reedy RC, Alley WM, McGuire VL, McMahon PB (2012) Groundwater depletion and sustainability of irrigation in the US High Plains and Central Valley. Proc Natl Acad Sci 109(24):9320–9325

Scanlon, B. R., Zhang, Z., Save, H., Wiese, D. N., Landerer, F. W., Long, D., Longuevergne, L., Chen, J., et al. (2016). Global evaluation of new GRACE mascon products for hydrologic applications. Water Resources Research, 52(12), 9412–9429. https://doi.org/10.1002/2016WR019494

Shen C (2018) A transdisciplinary review of deep learning research for groundwater and hydrology. Hydrol Earth Syst Sci 22(8):4087–4116

Singla N, Rani P, Tiwari MK (2023) Artificial intelligence-based approach to study the impacts of climate change and human activities on groundwater fluctuation. AQUA— Water Infrastruct, Ecosyst Soc 72(8):1404–1421

Smerdon BD (2017) A synopsis of climate change effects on groundwater recharge. J Hydrol 555:125–128

Sun AY, Green R, Osidele O (2021) Machine learning applications for hydrologic modeling. Water Resour Res 57(7):e2020WR028767

Taylor RG et al (2013a) Ground water and climate change. Nat Clim Chang 3(4):322–329

Taylor RG, Scanlon B, Doll P, Rodell M, van Beek R, Wada Y, Treidel H (2013b) Ground water and climate change. Nat Clim Chang 3(4):322–329

Terrett O, McGrath C, Byrne G (2020) Using synthetic aperture radar to map groundwater-dependent ecosystems: a machine learning approach. arXiv:2009.03129

The Groundwater Company (2024) Latest trends in groundwater management: How technology and innovation are changing the game. https://thegroundwatercompany.com/blog/latest-trends-in-groundwater-management-how-technology-and-innovation-are-changing-the-game/

Treidel H, Martin-Bordes JL, Gurdak JJ (2012) Climate change effects on groundwater resources: a global synthesis of findings and recommendations. CRC Press

Velis M, Conti KI, Biermann F (2017) Groundwater and human development: synergies and trade-offs within the context of the sustainable development goals. Sustain Sci 12:1007–1017

Wada Y et al (2010) Global depletion of groundwater resources. Geophys Res Lett 37(20)

Wang L, Tchelepi HA, You F (2022) Physics-informed Fourier neural operator for groundwater contamination under uncertain climate models. arXiv:2211.10884

Werner AD et al (2013a) Vulnerability indicators for seawater intrusion. Groundwater 51(3):316–328

Werner AD, Bakker M, Post VE, Vandenbohede A, Lu C, Ataie-Ashtiani B, Simmons CT (2013b) Seawater intrusion processes, investigation and management: recent advances and future challenges. Adv Water Resour 51:3–26

Winter TC, Harvey JW, Franke OL, Alley WM (1998) Ground water and surface water: a single resource. USGS Circular 1139

WWAP (World Water Assessment Programme) (2019) The United Nations World Water Development Report: Leaving No One Behind

Zhao J, Wang X, Li Y (2021) Blockchain technology for groundwater governance: potentials and challenges. Environ Res Lett 16(6):064026

Climate Change, Rising Groundwater Table and Flooding in Cape Town Informal Settlements: Impacts and Solutions Through Cutting-Edge Technologies

Johannes Bhanye and Ruvimbo Shayamunda

Abstract This chapter explores how climate change, rising groundwater tables, and frequent flooding affect informal settlements in Cape Town. It examines the socio-economic and environmental vulnerabilities that make these areas particularly susceptible to groundwater-induced flooding, especially during winter rainfall. Drawing on qualitative fieldwork and secondary data, the chapter unpacks how rising groundwater levels destabilize housing structures, compromise sanitation, and increase health risks. It highlights community experiences and coping strategies, revealing that while some residents deploy transient and adaptive mechanisms, others remain trapped in flood-prone areas due to poverty and limited options. The chapter evaluates innovative solutions including AI- and IoT-enabled early warning systems, flood-resilient infrastructure, and sustainable construction materials. It also emphasizes community-based adaptation and participatory planning as key to enhancing resilience. Finally, it critiques current governance and policy frameworks, proposing multi-scalar, integrated approaches to flood risk management that prioritize equity, sustainability, and long-term climate adaptation in informal settlements.

Keywords Climate change · Groundwater table · Flooding · informal settlements · Cape town · Smart technologies · Community-based approaches · Multi-scalar governance

J. Bhanye (✉)
African Climate and Development Initiative (ACDI), University of Cape Town, Cape Town, South Africa
e-mail: joebhanye@gmail.com

R. Shayamunda
Department of Urban and Regional Planning, University of the Free State, Bloemfontein, South Africa

A. Nath et al. (eds.), *Groundwater Depletion and Sustainability*, Advances in Geographic Information Science, https://doi.org/10.1007/978-3-032-09921-1_6

1 Introduction

Climate change poses significant threats to urban areas worldwide, affecting millions of lives through its impacts on weather patterns, sea levels, and natural habitats. Urban centers, with their high-density populations and economic activities, are particularly vulnerable to these changes (Zhou et al. 2022). The rise in global temperatures contributes to more frequent and severe weather events such as heatwaves, storms, and flooding (Putsoane et al. 2024). These phenomena not only disrupt urban infrastructure but also exacerbate existing social and environmental challenges. As cities continue to grow, the interplay between climate change and urbanization demands urgent attention to mitigate risks and enhance resilience.

Studies demonstrate that informal settlements are the most vulnerable to climate-induced disasters like flooding (Kikwasi and Mbuya 2019; Udo et al. 2024). For instance, experiences from cities like Dhaka, Jakarta, and Rio de Janeiro, reveal how governments grapple with significant populations living in high-risk informal settlements; with some of the cities experimenting with various approaches, such as community-led flood resilience programs and partnerships between local governments and non-governmental organizations to improve infrastructure and emergency response capabilities (Armond and Sant'Anna Neto JL. 2019; Haque 2021).

Globally, urban informal settlements are growing at an unprecedented rate as rural populations migrate to cities in search of better economic opportunities (Johannes et al. 2023; Bhanye 2024). This trend is particularly noticeable in developing countries where rapid urbanization outpaces the ability of local governments to provide adequate housing and infrastructure (Bhanye et al. 2023). As a result, a significant portion of the urban population ends up living in informal settlements, which are often under-equipped to handle the environmental challenges posed by climate change.

Scholars reveal that informal settlements share common vulnerabilities to climate change, including susceptibility to floods, landslides, and heatwaves, exacerbated by inadequate infrastructure and limited access to resources (Putsoane et al. 2024; Kikwasi and Mbuya 2019; Udo et al. 2024). These settlements are often built on land that is not only environmentally fragile but also ignored by urban planning regulations, leaving them highly exposed to the impacts of climate change. Cape Town's informal settlements, like those in other parts of the world, suffer from systemic neglect and a lack of resources. Cape Town faces unique challenges due to its geographical location at the tip of Africa, where it is exposed to both the Atlantic and Indian Oceans, making it particularly vulnerable to rising sea levels and coastal storms. Furthermore, the city's Mediterranean climate leads to dry, hot summers and wet, cold winters, a pattern that is increasingly disrupted by global climate change, resulting in unpredictable weather events devastating unprepared communities (Abiodun et al. 2017).

In Cape Town, rising groundwater table and increased frequency of flooding have become significant concerns. While previous research has extensively documented the socio-economic and infrastructural vulnerabilities of urban informal settlements to climate change (Rammile et al. 2024; Williams et al. 2019), there remains a critical gap in understanding the specific impacts of rising groundwater tables and their interaction with urban flooding in Cape Town. Most studies have focused on broader climatic effects, such as increased precipitation and sea-level rise, often overlooking how localized groundwater changes exacerbate flooding and disrupt the lives of those in informal settlements (Dube et al. 2021; Fitchett et al. 2016). Additionally, while there is substantial literature on the immediate impacts of flooding, such as displacement and health crises (Udo et al. 2024), there is less focus on long-term infrastructural and socio-economic challenges, leaving a gap in comprehensive risk assessment and mitigation strategies tailored to these communities. This study seeks to bridge these gaps by providing simplified, detailed analysis of the unique challenges faced by informal settlements in Cape Town due to climate change, rising groundwater levels, and frequent flooding. The chapter also proposes innovative solutions through cutting-edge technologies and community-based approaches as actionable insights for enhancing resilience and adaptive capacity in vulnerable urban populations leaving in informal settlements.

2 Methodology

The study employed a qualitative methodology to explore the impacts of climate change, rising groundwater levels, and flooding on informal settlements in Cape Town. The research was grounded in in-depth interviews with community members and leaders from various informal settlements in Cape Town including Khayelitsha, Zwezwe, Green Park, Los Angeles, and Covid-19. These interviews were designed to capture detailed, personal accounts and insights into the challenges and responses related to groundwater and flood-related issues within these communities.

In addition to interviews, the research methodology included observations through transect walks and drives across the various informal settlements in Cape Town. This approach allowed for the direct observation of environmental conditions and the ways in which residents adapted to and coped with their circumstances. It also provided a tangible sense of the spatial distribution of risks and the effectiveness of existing infrastructure against flooding.

The study also integrated desktop research to complement and contextualize findings from interviews and observations. Desktop research involved reviewing existing literature, reports, and case studies related to urban flooding, climate resilience, and informal settlement management. This component was crucial for understanding broader patterns and drawing comparisons with similar urban contexts globally.

Ethical considerations were paramount in this research. All participant interactions were conducted with informed consent, ensuring confidentiality and respecting the privacy of individuals. The research adhered to ethical guidelines to minimize any potential harm to participants, including emotional distress from discussing sensitive topics such as personal losses due to flooding. Figure 1 is the methodological flow diagram of the study.

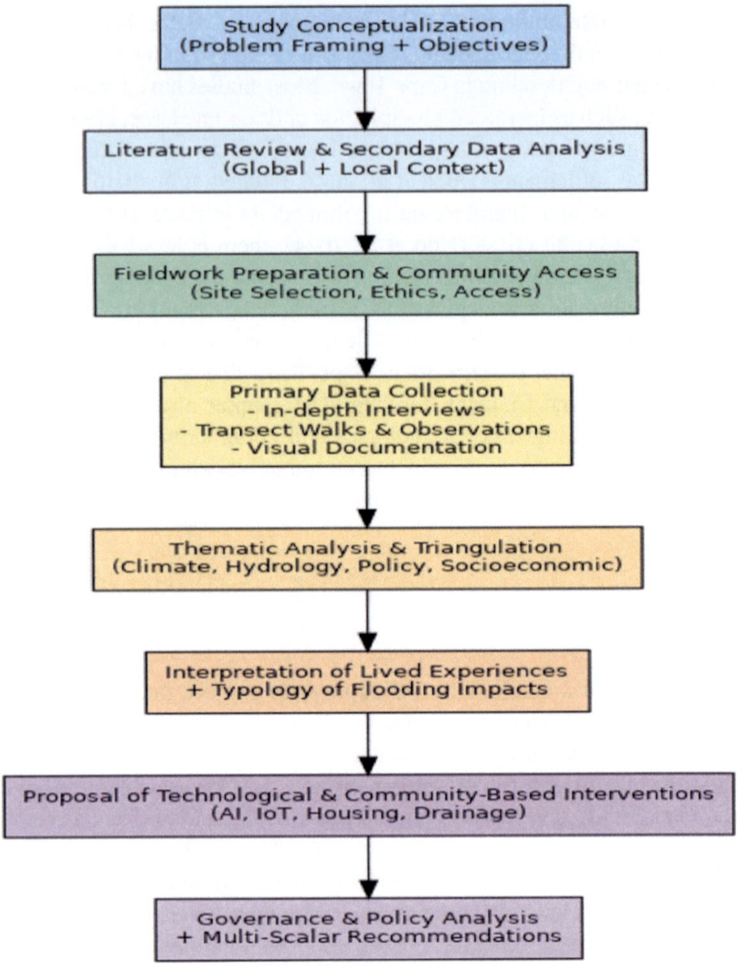

Fig. 1 Methodological flow diagram of the study

3 Climate Change and Urban Informal Settlements

Urban informal settlements in Cape Town, are typically characterized by high-density living conditions and makeshift housing structures. These areas, such as Khayelitsha, Philippi, and Gugulethu, are home to a substantial portion of the city's population. Residents of these settlements face numerous challenges including limited access to basic services such as clean water, sanitation, and reliable electricity (Ehebrecht 2015). The demographic profile of these communities is diverse, consisting predominantly of migrants from other parts of South Africa and neighbouring countries like Zimbabwe, Malawi and Zambia, seeking better economic opportunities in the city.

Fig. 2 Poor sanitation conditions in some informal settlements in Cape Town **a** Heap of garbage **b** Cluster of mobile toilets **c** Water tap next to a toilet. *Source* Fieldwork

Figure 2a–c show poor sanitation conditions in some informal settlements in Cape Town.

The socio-economic status of informal settlements is generally low (Bhanye et al. 2024). Unemployment rates are high, and many residents are employed in informal sectors or engage in day labour. The informal nature of housing in these areas often means that residents do not have legal ownership of the land on which their homes are built, which complicates their legal status and reduces their access to government services and protection. Education levels vary, but access to quality education and healthcare is typically below that of more formally developed areas of the city. This combination of high density, poor infrastructure, and low socio-economic status leaves residents particularly vulnerable to the impacts of climate change.

These vulnerabilities are increased by the physical location of many informal settlements. Often situated on marginal lands such as floodplains, riverbanks, and other environmentally sensitive areas, these communities are at an increased risk of natural disasters, including floods (Drivdal 2016). The non-permanent nature of many structures within these areas does not provide sufficient resistance to severe weather events, which are becoming more frequent and intense due to climate change. Additionally, the lack of formal drainage systems and adequate street paving often leads to significant water accumulation during rainstorms, further heightening the risk of flooding.

4 Vulnerability of Informal Settlements to Climate Change and Flooding

Urban informal settlements are uniquely vulnerable to the impacts of climate change due to a combination of socio-economic, environmental, and infrastructural factors. In Cape Town, like in many other urban centres, these vulnerabilities are particularly pronounced due to the city's geographic and climatic characteristics. These characteristics include its coastal location prone to sea level rise and storm surges, diverse microclimates influenced by surrounding mountainous terrain that exacerbates weather extremes, and a Mediterranean climate that brings wet winters and dry, hot summers, intensifying both flood and drought risks (Savelli et al. 2021).

The environmental setting of many informal settlements exacerbates their susceptibility to climate change. These areas are frequently located in zones prone to natural hazards such as floodplains or steep slopes, where land is often less desirable and thus more affordable for informal development (See Fig. 3a, b). The topography of these areas can lead to waterlogging and landslides, particularly during periods of heavy rainfall, which are becoming more common and intense with climate change. This geographical vulnerability is compounded by poor infrastructure. Most informal settlements lack adequate drainage and road systems, making it difficult to manage excess water during floods and to provide emergency services effectively.

The materials and construction methods used in informal settlements generally do not provide sufficient protection against extreme weather conditions. Homes are often built from salvaged materials like corrugated metal, wood, and plastic, which offer little resistance to high winds, heavy rains, and flooding (See Fig. 4a, b). The

Fig. 3 Informal settlements located in the flood plain in Cape Town **a** Shacks built on a flood plain **b** Rubble dumped on the flood plain in an attempt to lift the ground. *Source* Fieldwork

Fig. 4 Houses or shacks built with poor materials **a** Shacks built of zinc iron sheets **b** Shacks built on raised ground in a flood plain. *Source* Fieldwork

transient nature of these materials also means that they need frequent replacement, which can be costly and logistically challenging, further entrenching poverty.

The socio-economic conditions prevalent in informal settlements significantly heighten their vulnerability. High levels of poverty, limited access to education, and inadequate health services make it difficult for residents to prepare for, respond to, and recover from climate-related impacts. For example, the lack of economic resources can prevent individuals from investing in the improvement of their homes to withstand extreme weather or from relocating to safer areas. Additionally, these settlements often experience higher rates of disease due to overcrowded living conditions and poor sanitation, which can be exacerbated by the spread of waterborne diseases following flooding events. Figure 5 is a depiction of poor living conditions and poverty in some informal settlements in Cape Town. Figure 5a captures the spatial marginalization and lived experiences of residents in informal settlements, where basic services like sanitation, clean water, and secure housing remain inaccessible. Figure 5b shows accumulated solid waste and poor sanitation in an informal settlement in Cape Town. These conditions not only compromise human dignity but also increase vulnerability to disease outbreaks and climate-related risks like flooding and water contamination.

Lastly, there is often a lack of formal governance and institutional support in informal settlements, which hinders effective climate change adaptation strategies. The informal status of these areas frequently results in a minimal presence of government services and infrastructural development. This governance gap means that

Fig. 5 Depiction of poor living conditions and poverty in some informal settlements in Cape Town **a** A resident walks through a narrow footpath surrounded by precariously built shacks constructed from salvaged materials such as wood, metal sheets, and mesh wire. The ground is strewn with litter, indicating insufficient waste disposal systems **b** Piles of uncollected garbage, stagnant wastewater, and deteriorating infrastructure highlight the systemic neglect and lack of basic municipal services such as waste management and drainage. *Source* Fieldwork

responses to climate impacts are typically reactive rather than proactive, lacking in coordinated disaster preparedness and risk reduction measures that are essential in mitigating the effects of climate change.

Cape Town has experienced several significant flooding events over the years, often exacerbated by its winter rainfall pattern, which leading to intense storms and heavy downpours. Informal settlements, with their dense populations and inadequate infrastructure, are particularly vulnerable. Multiple, the winter storms over the past years are notable for the severe flooding they caused in areas like Philippi, Khayelitsha and Gugulethu, where hundreds of makeshift homes were washed away or severely damaged. These events highlight the critical deficiencies in drainage, emergency services, and community preparedness in the face of natural disasters.

Historical data on flooding in Cape Town's informal settlements offer valuable lessons on resilience, adaptation, and the gaps in current flood management practices. One of the primary lessons is the critical role of infrastructure in managing flood risks. Past events have repeatedly shown that areas with poor drainage systems and inadequate road networks suffer more severe impacts during flooding. Recent flooding events in informal settlements like Khayelitsha and Gugulethu, respectively, highlighted the devastation that can occur when water management systems are overwhelmed.

Historical flooding events also demonstrate the value of community engagement and preparedness. In instances where community members were involved in planning and response activities, the outcomes were noticeably better. Communities with established local networks and disaster response plans are able to mobilize more quickly and effectively, mitigating the impacts of flooding. This lesson points to the need for ongoing community education and involvement in disaster preparedness programs, ensuring that residents are not only aware of the risks but also equipped to respond effectively.

Another critical lesson is the need for integrated urban planning that incorporates flood risk management, especially in informal settlements. Many of Cape Town's informal settlements are situated in flood-prone areas due to historical and socio-economic factors that have confined poorer populations to less desirable land. While during the apartheid era, the government forcibly relocated people of colour to the Cape Flats, today, this area draws migrants from economically disadvantaged rural regions. Tens of thousands of residents now live in crowded conditions, either in low-cost, formally registered housing or in makeshift squatter camps.

Despite some progress, several challenges continue to recur with each flooding event (Table 1).

Table 1 Recurring challenges in flood management for informal settlements

Challenge	Description
Persistent Infrastructure deficiencies	Even with upgrades, the infrastructure in many informal settlements remains insufficient to handle severe weather events. The lack of durable roads, bridges, and proper drainage systems continues to exacerbate the vulnerability of these communities
Land tenure security	The informal nature of these settlements often means that residents lack legal claims to the land they occupy, which complicates the provision of municipal services and deters investment in permanent infrastructure improvements
Resource allocation	There is often a disparity in the allocation of resources for disaster management and recovery, with informal settlements typically receiving less than more formalized areas. This inequity affects the resilience capacity of these communities
Information and communication	Effective communication before, during, and after flood events remains a challenge. Ensuring that all community members receive timely and accurate information is crucial for minimizing risks and coordinating response efforts

5 Groundwater Dynamics and Flooding in Informal Settlements

5.1 Defining Ground Water Table

A water table, or groundwater table, is the level underground where the soil and rock are completely soaked with water. Imagine this section of ground as a sponge—it soaks up water and holds it there. The level of water can go up and down based on things like rain, snowmelt, and how much water people take from nearby wells and water supplies. Understanding the water table is important because it can affect the safety and condition of dwellings. The water table can rise when it rains a lot or when snow melts quickly. When the ground absorbs more water than it can handle, the water table goes up. It can also rise if there is a lot of water coming from higher places. On the other hand, the water table can drop during dry periods or if a lot of water is taken from nearby wells. The shape of the land also matters; water tables are usually higher in valleys and lower under hills.

In informal settlements, a high-water table presents significant risks to housing structures, particularly those built without formal planning or adequate foundations. When the water table rises, the increased pressure can undermine the stability of the soil, leading to ground saturation. This saturation exerts upward pressure on the foundations, potentially causing them to shift or crack. Over time, such pressure can result in uneven settling of the structure, leading to severe structural damage that is costly and complex to repair.

The presence of a high-water table elevates indoor humidity levels, creating an environment prone to various moisture-related problems. In informal settlements, where building materials often consist of impermanent or less durable components, increased humidity can accelerate the deterioration of materials through mold growth and wood rot. Additionally, metal components within these structures may rust more quickly due to the persistent dampness. This deterioration compromises the safety and longevity of homes, posing ongoing health and safety risks to residents.

Residents of informal settlements need to be vigilant about the signs of a high water table, as early detection can help mitigate some of its negative impacts (Table 2).

5.2 Groundwater Dynamics and Flooding: Understanding the Impact of Climate Change

Groundwater dynamics are significantly influenced by climate change, which affects both the quantity and quality of groundwater resources. The impact of climate change on groundwater levels is a complex interaction of various factors, including changes in precipitation patterns, temperature fluctuations, and sea-level rise (Taylor et al. 2013).

Table 2 Signs of a high water table in informal settlements

Sign	Description
Flooding	If multiple homes in the area experience flooding simultaneously, especially in lower-lying areas, it likely indicates a high water table
Standing water	Persistent puddles that do not drain after rainfall suggest that the ground is overly saturated, a common symptom of a high water table
Dampness in structures	Visible water stains or damp spots on floors and walls are clear indicators of excess moisture, possibly driven by water pushing upward from below the surface
Wastewater management issues	In many informal settlements, improper or insufficient wastewater systems can exacerbate problems with high water tables. If septic systems or informal drainage areas begin to back up, it might be due to the high water table preventing proper drainage and percolation

Firstly, climate change alters precipitation patterns, leading to variations in groundwater recharge rates. In regions where increased rainfall is observed, there might be a rise in groundwater levels due to enhanced recharge. Conversely, areas experiencing reduced rainfall and prolonged droughts face decreased groundwater recharge, leading to lower groundwater levels. These fluctuations are critical as they directly impact water availability for consumption, agriculture, and other uses, particularly in urban settings where demand is high.

Moreover, temperature increases due to climate change also play a crucial role. Higher temperatures lead to greater evaporation rates, which can reduce surface water bodies and indirectly decrease groundwater recharge. Additionally, warmer temperatures can increase water consumption, further straining groundwater resources. In urban areas like Cape Town, where heat islands are common due to dense construction and limited vegetation, the impact on groundwater can be intensified, exacerbating the challenges of managing urban water systems.

Sea-level rise is another critical factor affecting coastal groundwater systems. As sea levels rise, saltwater intrusion into coastal aquifers becomes more prevalent, leading to the salinization of groundwater resources. This not only reduces the availability of fresh water but also compromises water quality, posing serious challenges for urban water management in coastal cities like Cape Town. The intrusion of saltwater can render groundwater unsuitable for most uses without significant and costly treatment.

In Cape Town, these factors combine to create a precarious situation for groundwater management. The city has experienced both severe droughts and intense rainfalls in recent years, events that are becoming more common and severe with climate change (Savelli et al. 2021). These extreme weather conditions strain the city's groundwater systems, which are crucial for providing water to its sprawling informal settlements. During droughts, lowered groundwater levels can lead to water shortages, severely affecting communities that rely on groundwater for their daily needs. Conversely, intense and sudden rainfalls can lead to rapid increases in groundwater levels, which, when combined with poor drainage and the city's topography, lead to flooding, particularly in low-lying informal settlements.

5.3 Analysis of How Rising Groundwater Tables Contribute to Flooding

The relationship between rising groundwater tables and flooding is complex and significant, especially in urban environments where the consequences can be particularly severe. In Cape Town's informal settlements, understanding this connection is crucial for effective flood risk management and urban planning.

Rising groundwater tables occur when the balance between groundwater recharge and discharge is disrupted. This imbalance can be due to increased precipitation, reduced evaporation rates due to cooler local climates, or changes in land use that affect water absorption and runoff. In urban settings like Cape Town, impermeable surfaces such as pavement and buildings exacerbate this issue by preventing rainwater from naturally seeping into the ground, leading to higher runoff and reduced natural groundwater recharge. However, during periods of excessive rainfall or when changes in nearby land uses increase water inflow into the ground, groundwater levels can rise unexpectedly.

The impacts of groundwater flooding generated by Sea-Level Rise (SLR) are generally expected at lower elevations, in underground structures, close to tidal rivers and coastlines, where there is a direct pathway between oceans and exposed areas (Bosserelle et al. 2022) (Fig. 6).

When groundwater levels rise close to the surface, the capacity of the soil to absorb additional rainwater diminishes significantly. This condition is particularly critical in areas with poor drainage systems or where natural water paths have been blocked by urban development. For Cape Town's informal settlements, which are often situated in low-lying areas or on reclaimed land, this means that even moderate rainfall can lead to waterlogging and flooding. The elevated groundwater can saturate the soil, reducing its infiltration capacity and causing runoff to occur more rapidly than it can be managed by existing drainage infrastructure.

Moreover, the rise in groundwater levels can lead to the phenomenon of groundwater flooding, which occurs independently of river or surface water flooding. Groundwater flooding happens when the groundwater table rises to a point where water emerges on the surface, creating standing water without any apparent source like rainfall or overflowing water bodies. This type of flooding is particularly challenging to predict and manage because it does not necessarily follow patterns of storm events and can occur even during dry periods if groundwater levels continue to rise.

In addition to causing immediate flooding, elevated groundwater tables also pose long-term risks to urban infrastructure. The increased moisture can weaken building foundations and road beds, lead to the growth of mold and fungi, and exacerbate

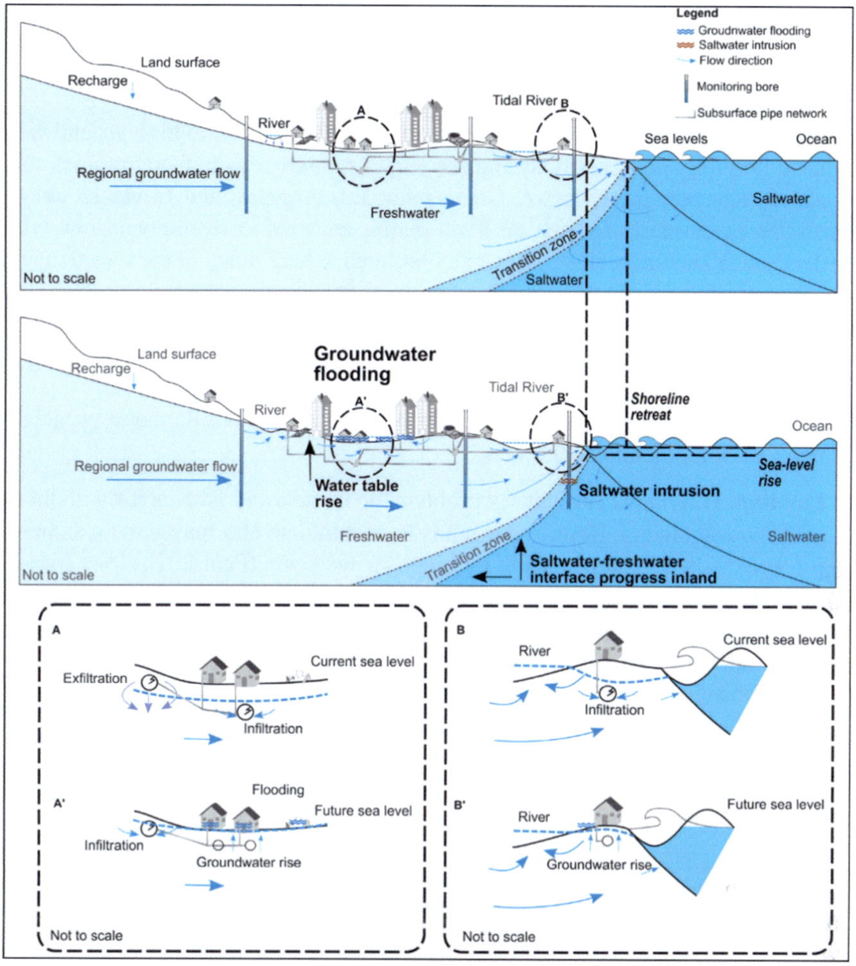

Fig. 6 Conceptual diagram showing the comparison of groundwater levels under current sea levels (top) and under sea-level rise (SLR) (middle) in urban to coastal settings. Bosserelle et al. 2022 *Source*

corrosion of pipes and other underground utilities. These effects can be particularly detrimental in informal settlements, where construction materials and techniques often are not equipped to handle prolonged exposure to moisture.

5.4 *Lived Realities of High Groundwater Induced Flooding in Cape Town Informal Settlements*

Informal settlements in Cape Town face severe challenges due to high groundwater induced flooding, particularly during the winter rainfall season. Communities such as Island, Khayelitsha, Zwezwe, Green Park, Los Angeles, and Covid-19 among others are precariously located on flood plains, adjacent to rivers, or in low-lying wetland areas known as vleis. The Cape Flats area, where many of these settlements are situated, epitomizes this challenge with its flat, sandy terrain that offers poor drainage and becomes a catchment for regular flooding. Each year, such conditions lead to the displacement of up to 8,000 households, significantly impacting the health, safety, and economic stability of thousands of residents.

The are different types of flooding in cape town's informal settlements especially related to rising ground water levels.

Ponding: Ponding is particularly problematic in vleis and other areas with inherently poor drainage. Here, water tends to accumulate and may remain standing for weeks, creating a persistent problem for residents (Feni 2018) (See Fig. 7). This situation is often worsened by a naturally high water table that becomes excessively waterlogged. Additionally, during periods of heavy rain, the existing stormwater drainage systems are overwhelmed, failing to manage the deluge. This results in significant water backup, which further exacerbates the flooding, especially when heavy runoff from paved surfaces flows into these already saturated areas.

Fig. 7 People still living in these shacks in Khayelitsha even though there is a large muddy pool right outside their homes. Feni 2018 *Source*

Water table or seepage flooding: Seepage flooding occurs when the water table rises above the ground surface, a common occurrence in the Cape Flats. This region, historically a wetland or vlei area, is particularly susceptible to this type of flooding due to its geographical makeup. Significant winter rainfall causes the underground aquifers to overflow, leading to widespread seepage that affects homes and public infrastructure. The rising groundwater not only floods homes but also seeps into the foundations, weakening structures and making them unsafe.
Riverine and estuarine flooding: Heavy rains and severe storms in the catchment areas can lead to riverine and estuarine flooding. The rivers and estuaries, unable to contain the rapid influx of water, often overflow their banks. This overflow can escalate to what is known as the 1 in 50-year or 1 in 100-year flood levels, creating catastrophic conditions for those living along these waterways. Figure 8a, b shows a boulder flood marker placed within the yard of an informal settlement resident in Island settlement, Cape Town. The stone markings starkly illustrate the proximity of extreme flood risk to daily life in informal settlements. The communities situated along these floodplains are subjected to sudden and severe flooding, posing risks to life and property and necessitating urgent and often life-saving responses.

The profound impact of high groundwater-induced flooding on the daily lives of residents in Cape Town's informal settlements is both devastating and complex.

Fig. 8 Large stone boulders marking the 1 in 100-year flood levels in Island informal settlements **a** and **b** Boulder flood marker placed within the yard of an informal settlement resident in Island settlement, Cape Town. *Source* Fieldwork

Fig. 9 RR Section, Site B, Khayelitsha badly flooded following the 2018 flooding. Feni 2018 *Source*

Personal accounts from community members vividly illustrate the struggles and hardships they endure during these frequent and severe flooding events (See Fig. 9). In Khayelitsha, a community leader describes the immediate impact on homes, stating, "Each winter, the floodwaters rush into our homes so quickly that many of us can't do anything to stop it. Last year, we tried elevating our floors with anything we could find—gravel, bricks, you name it. It was a temporary fix that barely held up against the relentless water." This sentiment is echoed across the settlements, where the rapid onset of flooding catches many residents off guard, forcing them to resort to makeshift solutions that offer little respite.

Tabeth, a resident and mother of three, shared her ongoing struggle with flooding: "As soon as the rain starts, water begins seeping from the ground. I spent hours using every bowl and dish I had to scoop out water, but it was useless. The flood was worse than anything I had imagined when moving here." Her experience highlighted the inadequate infrastructure that fails to protect residents from the rising waters, leaving them to fend for themselves in increasingly desperate conditions. The loss of personal belongings is a common and heart-wrenching outcome of these floods. Another resident recounted, "After the flood, returning home to find your life's possessions destroyed is a gut-wrenching experience. It is not just water. It is mud and debris that ruin everything they touch." This destruction extends beyond material losses, striking at the heart of their sense of security and home. Figure 10 shows a resident emptying water out of his flooded home in Khayelitsha informal settlement (Lali et al. 2019).

Health concerns are rampant, as a community leader in Green Park informal settlement explains, "Post-flood conditions are ripe for disease outbreaks. We see a spike in waterborne illnesses as contaminated water becomes a common problem. It is

Fig. 10 A resident empties water out of his flooded home in Khayelitsha informal settlement. Lali et al. 2019 *Source*

a public health emergency every time." The flooding exacerbates already challenging living conditions, making the spread of diseases more likely and more difficult to manage.

Displacement and the need for temporary shelter further complicate the lives of affected residents. A young mother from Khayelitsha shares her family's ordeal, "My family was displaced last winter, and we spent weeks in a temporary shelter. The crowded conditions were hardly ideal, especially with small children. It felt like we were escaping one hazard only to face another." This displacement disrupts lives and contributes to a cycle of instability and uncertainty. Figure 11 shows an abandoned shack in Siyahlala Town 2, Khayelitsha after the rain had flooded it too badly. Community solidarity often shines through in these tough times. A local NGO coordinator points out, "Despite the hardship, it is inspiring to see neighbours helping each other out during floods. People share their homes, food, and resources. But solidarity alone isn't enough; we need real solutions from those in power." This community spirit is a testament to the resilience of the residents but also underscores the inadequacy of current measures to effectively address the challenges they face.

Frustration with government response is palpable across the settlements. "The government's flood interventions feel like a band-aid on a gaping wound," a community leader expressed. "They come in, do a quick fix, and leave, without addressing the underlying issues. We're tired of temporary solutions to permanent problems." This sentiment reflects a broader disillusionment with the efficacy of governmental efforts in providing lasting solutions.

Educational disruptions due to flooding are another significant concern, as a father from Zwezwe explains, "Schools often close during major floods, and our kids fall behind every year. Educational disruption is one of the less talked about but critical

Fig. 11 Neighbours said this home in Siyahlala Town 2, Khayelitsha was abandoned because the rain had flooded it too badly. Feni 2018 *Source*

impacts of these floods." The interruptions in education threaten to have long-term effects on the community's youth, adding to the generational impact of these disasters.

Economic impact of flooding is severe, with a shop owner in Khayelitsha stating, "It is a complete standstill economically when the area floods. No one can go to work, and many lose their wages." The financial strain exacerbates the already challenging economic conditions faced by residents, highlighting the need for more robust economic support systems during and after flooding events.

Finally, the psychological toll of repeated flooding events cannot be underestimated. A community leader noted, "The repeated trauma of flooding leads to significant psychological stress. Many residents suffer from anxiety, sleeplessness, and even PTSD. The mental health aftermath is as severe as the physical destruction." Addressing the mental health impact is crucial for the community's overall recovery and resilience.

5.5 *Managing and Fixing High Water Table Issues in Informal Settlements*

In many informal settlements in Cape Town, managing a high-water table presents significant challenges that can severely impact the stability of homes and the well-being of residents. Effective management strategies are crucial to mitigate these impacts and ensure the safety and comfort of the community. From robust yard drainage systems that handle both surface and subsurface water to sophisticated waterproofing techniques and natural landscaping solutions, a comprehensive approach is needed (Table 3). These methods not only prevent structural damage by diverting water away from homes but also enhance the living environment, making them essential for long-term resilience in flood-prone areas.

Table 3 Strategies for managing high water table in informal settlements

Strategy category	Description
Integrated yard drainage solutions	Implementing comprehensive drainage systems is essential to manage high water tables effectively. Techniques like French drains, which are trenches filled with gravel or rock, help redirect both surface and groundwater away from homes. Additionally, constructing dry wells can control excess water by allowing it to percolate away from structural foundations. Proper grading of the settlement terrain to direct runoff away from homes is also crucial
Basement and foundation waterproofing techniques	In areas where formal basements are rare, ensuring the integrity of home foundations through waterproofing is vital. Utilizing sump pumps helps remove water accumulated in the ground, protecting against hydrostatic pressure. Interior drainage systems, like perimeter drains, are critical in keeping lower home sections dry. Applying waterproof coatings and sealing visible cracks are also effective measures to prevent moisture penetration
Utilizing natural drainage and landscaping	Employing natural drainage solutions through strategic landscaping helps manage soil moisture levels and reduces runoff. The roots of plants absorb water, while their canopies slow down rainfall impact. This dual role not only aids in water management but also improves the living environment, crucial in densely populated settlements. Integrating these natural techniques with structural solutions offers a holistic approach to water management
Engineering swales and customized water channels	Engineered solutions like swales, shallow vegetated ditches that capture and channel rainwater, are significant in alleviating high water table issues. Designed to follow the natural contours of the land, swales direct excess water toward designated drainage areas, preventing pooling around living spaces. Their placement and design should consider the overall landscape and drainage patterns of the settlement to maximize their effectiveness in water management

6 Technological Innovations for Flood Management and Resilience

Advancements in technology offer promising solutions for improving flood management and enhancing resilience in urban informal settlements. This section introduces various cutting-edge technologies that are currently being employed or have potential applications in the field of flood prediction and management in informal settlements. These technological innovations can significantly alter how communities predict, prepare for, and respond to flooding events.

6.1 Flood Prediction and Management Technologies

Effective flood management begins with accurate prediction and early warning systems. Advances in meteorological technology now allow for more precise weather forecasts, which are crucial for anticipating flood events. Technologies such as remote sensing and satellite imagery provide real-time data on weather patterns, water levels, and even soil saturation rates, which can be analyzed to predict flooding likelihood and severity.

Artificial Intelligence (AI) and machine learning are also transforming flood prediction. These technologies can process vast amounts of data from various sources, including weather stations, satellite images, and historical flood data, to identify patterns and predict future occurrences with high accuracy. For example, AI models can predict how rainfall will affect specific areas, allowing for targeted alerts to communities at risk. In Cape Town, implementing such AI-driven systems could provide crucial lead times for residents to prepare or evacuate, thereby reducing potential casualties and property damage.

Once a flood is predicted, effective management tools are essential. Geographic Information Systems (GIS) play a pivotal role in this regard, enabling city planners and disaster management teams to visualize flood data geographically. This helps in planning evacuation routes, identifying critical infrastructure at risk, and deploying resources more effectively. GIS can also assist in urban planning, helping to design flood-resilient urban spaces that incorporate natural water flow patterns and reduce impervious surfaces.

Internet of Things (IoT) devices, such as sensors placed in rivers, drains, and on critical infrastructure, can monitor water levels and flow rates in real-time. This information can be instantly communicated to central systems and warning networks, facilitating swift responses. For instance, smart drainage systems equipped with IoT sensors can automatically adjust to divert or store flood waters in less critical areas, thereby minimizing impact.

6.2 Flood Resilience Technologies with AI and IoT

Building resilience against flooding doesn't end with management; it also involves recovery and prevention. Technological innovations such as modular floating structures are being explored in some flood-prone regions. These structures rise with the water level, reducing the risk of water damage during floods. Additionally, construction technologies that utilize water-resistant materials and elevated building designs can help ensure that homes and infrastructure are less likely to be damaged in the event of a flood.

Another area of technological advancement is in community engagement platforms that utilize mobile and web-based applications to disseminate information quickly and gather community feedback. These platforms can be used for alerting residents about impending floods, providing information on safe zones, and facilitating post-flood recovery efforts by coordinating aid and assessing damage more efficiently.

The integration of Artificial Intelligence (AI) and the Internet of Things (IoT) into early warning systems represents a significant advancement in disaster management, particularly in the context of flood risk management. These technologies provide capabilities that enhance the accuracy, timeliness, and effectiveness of warnings, which are crucial for minimizing the impact of floods, especially in vulnerable urban areas like Cape Town's informal settlements.

6.3 Role of AI in Early Warning Systems

AI enhances flood forecasting and early warning systems through its ability to analyze large datasets quickly and accurately. By integrating data from various sources—such as satellite images, weather forecasts, historical flood data, and real-time sensors—AI algorithms can identify patterns and predict potential flood events with remarkable precision. For example, machine learning models can be trained to forecast the probability of flooding based on changes in weather patterns, river levels, and urban water flow dynamics. These predictions allow for timely warnings to be issued to at-risk communities, enabling early preparations and evacuations that can save lives and property. Moreover, AI can continuously learn and improve its predictions over time. As it processes more data, an AI system can adjust its models to better reflect the realities of a changing climate and urban development, thereby enhancing the accuracy of its forecasts. This adaptive capability is particularly important in areas experiencing rapid urbanization and environmental change, such as Cape Town.

IoT devices play a critical role in the operational aspects of early warning systems by providing real-time, actionable data. Networks of sensors placed throughout a city can monitor various environmental parameters that are indicative of flooding. These sensors can measure rainfall intensity, water levels in rivers and reservoirs, soil moisture, and even structural integrity of flood defenses. The data collected by these sensors are transmitted in real-time to centralized monitoring centers where AI systems analyze them to assess flood risks immediately. In practical terms, IoT enables a dynamic response to developing flood situations. For instance, sensors in storm drains can detect when water levels are rising unusually fast, triggering automatic alerts to both local authorities and the public. This immediate feedback loop allows for quicker mobilizations and more targeted emergency responses, reducing the time between detection and action.

6.4 Synergy of AI and IoT in Enhancing Early Warning Systems

The synergy between AI and IoT technologies enhances the functionality of early warning systems by enabling a proactive rather than reactive approach to flood management. AI's predictive analytics combined with IoT's real-time monitoring creates a powerful tool for disaster preparedness. For example, in Cape Town, such a system could preemptively identify areas that will be hardest hit by an upcoming storm and direct emergency services to prioritize these areas before the flood even begins. Furthermore, these technologies also support the post-event recovery phase. AI can analyze data collected during a flood to identify which strategies and responses were most effective, providing insights that can be used to improve future responses. IoT devices can continue to monitor affected areas for potential secondary impacts, such as landslides or further rising water levels, ensuring ongoing safety and support during the recovery process.

6.5 Sustainable Construction Materials and Flood-Resilient Infrastructure

In the context of increasing flood risks due to climate change, the importance of integrating sustainable construction materials and designing flood-resilient infrastructure cannot be overstated. These elements are critical for enhancing the sustainability and durability of urban environments, particularly in flood-prone areas like Cape Town's informal settlements.

Sustainable Construction Materials

The choice of materials used in construction plays a pivotal role in determining the resilience of buildings to flooding. Sustainable construction materials that are flood-resistant not only withstand the immediate impacts of water but also contribute to longer-term sustainability by reducing the need for frequent repairs and replacements, which can be both economically and environmentally costly.

One such material is autoclaved aerated concrete (AAC), which is lightweight, provides excellent insulation, and, most importantly, is highly resistant to water damage. Unlike traditional concrete or wood, AAC does not absorb water, which prevents the deterioration that often occurs in flood-exposed structures. Additionally, recycled plastic composites have emerged as a viable alternative to traditional wood, offering superior resistance to moisture and decay.

Green roofing systems, which involve the use of vegetative layers on rooftops, can also play a significant role in managing flood risks. These roofs absorb rainfall, reduce runoff, and help manage peak flow rates during heavy downpours. By retaining rainwater, green roofs can significantly mitigate the impact on the ground-level drainage system, thus easing the stress on urban sewer systems during storms.

Flood-Resilient Infrastructure

Beyond the use of sustainable materials, the design and implementation of flood-resilient infrastructure are crucial. This involves creating systems and structures designed to withstand, recover from, and adapt to flood conditions.

Elevated structures are one of the most effective means of flood-resistant architecture. Raising the foundation of buildings ensures that the critical structural and habitation areas are above typical flood levels. In Cape Town, adopting such elevation techniques in new constructions and retrofits in informal settlements can prevent water from entering homes during floods, thereby safeguarding both inhabitants and assets.

Flood walls and barriers are another essential component of resilient infrastructure. These can be permanent fixtures or temporary, deployable barriers that can be erected in anticipation of a flood event. Furthermore, the strategic placement of bioswales and rain gardens throughout urban areas can enhance the landscape's natural ability to absorb and filter runoff, reducing the volume of water that reaches critical areas.

Urban Planning and Flood Resilience

Integrating these materials and infrastructures into urban planning requires a coordinated and well-thought-out approach. Urban planners must consider not only the immediate cost and technical feasibility of these solutions but also their long-term sustainability and impact on the community. This includes ensuring that the materials are locally available or sustainable to produce, and that the infrastructure designs are culturally and contextually appropriate. For effective implementation, partnerships between local governments, community organizations, and private sector stakeholders are essential. These collaborations can facilitate knowledge sharing, provide necessary funding, and ensure that the infrastructural developments are aligned with the community's needs and preferences.

7 Community-Based Approaches and Participatory Planning

Beyond adopting cutting edge technologies, incorporating community-based approaches and participatory planning into urban development and disaster management strategies is essential for creating resilient and sustainable communities, especially in areas vulnerable to flooding like informal settlements in Cape Town. This section explores the importance of engaging community members in the planning and implementation processes of flood management initiatives, highlighting the benefits of such involvement.

7.1 Importance of Community Involvement and Participatory Planning Processes

Community involvement in urban planning and disaster management is crucial for several reasons. First, it ensures that the solutions developed are relevant and appropriately address the specific needs and concerns of the community. Residents often have intimate knowledge of their environment and can identify potential problems and propose practical solutions that outsiders might overlook. For example, in flood-prone areas, local inhabitants can provide valuable insights into historical flood patterns, effective evacuation routes, and safe havens that are not evident on maps or to external observers. Second, involving community members in the planning process helps build a sense of ownership and responsibility towards the implemented solutions. When people are actively involved in the decision-making process, they are more likely to support and maintain the initiatives, leading to better sustainability and effectiveness of the projects. This is particularly important in informal settlements where local governance structures may be weak, and community leadership plays a pivotal role in mobilizing and directing resident actions.

Participatory planning involves various methodologies that facilitate direct involvement from the community in the planning and implementation stages. Techniques such as community workshops, focus groups, and public consultations allow for the collection of diverse viewpoints and foster a collaborative environment where community members can contribute openly. For instance, participatory mapping is a powerful tool where residents use maps to identify key resources, risks, and areas needing attention in their community. This process not only gathers spatial data but also enhances community members' understanding of their environment and its challenges. In Cape Town, such mapping could be crucial for identifying flood-prone zones and planning appropriate interventions.

Implementing community-based projects effectively requires careful coordination and support from local authorities and NGOs. These entities can provide the necessary resources, training, and expertise that communities might lack in flood adaptation. For example, local governments can supply materials and technical assistance for building flood-resistant infrastructure, while NGOs can offer training on disaster preparedness and response. Furthermore, it is important to establish clear communication channels between the community, government agencies, and other stakeholders. Regular updates, meetings, and feedback sessions ensure that all parties remain informed and engaged throughout the project lifecycle. This ongoing dialogue helps adapt flood management projects as needed based on real-time feedback and changing conditions.

7.2 Strategies for Empowering Residents and Local Governance Structures

Empowering residents and strengthening local governance structures are pivotal steps toward improving resilience and adaptive capacity in flood-prone areas such as the informal settlements in Cape Town. This section outlines strategies that can facilitate community empowerment and enhance the effectiveness of local governance in managing flood risks (Table 4).

8 Gaps in Current Legislative and Regulatory Frameworks

Addressing flood risks effectively, especially in urban areas prone to frequent and severe flooding like informal settlements in Cape Town, requires a robust legislative and regulatory framework. However, current frameworks often exhibit significant gaps that can hinder efficient and equitable flood management. This section identifies critical gaps and outlines their implications for flood resilience.

8.1 Inadequate Coverage of Informal Settlements

A significant gap in many urban flood management frameworks is the inadequate coverage of policies concerning informal settlements. These areas often develop without formal planning or adherence to building regulations, making them particularly vulnerable to flooding. Existing laws and regulations may not recognize informal settlements officially, which means they often do not receive the necessary infrastructure investments, such as proper drainage systems, which are crucial for managing flood risks. Additionally, the lack of legal recognition for these communities can inhibit residents' ability to claim assistance or participate in decision-making processes related to flood management.

Table 4 Strategies for empowering residents and local governance structures

Strategy	Description
Community education and training	Education and training programs are vital for empowering residents by building local capacity to manage flood risks. These can range from teaching basic flood preparedness and emergency response to sustainable building techniques and community-based water management. Workshops might include instructions on constructing raised platforms for housing or creating rain gardens to absorb floodwaters, while local schools and community centers act as hubs for ongoing education about climate change impacts
Inclusive participatory planning	Inclusive participatory planning ensures that all community voices, especially those of vulnerable groups, are heard in managing flood risks. This involves regular community meetings, participatory mapping tools to identify risks, and integrating local knowledge into planning. Local governments can support these efforts by providing platforms for dialogue and ensuring the processes are transparent and accountable
Strengthening local governance capacities	Local governance structures need strengthening to effectively manage flood risks, requiring both resource allocation and decentralization of decision-making powers. Local authorities should have the resources to respond promptly to flooding, supported by dedicated funding for disaster risk reduction and local emergency response teams. Additionally, investing in GIS technologies and training enhances risk assessment and planning
Establishing community-based monitoring systems	Community-based monitoring systems allow residents to actively participate in observing and reporting flood risks, involving measures like marked poles for water levels or sensor-based systems accessible via smartphones. These systems provide real-time data, enhancing response efforts and promoting a culture of preparedness
Creating collaborative networks	Creating networks among residents, local governments, NGOs, and other stakeholders facilitates a coordinated response to flooding. These networks support the sharing of resources, information, and best practices and are instrumental in advocating for policy changes and securing funding. Partnerships with academic institutions and the private sector can also bring additional expertise and resources

8.2 Fragmented Responsibility Among Agencies

Effective flood risk management requires coordinated actions across multiple sectors, including water management, urban planning, disaster response, and public health. However, a common gap in the legislative framework is the fragmentation of responsibilities among different government agencies. This lack of integrated approach can lead to inefficient resource use, conflicting policies, and delays in response times during flooding events. Without a unified command or a clear delineation of responsibilities, it can be challenging to implement comprehensive and coherent flood management strategies.

8.3 Limited Focus on Adaptive and Resilient Infrastructure

While some regions have begun to incorporate climate change adaptation into their urban planning policies, many still lack a specific focus on developing adaptive and resilient infrastructure. Current regulations often emphasize short-term solutions rather than long-term resilience. This gap is particularly problematic given the increasing frequency and intensity of extreme weather events due to climate change. Legislation needs to prioritize investments in infrastructure that can adapt to changing climate conditions and mitigate the impacts of floods when they occur.

8.4 Insufficient Community Involvement in Policymaking

Another significant gap is the limited involvement of affected communities in the policymaking and planning processes. Community members often have first-hand knowledge of the challenges and risks associated with living in flood-prone areas and can offer valuable insights into effective solutions. However, regulatory frameworks frequently do not provide mechanisms for meaningful community engagement, which can lead to the implementation of policies that are not well-suited to the local context or that fail to gain public support.

8.5 Lack of Comprehensive Data and Research

Effective flood management policies must be based on accurate and comprehensive data regarding flood risks, vulnerabilities, and the effectiveness of different management strategies. A critical gap in many legislative frameworks is the lack of provisions for ongoing research and data collection. Without this data, it is difficult to assess the current risks accurately, predict future challenges, or evaluate the effectiveness of policies and interventions.

8.6 Recommendations to Address These Gaps

To address these gaps, several steps should be considered:

- **Legal recognition and support for informal settlements:** Updating the legislative framework to recognize and include informal settlements can ensure that these areas receive necessary infrastructure investments and legal protections.
- **Integrated policy approach:** Establishing inter-agency collaborations and creating integrated policies that span different sectors can enhance the efficiency and effectiveness of flood risk management.
- **Investment in resilient infrastructure:** Legislators should prioritize funding for infrastructure that can withstand extreme weather events and adapt to long-term environmental changes.
- **Formal channels for community participation:** Developing formal mechanisms for community involvement in planning and decision-making processes can enhance the relevance and acceptance of flood management policies.
- **Enhanced research and data collection:** Implementing policies that promote ongoing research and data collection related to flood risks and management strategies can provide a solid basis for informed decision-making.

9 Recommendations for Policy Enhancements and Multi-Scalar Governance Approaches

To effectively address the increasing risks of flooding in urban informal settlements, such as those in Cape Town, policy enhancements and multi-scalar governance approaches are essential. These strategies should aim to integrate national, regional, and local governance efforts while ensuring inclusivity, sustainability, and adaptability. This section outlines key recommendations for improving policies and governance frameworks to strengthen flood resilience.

9.1 Strengthen Local Governance Capacities

Local governments play a pivotal role in implementing flood risk management policies, yet they often face resource and capacity constraints. Strengthening local governance involves allocating adequate funding, providing technical training, and equipping local authorities with tools such as Geographic Information Systems (GIS) for flood risk mapping and monitoring. Enhanced local capacities can improve the planning, implementation, and maintenance of flood mitigation measures tailored to the unique challenges of informal settlements. Additionally, decentralizing decision-making powers can enable local authorities to act swiftly in response to flood risks. For example, giving municipalities control over local disaster response budgets and resources allows them to implement immediate and context-specific interventions. Establishing dedicated disaster management units within municipal governments can further enhance their ability to coordinate flood risk reduction efforts.

9.2 Foster Integrated Multi-Scalar Governance

Effective flood risk management requires a coordinated approach that aligns efforts across national, regional, and local levels. Policies must be designed to ensure vertical and horizontal integration, creating synergies between various governance scales and sectors. For instance, national governments can provide overarching frameworks and funding, while regional bodies facilitate cross-municipality collaboration, and local governments implement community-specific initiatives. Integrated governance can also improve resource sharing and knowledge exchange. For example, regional authorities can establish centralized databases to share real-time data on rainfall, water levels, and flood risks, enabling more coordinated responses across multiple jurisdictions.

9.3 Enhance Community Participation in Policy Formulation

Policies are more effective when they reflect the lived experiences and needs of the people they aim to serve. To achieve this, governance frameworks should incorporate mechanisms for meaningful community engagement. This includes participatory planning processes, public consultations, and the establishment of community advisory boards. Involving community members not only ensures that policies are relevant and practical but also fosters a sense of ownership and accountability. For example, residents can contribute local knowledge about historical flood patterns, vulnerable areas, and effective coping mechanisms, which can be invaluable for designing responsive policies.

9.4 Develop Comprehensive Climate Resilience Policies

Flood risk management should be integrated into broader climate resilience policies that address the underlying drivers of vulnerability. These policies should include measures to mitigate climate change, such as reducing greenhouse gas emissions, while also focusing on adaptation strategies, such as enhancing urban green spaces, improving drainage systems, and promoting sustainable land use practices. Policies must also address socio-economic vulnerabilities that exacerbate flood risks in informal settlements, including poverty, housing insecurity, and limited access to basic services. Comprehensive strategies that combine physical infrastructure improvements with social interventions can create more equitable and sustainable outcomes.

9.5 Strengthen Regulatory Frameworks and Enforcement

Strong regulatory frameworks are critical for ensuring compliance with flood risk management policies. Governments must enforce zoning laws that restrict construction in flood-prone areas, mandate the use of flood-resistant building materials, and ensure the maintenance of critical infrastructure such as drainage systems and levees. To enhance enforcement, governments can adopt digital tools and technologies. For instance, drone surveillance and remote sensing can be used to monitor compliance with zoning regulations and detect unauthorized developments in vulnerable areas.

9.6 Promote Public–Private Partnerships (PPPs)

Public–private partnerships can provide additional resources and expertise to support flood risk management. Private sector involvement can drive innovation, such as the development of smart water management technologies, and contribute funding for large-scale infrastructure projects. Governments can create incentives, such as tax breaks or grants, to encourage private sector investment in flood resilience initiatives. PPPs can also facilitate the delivery of affordable insurance products tailored to low-income households in informal settlements, helping residents recover more quickly after flood events.

9.7 Establish Monitoring and Evaluation Mechanisms

To ensure the effectiveness of policies and governance strategies, it is essential to establish robust monitoring and evaluation mechanisms. Regular assessments can identify gaps, measure progress, and provide evidence-based recommendations for policy adjustments. These mechanisms should include key performance indicators (KPIs) related to flood mitigation, community resilience, and governance efficiency. Monitoring systems should be transparent and accessible to the public, fostering accountability and trust between governments and communities. Governments can also leverage AI and data analytics to analyze trends and optimize policy interventions over time.

10 Conclusions and Future Directions

The impacts of climate change on urban informal settlements, particularly in Cape Town, highlight the pressing need for comprehensive flood risk management strategies. This chapter explored the intersection of rising groundwater tables, flooding, and the vulnerabilities faced by informal settlements, offering insights into innovative solutions, community engagement, and governance enhancements. The findings highlight several critical implications for Cape Town and similar urban areas globally.

The analysis reveals that informal settlements in Cape Town are disproportionately affected by flooding due to their location on marginal lands, inadequate infrastructure, and socio-economic vulnerabilities. Rising groundwater tables, exacerbated by climate change, further compound the risks, leading to frequent flooding that disrupts livelihoods, damages housing, and undermines health outcomes. The study also highlights the effectiveness of integrating cutting-edge technologies, such as Artificial Intelligence (AI) and the Internet of Things (IoT), into early warning systems, as well as the potential of sustainable construction materials and flood-resilient infrastructure to mitigate flood impacts.

Community-based approaches and participatory planning emerge as indispensable components of effective flood risk management. Engaging residents in decision-making processes fosters a sense of ownership and ensures that solutions are contextually relevant and sustainable. However, the evaluation of existing policies reveals significant gaps, including inadequate support for informal settlements, fragmented governance, and limited community involvement.

Given the increasing frequency and intensity of extreme weather events, ongoing adaptation and resilience-building efforts are essential. Climate change is a dynamic and evolving challenge, requiring policies and strategies that are flexible, forward-looking, and responsive to emerging risks. In Cape Town, this means prioritizing investments in resilient infrastructure, enhancing the capacity of local governance structures, and fostering a culture of preparedness within communities. Resilience-building efforts must also address the socio-economic drivers of vulnerability, including poverty and insecure land tenure. Policies should focus on empowering communities through education, economic opportunities, and equitable access to resources, ensuring that resilience is built not just into the physical environment but also into the social fabric of vulnerable populations.

Addressing the multifaceted challenges of flooding in informal settlements requires integrated efforts from policymakers, technologists, and community leaders. Policymakers must develop and enforce comprehensive, multi-scalar governance frameworks that align national, regional, and local efforts. These frameworks should promote collaboration across sectors, integrating urban planning, disaster management, and climate adaptation policies.

Technologists have a crucial role in advancing innovative solutions, such as AI-driven early warning systems, smart infrastructure, and community engagement platforms. These technologies must be designed with accessibility and scalability in mind, ensuring that even the most resource-constrained communities can benefit. Community leaders and organizations are essential partners in driving grassroots initiatives, mobilizing resources, and fostering social cohesion. Their involvement ensures that interventions are inclusive, equitable, and aligned with the needs and aspirations of local residents.

In conclusion, the challenges posed by climate change and flooding in informal settlements demand urgent and collaborative action. Cape Town, like many cities in the Global South, stands at a critical juncture where it can either continue to struggle with recurring disasters or lead the way in building resilient, inclusive, and adaptive urban environments. By fostering integrated efforts across all levels of governance and sectors of society, cities can not only mitigate flood risks but also enhance the overall quality of life for their most vulnerable populations. This chapter calls on stakeholders to act decisively, innovate boldly, and collaborate inclusively to shape a sustainable and resilient future for all.

References

Abiodun BJ, Adegoke J, Abatan AA, Ibe CA, Egbebiyi TS, Engelbrecht F, Pinto I (2017) Potential impacts of climate change on extreme precipitation over four African coastal cities. Clim Change 143:399–413

Armond NB, Sant'Anna Neto JL (2019) The urban climate system and the impacts of flooding on Rio de Janeiro, Brazil. Urban Climates in Latin America 259–80

Bhanye J (2024) Beyond informality:'Nimble peri-urban land transactions'-how migrants on the margins trade, access and hold land for settlement. Discov Glob Society 2(1):6

Bhanye JI, Mangara F, Matamanda AR, Kachena L (2023) COVID-19 Lockdowns and the urban poor in Harare. Springer International Publishing AG, Zimbabwe

Bhanye JI, Dzingirai V, Campbell MM (2024) Transactional and supplementary strategies for accessing land among migrants on the margins: an ethnographic study among Malawian migrants at Lydiate informal settlement, Zimbabwe. In: Secondary cities and local governance in Southern Africa. Cham: Springer International Publishing, pp 87–114

Bosserelle AL, Morgan LK, Hughes MW. (2022) Groundwater rise and associated flooding in coastal settlements due to sea-level rise: a review of processes and methods. Earth's Futur 10(7):e2021EF002580

Drivdal L (2016) Flooding in Cape Town's informal settlements: Conditions for community leaders to work towards adaptation. South African Geographical Journal Suid-Afrikaanse Geografiese Tydskrif 98(1):21–36

Dube K, Nhamo G, Chikodzi D (2021) Rising sea level and its implications on coastal tourism development in Cape Town, South Africa. J Outdoor Recreat Tour 33:100346

Ehebrecht D. (2015) The challenge of informal settlement upgrading: Breaking new ground in Hangberg, Cape Town? Universitätsverlag Potsdam

Feni M (2018) Shacks flooded after week of Cape rains. https://groundup.org.za/article/townships-left-flooded-after-cape-rains/ [Accessed 20 July 2025]

Fitchett JM, Grant B, Hoogendoorn G (2016) Climate change threats to two low-lying South African coastal towns: risks and perceptions. S Afr J Sci 112(5–6):1–9

Haque AN (2021) Climate risk responses and the urban poor in the global South: the case of Dhaka's flood risk in the low-income settlements. Int J Disaster Risk Reduction 64:102534

Johannes B, Dzingirai V, Chirisa I (2023) Transnational migrants on the margin: agency, aspirations, and perceptions for the future among Malawian migrants in Zimbabwe's Norton peri-urban. In: The Palgrave encyclopedia of urban and regional futures. Cham: Springer International Publishing, pp 1876–1886

Kikwasi G, Mbuya E (2019) Vulnerability analysis of building structures to floods: The case of flooding informal settlements in Dar es salaam, Tanzania. Int J Build Pathol Adaptation 37(5):629–656

Lali V, Feni M, GroundUp Staff. (2019) Cape Town residents battle spring floods. https://groundup.org.za/article/cape-town-flooded-spring/

Putsoane T, Bhanye JI, Matamanda A (2024) Extreme weather events and health inequalities: exploring vulnerability and resilience in marginalized communities. Dev Environ Science 15:225–248

Rammile S, Bhanye JI, Lehobo T, Hugo M. (2024) Building resilient urban Infrastructure and Innovation for Urban public health: a case study of Bloemfontein, South Africa. In: Sustainable development goals and urban health: strides, challenges and way forward for poor neighborhoods. Cham: Springer Nature Switzerland, pp. 69–96

Savelli E, Rusca M, Cloke H, Di Baldassarre G (2021) Don't blame the rain: social power and the 2015–2017 drought in Cape Town. J Hydrol 594:125953

Taylor RG, Scanlon B, Döll P, Rodell M, Van Beek R, Wada Y, Longuevergne L, Leblanc M, Famiglietti JS, Edmunds M, Konikow L (2013) Ground water and climate change. Nat Clim Chang 3(4):322–329

Udo F, Bhanye J, Daoudiallo B, Naidu M. (2024) Evaluating the sustainability of local women's climate change adaptation strategies in Durban, South Africa: a feminist political ecology and intersectionality perspective. Sustainable Development

Williams DS, Máñez Costa M, Sutherland C, Celliers L, Scheffran J (2019) Vulnerability of informal settlements in the context of rapid urbanization and climate change. Environ Urban 31(1):157–176

Zhou L, Kori DS, Sibanda M, Nhundu K (2022) An analysis of the differences in vulnerability to climate change: a review of rural and urban areas in south Africa. Climate 10(8):118

Climate Change Impact of Cutting-Edge Technology and Groundwater Sustainability

Deepashree Choudhury

Abstract Climate change is exerting unprecedented pressure on global water resources, particularly on groundwater, which plays a crucial role in meeting human, agricultural, and industrial needs. Groundwater is highly susceptible to changes in recharge patterns, increased extraction, and quality degradation due to the escalating effects of climate change. This chapter aims to examine the impact of climate change on groundwater and discuss how recent technological advancements can help manage and sustain this vital resource through a detailed literature study. Also, a brief case study of the National Capital Region (NCR), India, with a focus on Sanjay Lake in East Delhi, is included to discuss groundwater recharge strategies with reference to its role in sustainable groundwater management.

Keywords Groundwater sustainability · Climate change · Cutting-edge technology · Groundwater recharge

1 Introduction

Climate change poses significant challenges to water resources worldwide, with groundwater being particularly susceptible to its impacts (Dao et al. 2024). Groundwater, which constitutes nearly 30% of the world's available freshwater, is essential for drinking water, agriculture, and industrial uses (IGRAC—International Groundwater Resources Assessment Centre 2021). It is a key resource, especially in regions prone to water scarcity, where it serves as a buffer against surface water shortages (Scanlon et al. 2023). However, climate change is affecting groundwater systems through altered recharge rates, increased demand due to changing weather patterns, and potential quality degradation. These impacts threaten the sustainability of groundwater, a critical component of global water security (Mukherjee et al. 2021). Addressing these issues requires the application of innovative approaches to enhance groundwater sustainability. Cutting-edge technologies, such as remote

D. Choudhury (✉)
Department of Interior Design, Budha Group of Institutions, Karnal, India
e-mail: ar.deepachoudhury@gmail.com

A. Nath et al. (eds.), *Groundwater Depletion and Sustainability*, Advances in Geographic Information Science, https://doi.org/10.1007/978-3-032-09921-1_7

sensing, artificial intelligence, and advanced recharge systems, have emerged as promising tools for monitoring and managing groundwater more effectively. The deployment of these technologies aims to better understand groundwater dynamics, enhance recharge processes, optimize groundwater use, and mitigate risks associated with over-extraction and contamination (Shaikh and Birajdar 2024b).

This chapter aims to examine the impact of climate change on groundwater and discuss how recent technological advancements can help manage and sustain this vital resource. We will explore different aspects of these technologies, their effectiveness, and their limitations in promoting sustainable groundwater management.

Groundwater contributes to a large extent to maintain the equilibrium of freshwater resources, with its usage spanning agricultural, industrial, and domestic sectors. Groundwater availability and quality are influenced by various natural and anthropogenic factors, including climate patterns, population growth, and economic activities (Patel et al. 2020). With climate change, these dynamics have become increasingly complex. Changing rainfall patterns, prolonged droughts, and more frequent extreme weather events are all altering the natural recharge of aquifers and increasing the dependence on groundwater as a water source (Akhtar et al. 2021). Additionally, increased temperatures lead to greater evaporation, reducing the effectiveness of recharge processes, while rising sea levels can exacerbate saltwater intrusion into coastal aquifers (Abanyie et al. 2023).

These challenges necessitate the development of advanced solutions for managing groundwater sustainably. Emerging technologies hold the potential to transform groundwater monitoring, management, and policy development (Wang et al. 2022). Remote sensing and Geographic Information System (GIS) technologies, for example, have provided enhanced capabilities for real-time monitoring of groundwater levels and recharge areas. Machine learning models are being used to predict future water availability based on climate scenarios and groundwater extraction patterns. Moreover, artificial recharge techniques and managed aquifer recharge (MAR) systems offer a means to increase the resilience of aquifers against climate-induced stress (Arabameri et al. 2021).

While these technologies present new opportunities, their adoption faces several challenges, such as high costs, lack of technical expertise, and varying policy frameworks (Shaikh and Birajdar 2024a). Thus, it is essential to not only explore the technical capabilities of cutting-edge technology but also evaluate their feasibility, limitations, and the enabling conditions required for their widespread implementation (Ministry of Jal Shakti, Government of India 2018). This chapter will provide a comprehensive overview of these technologies, their role in addressing climate-induced impacts on groundwater, and the pathways toward ensuring groundwater sustainability amidst climate change.

1.1 Overview of the State of Research on Climate Change on Groundwater Resources

During the past decades, the impacts of climate change on groundwater resources have garnered increasing attention, with researchers employing various hydrological models to evaluate these effects. Groundwater recharge, a critical component of these studies, is influenced by hydrological processes, surface structures, and soil characteristics (Bates et al. 2008).

An initial study conducted in the Coastal Plain of Western Australia utilized a one-dimensional model of the unsaturated zone, employing Richard's equation as its foundation, to analyze how changes in rainfall affect recharge. The study revealed that factors beyond rainfall, such as vegetation cover, play a significant role in determining recharge dynamics (Anandhi et al. 2011).

Groundwater models have proven indispensable for understanding the complexities of groundwater systems and supporting effective management and decision-making. These models provide a theoretical framework to explore system dynamics and the influences of human activities, making them invaluable tools for water resource assessment, conservation, and restoration. They offer cost-effective insights for developing, evaluating, and refining strategies, legislation, and designs related to groundwater management. The availability of diverse groundwater modeling codes, each with unique capabilities and limitations, enhances their applicability across various scenarios (Arnold et al. 2007).

Incorporating climate change into groundwater studies often involves using outputs from general circulation models (GCMs) combined with techniques such as delta methods or stochastic weather generators to simulate changes in recharge. Downscaling methods are frequently applied to refine the spatial resolution of GCMs, enabling more precise climate projections at the watershed scale. Numerical models are also widely used to simulate flow in the vadose zone, groundwater systems, and watersheds, projecting responses to changes in recharge or surface water interactions. In coastal areas, models that account for saltwater density effects are critical for assessing the potential impacts of sea level rise on aquifers (Candela et al. 2009).

Despite significant advancements, research on the effects of climate change on groundwater quality remains limited. Some studies have examined how changes in recharge affect the flushing of salts or contaminants from the vadose zone, while others have explored the impact of altered flow regimes on pollutant concentrations from point and nonpoint sources. However, the ability to compare findings across studies is constrained by differing assumptions regarding climate inputs, greenhouse gas (GHG) emissions, evaluation timeframes, and hydrological modeling methodologies.

The site-specific nature of hydrogeological settings further complicates extrapolation. Factors such as recharge characteristics, soil and aquifer properties, snow presence, and groundwater-surface water interactions significantly influence study outcomes. Consequently, while groundwater modeling and climate projections offer powerful tools for assessing climate impacts and informing groundwater management, care must be taken when generalizing findings across different regions or conditions.

1.2 Dataset and Methodology

This chapter presents a comprehensive literature review of the key topics under investigation. The review involved an extensive search across various academic and institutional databases, including Google Scholar, Scopus, Web of Science, SAGE Journals, the Global Groundwater Information System (GGIS), and the India Water Resource Information System (WRIS). The WRIS, established by the Ministry of Jal Shakti (MoJS) in 2018 as part of the National Hydrology Project, serves as a centralized repository for nationwide water resources data, providing a "single window" access to updated information on water resources and related themes. Additionally, specialized resources like the Developing Library Network (DELNET) in New Delhi and ShodhGanga (a repository of Indian theses and dissertations) were explored for insights, particularly regarding the impacts of climate change on groundwater management. From an initial screening of 100 research papers, 30 were selected for detailed review based on their relevance to the chapter's objectives. Supplementary materials, including working papers, reports, and books, were also analyzed to provide a broader context and strengthen the study's framework. Furthermore, a brief case study of the National Capital Region (NCR), India, with a focus on Sanjay Lake in East Delhi, is included to discuss groundwater recharge strategies.

2 Impact of Climate Change on Groundwater

Climate change significantly impacts groundwater resources, which are vital for agricultural, industrial, and domestic uses. The effects are felt through alterations in recharge patterns, increased demand and overexploitation, and quality degradation. These changes threaten the sustainability and availability of groundwater. This section discusses these impacts in detail.

2.1 Changes in Recharge Patterns

Groundwater recharge is the process by which water from precipitation and other sources infiltrates the ground and replenishes aquifers. Climate change affects this process in multiple ways, primarily through altered precipitation patterns, increased evaporation rates, and the occurrence of extreme weather events (Hughes et al. 2021).

Altered Precipitation Patterns and Increased Evaporation Rates: Climate change leads to shifts in precipitation patterns, affecting the amount, intensity, and timing of rainfall. In some regions, prolonged dry periods are becoming more common, reducing the amount of water available for infiltration and recharge. Conversely, some areas are experiencing more intense rainfall over shorter periods. While these heavy rain events may seem beneficial, they often lead to increased surface runoff rather than groundwater recharge, as the ground may not absorb the water effectively during such rapid events. Moreover, rising global temperatures are increasing evaporation rates, which directly affects groundwater recharge. Higher temperatures lead to greater evaporation of surface water and moisture in the soil, reducing the overall availability of water that can infiltrate into the ground. This phenomenon is particularly pronounced in arid and semi-arid regions, where the balance between evaporation and precipitation is already precarious (McCallum et al. 2020).

Extreme Weather Events:

Extreme weather events, such as floods and droughts, are becoming more frequent and intense as a result of climate change. Droughts can lead to reduced groundwater recharge by limiting the availability of surface water to infiltrate aquifers. During prolonged droughts, the water table may decline, making it more challenging for the aquifer to recover, even after the drought ends. Floods, on the other hand, can have complex effects on groundwater recharge. In some cases, floodwaters may contribute to recharge; however, in many situations, the rapid influx of water leads to increased runoff rather than infiltration. Additionally, floods can carry contaminants from surface sources, such as agricultural fields and urban areas, into aquifers, affecting groundwater quality (Davamani et al. 2024).

Type of event	Region	Event description	Impact on groundwater	Source/References
Drought	California, USA	Prolonged drought (2012–2016)	Severe depletion of groundwater; overdrafting led to land subsidence and long-term aquifer damage	NASA, USGS Reports (California Water Science Center, 2023; NASA Scientific Visualization Studio, 2014)
Flood	Queensland, Australia	2010–2011 floods after long drought	Initial recharge benefit, but most water ran off; also caused groundwater contamination in some areas	Commonwealth Scientific and Industrial Research Organisation (CSIRO), Australian Bureau of Meteorology (Floods, 2022; Chap. 1—Australian Bureau of Meteorology Flood Forecasting and Warning, 2016)

(continued)

(continued)

Type of event	Region	Event description	Impact on groundwater	Source/References
Drought	Marathwada, India	Recurring droughts (2014–2016)	Decreased recharge and increased dependence on groundwater for agriculture, causing aquifer stress	Central Water Comission-CWC, Indian Meteorological Department-IMD (M. P. Shewale & Kumar, 2005)
Flood	Kerala, India	Catastrophic floods in 2018	Recharge in some aquifers noted, but pollutants and septic overflows affected groundwater quality	Central Pollution Control Board-CPCB (Hydrological Studies Organisation, Hydrology (S) Directorate, 2018), Kerala State Groundwater Board (Ground Water Department & Central Ground Water Board, Government of Kerala, Government of India, 2022)
Drought	Cape Town, South Africa	Day Zero crisis (2017–2018)	Heavy reliance on groundwater due to surface water scarcity; led to over-extraction of coastal aquifers	City of Cape Town Water Reports (Brühl & Visser, 2021; Water and Sanitation Directorate, City of Capetown, 2019)
Flood	Bihar, India	Seasonal floods in the Ganga Basin	Limited recharge due to high runoff; contamination from agricultural runoff and household sewage	National Disaster Management Authority-NDMA (National Remote Sensing Centre, Government of India, 2020)
Drought	Rajasthan, India	Chronic drought-prone conditions	Led to adoption of Managed Aquifer Recharge (MAR) to counter low natural recharge rates	Groundwater Year Book (CGWB), 2022 (Central Ground Water Board, Rejuvenation, Ministry of Jal Shakti, 2023)

(continued)

(continued)

Type of event	Region	Event description	Impact on groundwater	Source/References
Flood	Jakarta, Indonesia	Regular Urban flooding due to monsoons and land subsidence	Poor recharge due to paved surfaces; floodwaters infiltrated aquifers with high contamination risks	United Nations Climate Change Technology Executive Committee-UNFCCC (United Nations Climate Change Technology Executive Committee, 2023; Nasution et al. 2022; Asdak et al. 2018)

2.2 *Increased Demand and Overexploitation*

The increasing frequency and severity of climate change-related impacts on surface water resources are contributing to a growing reliance on groundwater. This shift has led to increased demand and subsequent overexploitation, especially in water-scarce regions.

- Rising Temperatures and Changing Precipitation Patterns:

Climate change has led to rising temperatures, which in turn increases evapotranspiration from crops, resulting in higher water requirements for agriculture. As surface water availability becomes more unreliable due to altered precipitation patterns, there is a growing dependence on groundwater for irrigation purposes. In many regions, groundwater has become the primary source of water for agriculture, particularly in dry seasons when surface water is insufficient. For example, in the Indian States of Punjab and Haryana, the Green Revolution-induced reliance on high-yielding crop varieties like paddy has intensified groundwater withdrawal due to the high water requirements of these crops. Rising temperatures have further compounded the irrigation demand. The Central Valley of California, United Sytes responsible for a significant share of U.S. agriculture, has seen severe depletion of groundwater during recurring droughts, especially in the 2012–2016 period. With surface water reservoirs low, groundwater pumping increased drastically, resulting in overdraft.

Figure 1 illustrates a temporal comparison between average annual temperature rise and the volume of groundwater extracted, highlighting the influence of climate change on groundwater dependence in two highly stressed agricultural regions.

Additionally, the increasing variability in precipitation patterns has created uncertainty in water availability, further driving the dependence on groundwater for both drinking water and irrigation as a stable water source. During prolonged dry spells, groundwater often serves as the primary or only reliable water source, leading to increased extraction to meet the needs of communities, agriculture, and industries

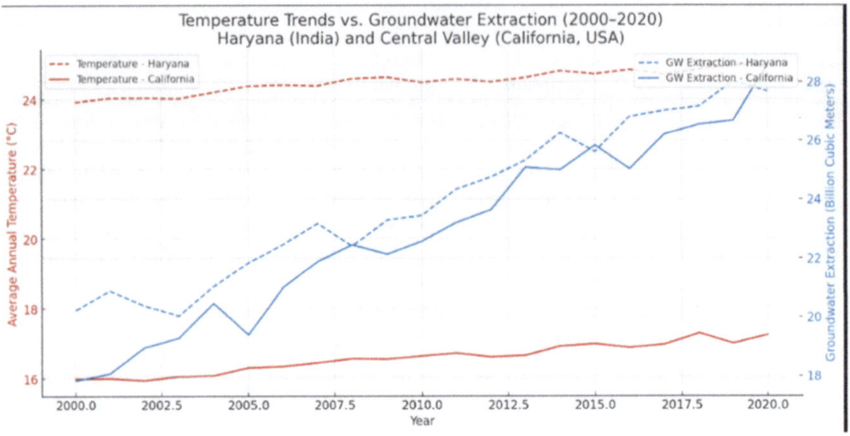

Fig. 1 Graph Depicting the Correlation Between Rising Temperatures and Increasing Groundwater Extraction in Haryana (India) and Central Valley (California, USA). (Central Ground Water Board, North Western Region, Chandigarh and Ground Water Cell, Irrigatiin and Water Resource Department, Haryana, 2023; Haryana Ground Water Department, 2025; California Department of Water Resources, 2023)

(Wu et al. 2020). In Marathwada, Maharashtra, an erratic monsoon rains over the last decade have led to increased dependence on groundwater. Over 80% of irrigation in the region is now groundwater-based, resulting in severe aquifer stress (Khetwani and Singh). In Sub-Saharan Africa, irregular rainfall and prolonged dry seasons have made groundwater the only dependable source of water in many rural communities (Nyika and Dinka 2023).

- Groundwater Depletion and Land Subsidence:

Overexploitation of groundwater resources has led to a significant decline in groundwater levels in many parts of the world. In areas where recharge rates are not sufficient to replenish the extracted water, groundwater depletion occurs. This is particularly evident in agricultural regions where groundwater is extracted at a much higher rate than it is recharged, leading to a gradual decline in the water table.

One of the consequences of groundwater depletion is land subsidence, which occurs when the water removed from underground aquifers results in the compression of the soil structure. This can lead to damage to infrastructure, such as roads, buildings, and pipelines, and can also reduce the aquifer's capacity to store water in the future. Land subsidence is a significant concern in urban areas, where the structural integrity of buildings and infrastructure is compromised due to the sinking of the ground (Hasan et al. 2023).

Aquifer depletion and the resulting land subsidence are critical challenges in several parts of the world, leading to profound environmental, infrastructural, and socio-economic consequences. In Mexico City, overextraction of groundwater has caused the city to sink at an alarming rate of 20–30 cm annually, resulting in tilted

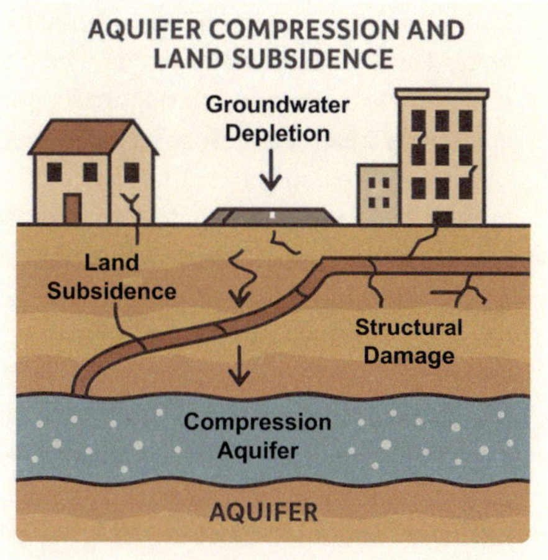

Fig. 2 Aquifer compression due to excessive groundwater extraction. (*Source* Author)

buildings, cracked roads, and damaged water supply systems, thereby imposing significant economic burdens and public safety concerns (Chaussard et al. 2014). Similarly, in India's Northwest region, particularly around Delhi and western Uttar Pradesh, aggressive groundwater pumping for agriculture and urban needs has led to a sharp decline in water tables—exceeding 1 m per year in some areas (Central Ground Water Board [CGWB], 2022). The situation is no less dire in Jakarta, Indonesia, where certain northern regions have experienced subsidence of over 2.5 m within the last decade, forcing the government to consider relocating the capital city due to the increasing vulnerability of the existing urban fabric (Abidin et al. 2015). These examples underline the urgent need for sustainable groundwater management and integrated land-use planning to mitigate the risks associated with excessive groundwater extraction (Fig. 2).

2.3 *Quality Degradation*

Climate change affects both the quantity and quality of groundwater. Rising sea levels, coastal flooding, and changes in agricultural practices contribute to the degradation of groundwater quality, posing a significant threat to the sustainability of this vital resource.

- Sea-Level Rise and Coastal Flooding Sea-level rise, driven by global warming, is a critical concern for coastal aquifers. As sea levels rise, saltwater intrusion into freshwater aquifers becomes more common. This process leads to the contamination of groundwater supplies, rendering them unsuitable for drinking, irrigation, and other uses. Coastal flooding also exacerbates saltwater intrusion by pushing saline water further inland, compromising the quality of groundwater in coastal regions.

- Saltwater intrusion has severe implications for communities that rely on groundwater for their freshwater needs. Once an aquifer is contaminated with saltwater, it becomes challenging to restore its quality, and the water may require extensive and costly treatment before it can be used again. The long-term effects of saltwater intrusion threaten the availability of potable water in coastal areas, making it a critical concern in the context of climate change adaptation (Yu et al. 2022).
- Increased Use of Fertilizers and Pesticides: The impact of climate change on agriculture, including prolonged dry periods and changing growing conditions, has led to changes in agricultural practices, such as increased use of fertilizers and pesticides to maintain crop yields. This increased usage can have detrimental effects on groundwater quality. Excess fertilizers and pesticides applied to agricultural fields can leach into the soil and eventually contaminate groundwater. Nitrate contamination, in particular, is a growing concern in many agricultural regions, as elevated nitrate levels in drinking water can pose serious health risks. Pesticides and herbicides also pose a threat to groundwater quality, as these chemicals can persist in the environment for extended periods and accumulate in groundwater, affecting its suitability for consumption and irrigation.
- Moreover, the higher frequency of extreme rainfall events can lead to increased runoff from agricultural fields, further exacerbating the risk of groundwater contamination. Runoff during heavy rain events often carries fertilizers, pesticides, and other pollutants into aquifers, threatening the safety of groundwater supplies (Shishaye et al. 2021).

3 Modern Artificial Recharge Methods

This section examines contemporary artificial recharge techniques, focusing on both direct and indirect approaches to increasing groundwater levels. With global water demand rising, artificial recharge has emerged as a crucial tool for sustainable water management.

3.1 Types of Artificial Recharge Methods

Direct Methods: Direct recharge methods involve the application of water directly to the surface, where it percolates into the ground and replenishes the aquifers below. These methods are designed to enhance recharge rates in areas where natural recharge is insufficient or where groundwater levels have been depleted due to over-extraction (Bhattacharya 2010).

Flooding: Controlled flooding is employed in areas where surface runoff can be diverted into large basins or low-lying lands. This allows water to infiltrate the soil over time. In Israel, for instance, managed aquifer recharge through surface flooding

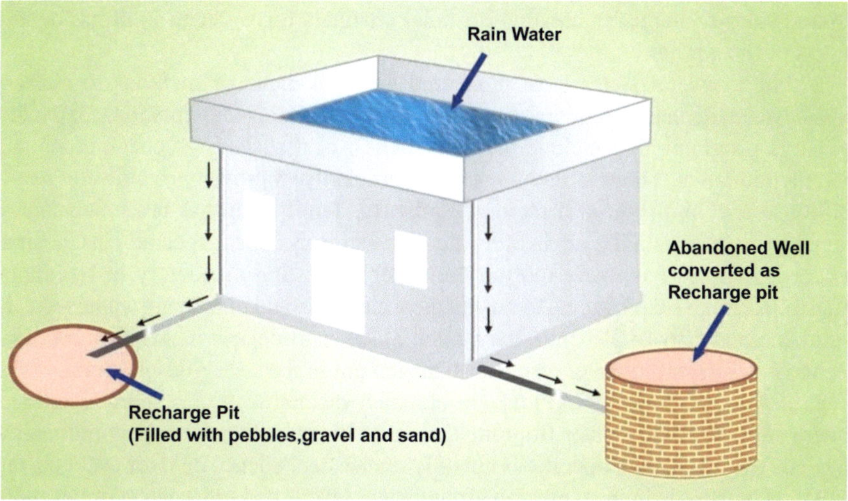

Fig. 3 Recharge structures for groundwater recharge

has successfully enhanced water storage and reduced the dependency on desalinated water (Fig. 3).

Ditch and Furrow: This method involves creating small ditches or furrows in areas with suitable soil conditions to guide water from rivers or streams into the ground. California has implemented ditch and furrow systems in its agricultural regions to recharge over-extracted aquifers, particularly in the Central Valley.

Recharge Basins: Recharge basins are shallow, man-made depressions designed to hold water, allowing it to infiltrate into the underlying aquifer. Widely used in India and the United States, recharge basins are an effective way to store rainwater or treated wastewater. The Florida Aquifer Storage and Recovery Project employs recharge basins to capture excess surface water and store it underground for later use during dry periods.

Induced Recharge Methods: Induced recharge methods are aimed at modifying aquifers or the landscape to improve their ability to absorb and retain water. These approaches often involve altering the subsurface structure or implementing engineering techniques to increase recharge efficiency.

Aquifer Modification: This method involves altering the physical structure of an aquifer to increase its capacity to hold water. The Netherlands has used subsurface aquifer storage systems, where treated water is injected into confined aquifers, increasing both water storage and availability during dry seasons.

Bore Blasting: Bore blasting, or well stimulation, involves drilling or blasting wells in aquifers to create fractures or voids, which improve water percolation rates. In India, bore blasting is employed in regions like Gujarat and Karnataka, where natural percolation is slow due to hard rock formations. The technique improves groundwater recharge by creating artificial channels for water to infiltrate deeper layers of the aquifer.

In India, several states have pioneered artificial recharge methods to address groundwater depletion. Gujarat has been at the forefront, implementing extensive recharge programs that include the construction of thousands of check dams and percolation tanks. These structures have significantly improved groundwater availability in arid and semi-arid regions. Similarly, Tamil Nadu has taken substantial steps by institutionalizing urban rainwater harvesting systems, especially in Chennai, where legislation has made rooftop rainwater harvesting mandatory in buildings. These efforts have contributed to raising the water table and improving water security in urban areas (Board 2010). In the United States, particularly in Arizona, artificial recharge is integral to water resource management in the state's desert climate. The Central Arizona Project (CAP) has developed a vast network of recharge facilities, where surplus surface water from the Colorado River is diverted and stored underground. These recharge operations not only reduce dependence on river water during droughts but also help in stabilizing groundwater levels, making water supplies more resilient for future needs (Area 2019). In Australia, artificial recharge has become an essential part of adapting to the effects of climate change. In Western Australia, stormwater harvesting is commonly practiced, with recharge basins designed to capture runoff from urban areas. The harvested stormwater is allowed to percolate into the ground, replenishing aquifers and providing a reliable source of water for irrigation and municipal use during dry spells. This strategy has gained prominence due to the region's declining rainfall patterns and increasing water demand (Hu et al. 2023). Germany presents an innovative example from Europe, where artificial recharge has been integrated with advanced wastewater treatment. In Berlin, treated wastewater is filtered through natural soil systems into underlying aquifers. This method not only replenishes groundwater reserves but also maintains water quality, ensuring a sustainable and safe water source for urban use. The process demonstrates how artificial recharge can align with environmental standards and urban infrastructure to enhance water security (Valhondo et al. 2020).

Modern artificial recharge techniques have advanced significantly in recent years, offering innovative ways to replenish depleted aquifers. By combining traditional methods with modern engineering approaches, regions worldwide can address the growing challenges of water scarcity and ensure sustainable groundwater management for future generations (Figs. 4, 5 and 6).

3.2 Urban Groundwater Recharge Solutions

Rooftop rainwater harvesting: It is gaining popularity as an urban recharge solution, where rainfall collected on rooftops is directed into recharge pits or wells to augment groundwater levels. The first Indian state to make rooftop rainwater harvesting compulsory for all buildings is Tamil Nadu. This was done in 2001 to

Fig. 4 Percolation tank for water storage (Jeet and Veeranna 2020)

Fig. 5 Check dam for surface water harvesting (Jeet and Veeranna 2020)

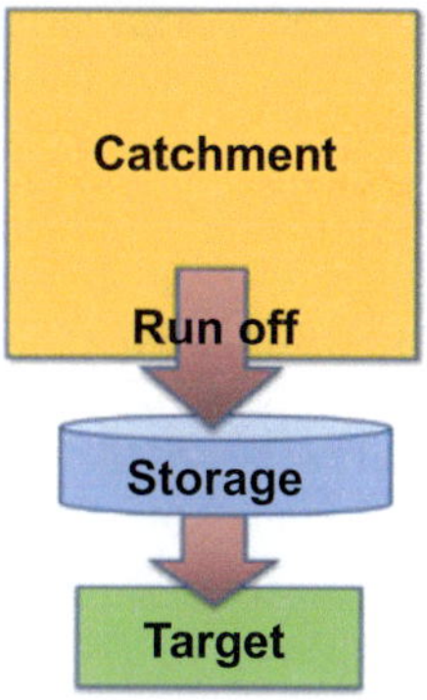

Fig. 6 Components of rainwater harvesting

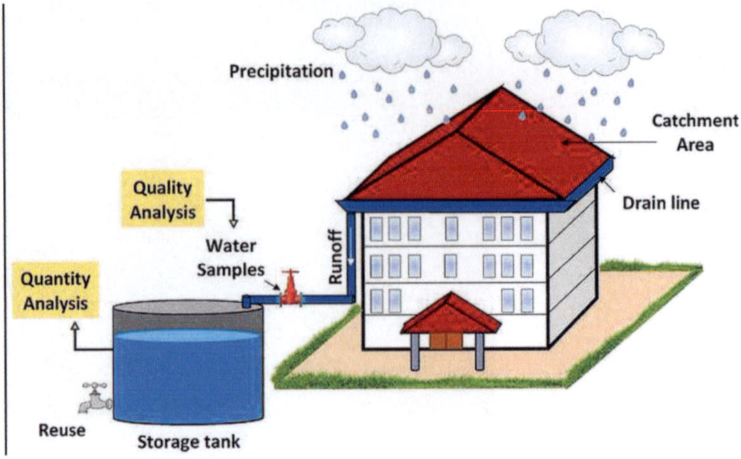

Fig. 7 Rooftop rainwater harvesting

avoid groundwater depletion. Tamil Nadu also launched a "Rainwater Harvesting" ordinance in 2003 to further promote the practice (Fig. 7).

Groundwater recharge from urban megastructures: Urban megastructures—such as malls, IT parks, high-rise residential towers, and expansive commercial complexes—significantly alter natural hydrological cycles. These developments often feature large impervious surfaces, including rooftops, parking lots, and paved open spaces, which drastically reduce groundwater infiltration and increase surface runoff. As rainfall cannot percolate into the ground naturally, it is swiftly drained away, leading to lower aquifer recharge and increased urban flooding risks. To counteract this, integrated groundwater recharge solutions must be embedded within the design of these structures. In Hyderabad's HITEC City area, dedicated recharge pits and stormwater collection systems have been constructed in large IT parks, allowing runoff from expansive roof and pavement areas to be diverted into engineered trenches and recharge wells. The Infosys Campus in Pune incorporates rooftop rainwater harvesting, stormwater collection, and infiltration trenches to retain most of its rainwater on-site. Likewise, Phoenix Market City in Mumbai uses recharge pits and percolation tanks to manage runoff from its vast built-up space. These examples demonstrate how embedding recharge structures in megastructures can effectively address issues like aquifer depletion, urban flooding, and the urban heat island effect, ultimately supporting sustainable water management and urban resilience (Singh et al. 2014).

Protection and enhancement of ground water recharge capacity of ponds, lakes, and other natural or artificial water bodies. These play a crucial role in supporting ecosystems and maintaining the hydrological balance and act as major recharge zones for groundwater. These water bodies allow rainwater and surface runoff to seep into the ground, replenishing underground aquifers. In regions facing water scarcity, their protection becomes even more essential as they serve as sources of drinking water, irrigation for agriculture, and microclimate regulation. Besides groundwater recharge, they help control floods by absorbing excess water, support biodiversity by hosting various aquatic plants and animals, and improve the aesthetic and recreational value of an area. Unfortunately, rapid urbanization, pollution, and neglect have led to the degradation of many such ecosystems. Hence, systematic efforts are required for their reclamation and restoration. Several techniques have been developed and adopted to restore and manage ponds and lakes effectively. One of the primary methods is desilting and deepening, where the accumulated silt is removed to restore the original depth and increase water holding capacity. This helps improve percolation and recharge of the groundwater table. Another essential technique is catchment area treatment. This involves afforestation, the construction of small check dams or bunds, and the reduction of erosion through vegetative cover. These measures not only reduce sedimentation in the water body but also promote rainwater infiltration into the soil. Effective inlet and outlet management is also crucial to maintaining the health of water bodies. Sedimentation tanks or filters are installed at the inlets to prevent polluted water or debris from entering the lake. Constructed wetlands are another innovative solution—using aquatic plants, these wetlands filter incoming water and enhance biodiversity. In addition to structural methods, rainwater harvesting systems can be integrated to direct surface runoff into these water bodies, thereby increasing their recharge potential. Recharge pits and trenches around lakes can also serve to channel water directly into aquifers. Lastly, community involvement is key to ensuring long-term sustainability. When local residents are engaged in the restoration process through education and stewardship programs, the chances of success and maintenance increase significantly. Several case studies demonstrate the successful restoration of water bodies with direct positive impacts on groundwater levels. One such example is the Kundalahalli Lake in Bengaluru, India. Once heavily polluted and encroached upon, the lake was restored through desilting, bund construction, inlet treatment using sedimentation tanks and wetlands, and replanting of native species. The project was led by both municipal authorities and local residents, who now help monitor the lake's health. A diagram of this lake shows distinct zones for inflow, sedimentation, wetland filtration, and recharge wells, creating a self-sustaining ecosystem. Another successful example is Sukhna Lake in Chandigarh, which faced issues of siltation due to erosion from the catchment area. Restoration involved afforestation of the catchment, construction of check dams upstream, and periodic desilting of the lakebed. These efforts significantly reduced sediment inflow and improved water storage and percolation. The lake diagram typically includes upstream forested areas, silt traps, and clearly defined recharge zones. A notable traditional example comes from the Alwar district in Rajasthan, where local communities revived old water harvesting structures called Johads. These small earthen check dams capture and store rainwater, allowing it to seep into the ground. Over time, they have helped raise the water table, rejuvenate dried-up wells, and support agriculture year-round. The diagram of a Johad typically features an earthen embankment, overflow channel, and percolation zone. These case studies highlight that the reclamation and restoration of water bodies is not only feasible but also vital for sustaining groundwater resources and ecological health. Integrating modern engineering with traditional knowledge, and involving the community, ensures that these efforts are both scientifically sound and socially inclusive.

Village Tanks as Recharge Structures are traditionally used in India and many other parts of the world for irrigation and domestic purposes, hold immense potential in addressing the growing problem of groundwater depletion. These tanks, often built centuries ago by communities, were designed to collect rainwater and regulate local water use throughout the year. However, due to neglect, encroachment, and sedimentation, many of these systems have lost their effectiveness. With proper restoration and scientific enhancement, these tanks can be revitalized and transformed into powerful groundwater recharge systems. The process of modifying existing village tanks involves several key interventions. First and foremost is desilting—removing layers of sediment that have accumulated over the years. Desilting restores the tank's storage capacity and improves water percolation through the tank bed. This allows more rainwater to seep into the ground rather than evaporating or overflowing. In many cases, the silt removed from the tanks is nutrient-rich and can be reused on agricultural fields to enhance soil fertility. Another effective method is the construction of recharge wells or percolation pits within or near the tank. These structures allow stored surface water to infiltrate rapidly into the deeper layers of the earth, reaching aquifers and recharging wells and boreholes in the vicinity. Stone-pitched percolation trenches can also be built around tanks to channel water directly to the recharge zone. Furthermore, check dams and contour bunds in the upstream catchment can reduce runoff speed, allowing more water to be captured and stored in the tank. The treatment of inflow channels is equally important. These channels should be cleared and realigned to ensure that clean water flows into the tank. In some cases, bio-filters or sedimentation tanks are installed to remove contaminants and debris. Additionally, planting native vegetation around the tank helps stabilize the embankments, reduces erosion, and enhances water quality. A successful example of such a transformation is seen in the Hiware Bazar village in Maharashtra, India. Once facing drought and migration, the village revived its traditional water harvesting systems. The tanks were desilted, new percolation tanks and check dams were constructed, and recharge shafts were drilled. As a result, groundwater levels rose dramatically, and the village became water-secure. Farmers could grow multiple crops, and the village even reported reverse migration. Another remarkable case is in Ralegan Siddhi, also in Maharashtra. Under the leadership of social activist Anna Hazare, village tanks were desilted, and a series of contour trenches and check dams were built. These interventions significantly improved groundwater recharge and turned the once-barren village into a model of sustainable water management. In Andhra Pradesh, the Mission Kakatiya project restored over 45,000 village tanks. Each tank was scientifically surveyed and restored with desilting, bund strengthening, and improved inlet/outlet management. The project led to significant improvements in groundwater levels, crop yields, and farmer incomes across the state.

Groundwater dams and subsurface dykes are innovative and effective structures used to conserve groundwater by intercepting and storing subsurface flows in aquifers. These structures function by retarding the natural underground flow of water in porous soil and rock formations, thereby increasing the water table in the upstream area and making groundwater more readily available through wells and boreholes. These methods are particularly useful in arid and semi-arid regions where surface water is scarce and rainfall is seasonal. A groundwater dam is typically constructed across a seasonal stream where there is alluvium or weathered rock that allows groundwater to flow underground. The dam consists of an impermeable barrier—often made of concrete, clay, or stone masonry—built below the ground level. This barrier intercepts the natural subsurface flow and causes water to accumulate behind it in the aquifer itself, rather than on the surface. The stored groundwater can then be accessed through wells in the upstream area. A subsurface dyke (or underground dyke) works on a similar principle. It involves constructing an underground wall across the flow direction of groundwater. The wall is buried completely beneath the surface, leaving no visible structure above ground. This method is advantageous because it does not result in land submergence, avoids evaporation losses, and does not interfere with land use, making it ideal for agricultural zones. Both groundwater dams and subsurface dykes improve groundwater retention without the need for large-scale land acquisition or displacement. They increase the availability of water in wells and hand pumps, especially during dry periods, and can contribute to drought-proofing rural areas. A notable example of a successful groundwater dam project comes from the Karnataka watershed programs, where groundwater dams were constructed across seasonal streams with sandy beds. These structures resulted in a significant rise in the water table, increased the duration of stream flow, and improved access to irrigation water. Farmers in the area reported increased crop yields and reduced dependency on erratic monsoon rains. Another successful case is the subsurface dyke constructed at Kothapally village in Telangana, India. A stone masonry wall was built 3 m below the surface across a stream bed. Post-construction, water levels in nearby wells rose substantially, and even during dry months, water remained accessible. The intervention helped stabilize farming activities in the region and supported livestock during drought seasons. In Japan and Africa, subsurface dams have long been used to store underground water efficiently. In Kenya's Kitui region, a series of sand and subsurface dams were built to harvest seasonal flow in sandy riverbeds. These projects led to improved water availability for both domestic use and small-scale irrigation.

In the context of increasing water scarcity and rapid urbanization, reusing treated wastewater for groundwater recharge through artificial means has emerged as a sustainable and efficient solution. Instead of discharging treated wastewater into rivers or letting it go to waste, it can be directed into aquifers using various recharge techniques. This method not only reduces the pressure on freshwater sources but also contributes to maintaining groundwater levels in urban and semi-urban areas where demand is high and recharge is limited. Artificial recharge using treated wastewater involves treating municipal sewage or industrial effluent to an acceptable standard and then introducing the water into the ground through structures such as recharge pits, percolation tanks, infiltration basins, or recharge wells. In some advanced systems, aquifer storage and recovery (ASR) is also used, where treated water is injected into deep aquifers and later recovered when needed. The treatment process usually involves primary (removal of solids), secondary (biological treatment), and tertiary (filtration, chlorination or UV disinfection) stages to ensure the water is safe for recharging aquifers. The advantages of this approach are numerous. Firstly, it provides a reliable, year-round source of recharge water, especially in urban areas where wastewater is continuously generated. Secondly, it reduces environmental pollution by preventing untreated or partially treated sewage from entering water bodies. Thirdly, it supports the sustainability of groundwater-dependent agriculture and drinking water supplies, particularly during dry seasons. Additionally, this method helps in improving groundwater quality, especially in over-extracted and saline aquifers, by diluting contaminants with cleaner treated water. A pioneering example is the National Capital Region (NCR) of Delhi, India, where treated wastewater from sewage treatment plants (STPs) is being reused for groundwater recharge in areas like Dwarka and Palla. The treated water, after undergoing tertiary treatment, is routed through recharge wells and infiltration basins, helping raise the groundwater levels in severely depleted zones. Another example is from Chennai, Tamil Nadu, where the Koyambedu Sewage Treatment Plant treats nearly 45 ML per day (MLD) of wastewater. A portion of this water is used to recharge aquifers through an advanced system of recharge wells and percolation ponds. This has contributed significantly to stabilizing groundwater levels in adjoining residential zones. In Israel, a world leader in wastewater reuse, treated wastewater is used for aquifer recharge via spreading basins in the Dan Region Reclamation Project. The system combines high-level treatment with soil-aquifer filtration (SAF), allowing for natural polishing of the water as it percolates through the soil. This process has enabled safe, large-scale reuse of wastewater in agriculture and urban water supply.

4 Cutting-Edge Technologies in Groundwater Management

In response to the growing pressures on groundwater resources, technological advancements are playing a critical role in managing and sustaining groundwater. These cutting-edge technologies—including remote sensing, AI, IoT, and advanced water treatment systems—offer innovative tools for monitoring, prediction, and quality improvement. Below is a detailed exploration of these major technologies and their contributions to sustainable groundwater management.

Remote Sensing and Geographic Information Systems (GIS), Satellite Imagery, and GIS Tools for Groundwater Monitoring-Remote sensing and GIS are powerful tools for large-scale groundwater monitoring and management-Satellite imagery provides extensive data on land surface conditions, which can be correlated with subsurface groundwater levels. Remote sensing can detect changes in land deformation linked to groundwater extraction, while GIS offers spatial data analysis and visualization, helping identify recharge areas and manage aquifers more efficiently. Satellite imagery, combined with GIS tools, provides valuable data on land surface conditions, vegetation health, and water bodies. This data is used to infer subsurface groundwater levels and identify areas for recharge. By integrating environmental, social, and economic factors, decision-makers can optimize resource allocation, improve conservation efforts, and ensure groundwater supplies meet long-term sustainability goals.

The Bhujal Monitoring Program uses satellite-based remote sensing to track groundwater extraction and recharge patterns in various states in India. In the USA, The Gravity Recovery and Climate Experiment (GRACE) satellite mission has been instrumental in monitoring global groundwater depletion, including in areas like California (Shaikh and Birajdar 2024).

Artificial Intelligence (AI) and Machine Learning (ML)-AI and ML technologies are transforming how data is analyzed for groundwater management, offering predictive capabilities that traditional methods lack. AI-powered adaptive management systems can optimize groundwater extraction and recharge activities in real time. By predicting periods of scarcity and adjusting resource allocation accordingly, these technologies enhance the sustainability of groundwater use, particularly in regions facing chronic water shortages. In Australia, CSIRO's AI models are used to forecast groundwater availability in the Great Artesian Basin, ensuring water security in one of the driest continents (Bahri and Dasarathy 2024).

Predictive Models for Groundwater Availability and Demand-AI processes vast datasets on climate, land use, and groundwater extraction, enabling ML algorithms to predict future groundwater levels and recharge rates. These predictive models help in forecasting trends in groundwater availability and demand, offering a proactive approach to preventing overexploitation. Machine learning algorithms process vast datasets on climate, land use, and groundwater extraction to forecast future trends in groundwater levels and demand. In India, The Groundwater Information System developed by the Indian government integrates ML models to predict water scarcity trends in agriculture-heavy regions like Punjab. In the USA and Texas, AI-powered predictive models are used to forecast groundwater depletion in the state's major aquifers (Tao et al. 2022).

Internet of Things (IoT) and Sensor Networks-IoT devices and sensor networks provide real-time monitoring of groundwater conditions, offering a dynamic approach to water management. IoT devices and sensor networks offer real-time monitoring of groundwater levels, quality, and flow rates. These sensors can be deployed across aquifers, wells, and boreholes, providing continuous data to manage groundwater more dynamically. This real-time monitoring allows for quick detection of issues like depletion or contamination, enabling timely corrective actions. In India, the Central Groundwater Board is using IoT-based sensors in states like Gujarat to monitor water quality and groundwater depletion. In the Netherlands and Friesland, IoT networks are being used to monitor groundwater extraction by farmers, ensuring compliance with sustainable water use regulations. With real-time data, water managers can rapidly adjust groundwater extraction rates or implement artificial recharge methods when necessary. For example, when sensors detect declining water levels or deteriorating water quality, immediate measures can be taken to mitigate the impact and ensure the sustainability of groundwater resources (Pradeep et al. 2017).

Advanced Water Treatment Technologies-Advanced water treatment technologies, such as membrane filtration and advanced oxidation processes (AOPs) improve the quality of groundwater, ensuring it remains safe for consumption and other uses. Membrane filtration techniques like reverse osmosis and nanofiltration remove contaminants such as heavy metals and chemicals from groundwater. Advanced oxidation processes (AOPs), which use hydroxyl radicals to degrade organic pollutants, offer another layer of protection, particularly in regions affected by agricultural and industrial contamination. In India, Gujarat has implemented reverse osmosis plants to treat groundwater contaminated by high salt content, making it potable for rural populations. In California, US, advanced membrane filtration and AOP systems are being used to treat groundwater contaminated by industrial waste, making it safe for irrigation and drinking.

5 Case Studies

Table 1 highlights case studies showcasing innovative and advanced technologies employed in groundwater management. These examples illustrate how cutting-edge tools and strategies are effectively addressing challenges such as groundwater depletion, quality monitoring, and recharge optimization in diverse contexts.

The studies reveal that technological advancements play a transformative role in sustainable groundwater management. Remote Sensing and GIS have emerged as essential tools for large-scale monitoring and the identification of suitable recharge zones, particularly in regions facing excessive groundwater extraction or drought. Successful applications of these technologies in places like California and Gujarat have led to more efficient water resource planning. Similarly, Artificial Intelligence (AI) and Machine Learning (ML) are increasingly being used to develop predictive models that guide the optimized use of groundwater. This is particularly evident in agricultural states such as Karnataka and Punjab, where these tools have significantly improved water-use efficiency and minimized wastage. Additionally, the integration of Internet of Things (IoT) technologies has facilitated real-time monitoring of groundwater levels and quality. Australia's implementation of IoT systems has enabled timely responses to emerging issues, and Gujarat has also benefited from their use in preventing over-extraction and monitoring pollution. Together, these technologies provide a robust framework for managing groundwater sustainably, offering effective solutions to address water scarcity both globally and within India.

Table 1 Case studies describing cutting-edge technologies in groundwater management

	Case study	Description	Outcomes
1	Remote sensing in groundwater management: The Case of California, USA (Chaussard et al. 2017)	California has faced persistent droughts, leading to a significant decline in groundwater levels. Satellite-based remote sensing and GIS tools were deployed to monitor groundwater extraction and identify recharge areas. The Gravity Recovery and Climate Experiment (GRACE) satellite provided crucial data to assess groundwater depletion	These tools allowed water managers to identify critical groundwater depletion zones and direct resources for targeted interventions like artificial recharge and water conservation projects. Remote sensing enables large-scale, real-time monitoring of groundwater resources, especially in regions where direct access to aquifers is difficult. It supports more informed decision-making for sustainable groundwater use
2	AI and ML in Predictive Groundwater Modeling: The Case of India (Nordin et al. 2021)	In the agricultural regions of Punjab and Karnataka, AI-based predictive models have been developed to assess groundwater availability and forecast future demand. These models analyze vast datasets on rainfall, water extraction, and land use to generate predictions on groundwater trends	Significant improvements in water use efficiency were achieved, especially in areas facing chronic water shortages. The models enabled the government to allocate water resources more effectively and plan recharge efforts. AI and ML provide powerful tools for predictive modeling in groundwater management. They allow for the optimization of groundwater use, mitigating the risks of overexploitation and drought
3	IoT-Based Groundwater Monitoring: The Case of Australia (Essamlali et al. 2024)	Australia, especially in the Great Artesian Basin, implemented IoT sensor networks to provide real-time monitoring of groundwater conditions, including water levels, flow rates, and quality. These IoT devices offered continuous updates to water managers and allowed them to respond promptly to changes	The real-time data collected through IoT networks enabled timely intervention in case of water depletion or contamination. It also facilitated better planning for future groundwater extraction and conservation efforts. IoT sensor networks allow for dynamic groundwater management, offering a proactive approach to managing aquifers. They help in rapid response and adjusting water extraction practices when issues arise

(continued)

Table 1 (continued)

	Case study	Description	Outcomes
4	Remote Sensing in Groundwater Management: The Case of Gujarat, India (Chinnasamy et al. 2013)	Gujarat's Bhujal Monitoring Program used remote sensing and GIS tools to assess groundwater depletion in regions affected by industrial and agricultural water extraction. By integrating remote sensing data with on-ground surveys, the program aimed to identify recharge zones and direct artificial recharge efforts	Remote sensing improved groundwater monitoring and enabled the implementation of targeted recharge projects like check dams and recharge wells. GIS tools combined with remote sensing can effectively identify groundwater depletion zones, ensuring efficient allocation of recharge efforts in water-scarce regions
5	IoT-Based Groundwater Monitoring: The Case of Gujarat, India (Kabuya et al. 2021)	Gujarat deployed IoT devices across key agricultural areas to monitor groundwater extraction and prevent over-pumping. These sensors tracked groundwater levels and provided real-time data to water resource managers	IoT networks helped reduce over-extraction by 15% and improved the monitoring of groundwater contamination in heavily industrial areas. Real-time monitoring via IoT sensors supports better regulation of groundwater extraction and improves early detection of water quality issues, making it a critical tool in urban and agricultural areas alike
6	Managed Aquifer Recharge (MAR) in Rajasthan, India (Alam et al. 2022)	In Rajasthan, the Pali Groundwater Restoration Project implemented Managed Aquifer Recharge (MAR) techniques using check dams, recharge wells, and percolation tanks. This initiative aimed to address declining groundwater levels in the water-scarce Thar Desert region	MAR helped increase the groundwater table by 2–3 m in the affected areas, benefiting both agriculture and local communities. Area-specific recharge practices, such as check dams and percolation tanks, are effective in arid regions where water is a limited resource. MAR can play a key role in restoring groundwater levels in drought-prone zones
7	Groundwater Recharge and Sewerage Management in Tamil Nadu, India (Natarajan et al. 2017)	Tamil Nadu's Chennai City Water Supply Augmentation project integrated groundwater recharge with treated wastewater reuse. Artificial recharge pits and infiltration wells were established alongside the treatment of wastewater for irrigation and aquifer recharge	Groundwater levels in Chennai were stabilized, and wastewater was effectively reused, reducing the city's freshwater demand. Combining groundwater recharge with treated wastewater reuse offers a sustainable solution in urban settings with limited freshwater availability, especially in water-stressed cities like Chennai

(continued)

Table 1 (continued)

	Case study	Description	Outcomes
8	Artificial Recharge with Stormwater Management: Singapore (Ramovha et al. 2024)	Singapore, with limited natural freshwater sources, developed an extensive stormwater management and artificial recharge system as part of the Marina Barrage Project. Excess stormwater is collected, treated, and directed to recharge groundwater aquifers	Singapore's approach helped secure sustainable freshwater resources while protecting against urban flooding. Combining stormwater management with artificial recharge is an effective method in urbanized regions with limited groundwater reserves. It also reduces the risk of flooding in densely populated areas
9	Rainwater Harvesting for Groundwater Recharge: The Case of Israel (Nachson et al. 2022)	Israel's Negev Desert, known for its arid conditions, uses large-scale rainwater harvesting systems combined with artificial recharge wells. Collected rainwater is stored in underground cisterns and percolation ponds, allowing gradual recharge of local aquifers	These initiatives have allowed Israel to overcome severe water scarcity and maintain sustainable groundwater levels in the desert. Rainwater harvesting and artificial recharge provide critical solutions in desert regions with minimal precipitation, ensuring long-term water security
10	IoT-Enabled Groundwater Management: The Case of Texas, USA (Aderemi and Olwal 2021)	Texas implemented IoT sensor networks under the Edwards Aquifer Authority to monitor groundwater levels, flow, and quality across agricultural and urban zones. Data from these sensors was integrated into an automated decision-making system to regulate water extraction and recharge activities	Texas improved water conservation and significantly reduced the risk of aquifer depletion through real-time monitoring and responsive actions. The combination of IoT and automated decision-making systems enhances the sustainable use of aquifers, especially in regions facing over-extraction challenges

5.1 Groundwater Situation in NCR

Water is essential for human development and economic growth, with groundwater accounting for 97% of the world's accessible freshwater and providing drinking water for approximately 1.5 billion people. The connection between groundwater recharge, conservation efforts, and sustainable development is crucial. Yet, in the National Capital Region (NCR), this resource is increasingly under threat. As a water-scarce region, NCR relies heavily on external surface water sources to meet its growing demand, which was estimated at 33.39 BCM annually in 2005 (Central Ground Water Board 2010). Groundwater has emerged as a critical supply, contributing around 6.35 BCM per year. However, the growing imbalance between recharge and extraction is depleting reserves rapidly, with the water table falling by an average of

2 m annually. This unsustainable extraction requires immediate action to enhance groundwater recharge through artificial methods such as rainwater harvesting and restoration of water bodies, to secure the region's long-term water supply. Over the past decade (1993–2003), nearly all areas in NCR have experienced significant declines in groundwater levels, especially where fresh groundwater reserves are present (Ministry of Housing and Urban Affairs 2018). Alwar district in Rajasthan and parts of the National Capital Territory (NCT) of Delhi have recorded the steepest declines (Central Ground Water Board 2010; Candela et al. 2009). While natural groundwater replenishment occurs gradually, it cannot match the rate of exploitation, resulting in a critical depletion of reserves and worsening water scarcity in many regions (Motagh et al. 2022).

5.2 Groundwater Recharge Strategies for National Capital Region (NCR), India

The National Capital Region (NCR) is a planning area that is centered around the National Capital Territory (NCT) of Delhi, India, and includes Delhi and adjoining districts from the states of Haryana, Uttar Pradesh, and Rajasthan in its gambit (Fig. 8). NCR and the National Capital Region Planning Board (NCRPB) were established in 1985 to spearhead the development of the region with coordinated planning activities and formulate policies for land-use management and infrastructure development which considers the region as a whole. Key urban centers in the NCR include Delhi, Faridabad, Ghaziabad, Gurgaon, and Noida (Ministry of Housing and Urban Affairs 2018). The NCR consists of a large inventory of rural–urban areas with a population of more than 46 million citizens and is reported to have an urbanization rate of 62.6%. The region also boasts ecologically important resources like the Aravalli Ridge, forests, and numerous sanctuaries (Mishra et al. 2024).

A review of the literature on groundwater recharge in the NCR region highlights several strategies that align with the broader water management objectives of the region.

Groundwater as a Common Pool Resource and the Rule of Capture: Groundwater in the National Capital Region (NCR) is traditionally treated as a common pool resource, managed under an open access framework where ownership follows the "rule of capture." This approach implies that individual users exploit the resource without formal ownership, leading to over-extraction. As no single entity has an incentive to conserve groundwater for the future, unchecked self-interest results in excessive usage. This situation has contributed to the overexploitation of groundwater, posing significant challenges to long-term sustainability in the region (Ministry of Housing and Urban Affairs 2018).

Rainwater Harvesting and Artificial Recharge Techniques: To address the alarming depletion of groundwater, rainwater harvesting, and artificial recharge techniques

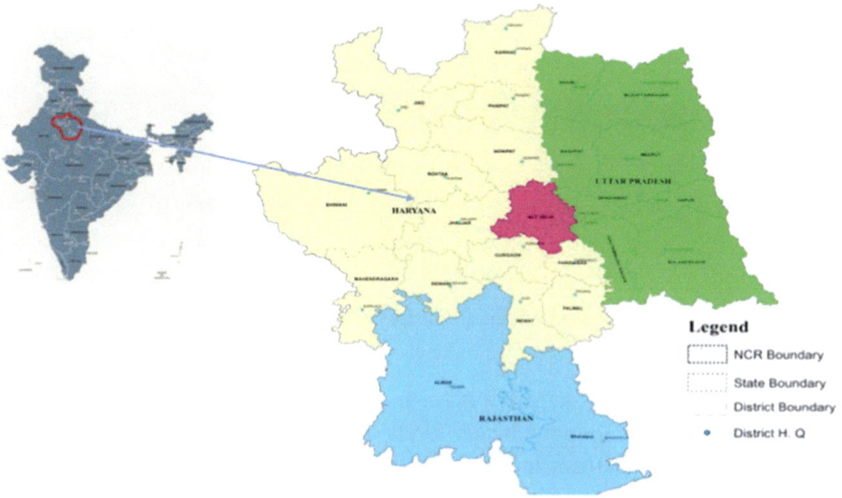

Fig. 8 Location of National Capital Region in India. (Ministry of Housing and Urban Affairs 2018)

have been promoted as viable solutions for augmenting groundwater levels. These efforts focus on enhancing the natural movement of surface water into groundwater reservoirs using engineered structures. Such methods depend on soil conditions, topography, and hydrogeology, making them context-specific and essential for regions like NCR that experience rainfall only for three months a year. The number of rainy days, ranging from 80 to 100 annually, limits the natural recharge period to a few months, underscoring the need for extended recharge techniques post-monsoon to ensure water availability during lean seasons (Khan et al. 2021).

Artificial Recharge Structures and Potential for Water Storage: Artificial recharge structures offer a technically feasible option for harvesting rainwater and excess monsoon runoff that otherwise goes unutilized. Subsurface reservoirs can store substantial amounts of water, which can be tapped into during periods of scarcity. By inducing the natural replenishment of groundwater, these recharge methods—whether planned (e.g., pits, tanks) or unplanned (e.g., irrigation seepage, leakage from pipes)—help conserve surface runoff and boost groundwater supplies. Moreover, incidental recharge from human activities, such as irrigation and leakages, also contributes to replenishing the groundwater table. Constructing basin recharge structures along floodplains of major rivers like the Yamuna could be an effective means of groundwater recharge.

Scope for Reviving Water Bodies and River-Induced Recharge: In NCR, numerous small and medium-sized water bodies, lakes, and ponds offer untapped potential for groundwater recharge through catchment treatment and de-siltation efforts. Additionally, floodplain areas present opportunities for induced recharge from rivers, providing another vital source of groundwater replenishment. Areas surrounding ridges can be treated with trenches, gully plugs, and gabion structures to promote

Fig. 9 Location of Sanjay Lake in Delhi. (Ministry of Housing and Urban Affairs 2018)

subsurface recharge. The revival of these natural and engineered systems can play a crucial role in bolstering groundwater levels in the region.

Community Involvement in Urban Rainwater Harvesting: Sparking a people's movement for rainwater harvesting is critical for achieving sustainable groundwater management in urban areas. Encouraging rooftop rainwater harvesting systems can significantly contribute to enhancing groundwater recharge, especially in densely populated urban centers where water demand continues to rise. Engaging communities in conservation efforts not only mitigates groundwater depletion but also fosters a culture of water stewardship, crucial for the long-term viability of groundwater resources in NCR.

Urban recharge structures and abandoned quarries: Park-type recharge facilities and rooftop rainwater harvesting for large institutional buildings. Over 300 identified quarries in the region could be revitalized for groundwater recharge (Fig. 9).

Incentives and awareness: Governments should encourage public involvement with incentives and financial aid to implement water conservation measures (Motagh et al. 2022).

5.3 *The Sanjay Lake Project, Delhi, India*

Sanjay Lake, an artificial water body developed by the Delhi Development Authority (DDA), is located in Trilokpuri, East Delhi, India, near the Mayur Vihar II residential area (Fig. 10). Spanning approximately 17 ha (42 ac), the lake is nestled within a 69-ha (170-ac) forest area known as Sanjay Lake Park. Constructed in the 1970s and inaugurated in 1982, the lake is a habitat for some migratory birds and features diverse native tree species. It also boasts a well-maintained fitness track, which is a favorite among walking enthusiasts (Fig. 8) (Davamani et al. 2024).

The Sanjay Lake project in East Delhi highlights the challenges of managing a water body located in a low-lying depression in the floodplain area. This lake, historically receiving rainwater but with no natural outlets, became heavily polluted due to untreated sewage inflows through multiple drains. Over time, with urbanization and a lack of sewer systems in East Delhi, sewage contaminated the lake, drastically

Fig. 10 Sanjay Lake, New Delhi,management of a water body (Ministry of Housing and Urban Affairs 2018)

reducing its dissolved oxygen levels and killing aquatic life. Efforts to divert many of these sewage inflows have reduced this problem, but some residual inflows and evaporation continue to affect the lake's health (Joshi et al. 2020).

The Sanjay Lake in East Delhi was developed with multiple objectives in mind, reflecting both ecological and societal needs. One of the primary aims is to maintain sufficient water levels throughout the year to support recreational activities, especially during the dry summer months. Additionally, the lake is intended to serve as a habitat for diverse avian and aquatic species, thus promoting local biodiversity and maintaining ecological balance. Another crucial objective is to facilitate groundwater recharge, helping to counteract groundwater depletion in the surrounding areas.

However, the lake faces several pressing issues that hinder the fulfillment of these goals. A major concern is the high rate of evaporation during peak summer months, particularly in May and June, which not only lowers water levels but also increases the concentration of dissolved solids. This change in water chemistry promotes the growth of invasive species such as water hyacinths. Compounding the problem is the absence of natural inflows; aside from direct rainfall, making the lake heavily dependent on artificial water inputs. Despite efforts to divert several sewage drains, untreated wastewater from a few remaining sources continues to enter the lake, deteriorating water quality.

To address these challenges, a set of targeted recommendations is proposed. Technologies should be introduced to reduce the inflow of organic nutrients, thereby improving overall water quality and curbing weed growth. Ensuring tertiary treatment of wastewater before it reaches the lake will prevent raw sewage contamination. Furthermore, excavating parts of the lakebed and installing recharge shafts can help store excess water during the monsoon season and promote the percolation of water into the aquifers, thereby contributing to the dual goals of lake rejuvenation and groundwater sustainability.

6 Conclusion

Groundwater, a critical lifeline for millions globally, is under increasing threat from the compounded impacts of climate change and unsustainable extraction practices. As this research underscores, the urgency to address groundwater sustainability has never been greater. Climate-induced changes in precipitation patterns, rising temperatures, and the occurrence of extreme weather events are altering recharge dynamics while escalating demand further exacerbates groundwater stress. However, advancements in technology present a silver lining. From remote sensing and GIS to AI-driven predictive models and IoT-based monitoring, cutting-edge technologies are revolutionizing groundwater management, enabling more accurate monitoring, efficient recharge methods, and sustainable use practices. Case studies from India and globally demonstrate the feasibility and success of such approaches, particularly in addressing region-specific challenges like those in the National Capital Region (NCR). Moving forward, it is essential to integrate these technological innovations with robust policy frameworks, community participation, and equitable resource governance. Only by adopting a holistic and adaptive approach can we ensure the long-term sustainability of groundwater resources in the face of evolving climate and societal pressures.

The case study of groundwater management in the National Capital Region (NCR), with a focus on Sanjay Van, highlights both the challenges and potential solutions for sustaining aquifers in densely populated urban areas. NCR's rapid urbanization, compounded by climate-induced variability in rainfall, has led to significant depletion of groundwater reserves. Sanjay Van, a critical green buffer, demonstrates the role of natural ecosystems in aiding groundwater recharge and mitigating urban heat island effects. Restoration efforts, including the creation of recharge zones and sustainable water management practices, have showcased the tangible benefits of integrating natural ecosystems with urban planning strategies.

The case studies demonstrate the effectiveness of various technologies and approaches in groundwater management, highlighting the role of innovation and interdisciplinary integration. Remote sensing tools like satellites and GIS were essential for monitoring groundwater depletion and identifying recharge zones in regions such as California and Gujarat. AI and ML in India helped optimize resource allocation and forecast groundwater availability, particularly in agricultural areas. IoT-enabled sensor networks in Australia, Gujarat, and Texas provided real-time data for groundwater levels, flow, and quality, allowing for responsive management and regulatory oversight. Managed Aquifer Recharge (MAR) techniques, such as those used in Rajasthan, significantly increased groundwater levels in arid areas. Integrated groundwater recharge systems combining wastewater reuse and stormwater management, as seen in Chennai and Singapore, offered sustainable solutions for urban areas with limited freshwater resources. Rainwater harvesting in Israel's Negev Desert provided a critical approach to maintaining groundwater levels in arid conditions. Together, these case studies demonstrate that combining innovative technologies with traditional water management techniques can enhance groundwater sustainability, reduce water scarcity, and improve overall resource management.

The findings from Sanjay Van emphasize the importance of protecting and revitalizing green spaces as natural recharge systems while incorporating advanced technological interventions, such as GIS mapping and IoT-based monitoring, to assess and manage groundwater levels effectively. This site-specific analysis underscores that urban ecosystems can serve as critical allies in the fight against groundwater depletion when combined with active community involvement and policy support. The success at Sanjay Van serves as a replicable model for other urban areas grappling with similar challenges, highlighting the synergy between nature-based solutions and technological advancements for resilient groundwater management.

7 Challenges and Future Directions

Sustainable groundwater management faces a range of challenges that hinder effective conservation and equitable access. The following sections highlight key obstacles and the steps needed to overcome them:

Data Availability and Quality: Ensuring the availability of high-quality data for accurate analysis and decision-making remains a critical challenge in sustainable groundwater management. Reliable and consistent data are essential for understanding groundwater recharge rates, depletion patterns, and potential areas for intervention. Addressing gaps in data collection and integration across different regions is necessary. In regions like NCR, data disparities between urban and rural areas, as well as variations in monitoring systems, make it difficult to form a holistic view of groundwater dynamics.

Technological Integration and Adoption: Overcoming barriers to the adoption of advanced technologies, particularly in developing regions, is vital for effective ground water management. Tools like remote sensing, geospatial mapping, and real-time monitoring systems offer promising solutions, but their widespread implementation faces logistical and financial constraints. Promoting capacity-building and training is essential to enable effective use of these technologies. Stakeholders, including local authorities, communities, and industries, must be equipped with the necessary knowledge and skills to leverage modern tools for sustainable groundwater management.

Policy and Regulatory Frameworks: Developing robust policies and regulatory frameworks is critical to supporting the sustainable management of groundwater resources. These frameworks should incorporate principles of equitable access, conservation, and environmental protection, ensuring that over-exploitation is curbed. Encouraging collaboration between governments, industries, and communities can foster a cooperative approach to groundwater management. Joint efforts in formulating policies, implementing recharge strategies, and raising awareness can drive more sustainable outcomes for all stakeholders involved.

References

Abanyie SK, Apea OB, Abagale SA (2023) Sources and factors influencing groundwater quality and associated health implications: a review. Emerg Contam 9(2):100207. https://doi.org/10.1016/j.emcon.2023.100207

Aderemi BA, Olwal TO (2021) A review of groundwater management models with a focus on IoT-based systems. Sustainability 14(3):1–10. https://doi.org/10.20944/preprints202107.0227.v3

Akhtar N, Izzuddin M, Ishak S (2021) Various natural and anthropogenic factors responsible for water quality degradation: a review. Water 13(2):1–33. https://doi.org/10.3390/w13192660

Alam MF, Pavelic P, Villholth KG (2022) Impact of high-density managed aquifer recharge implementation on groundwater storage, food production, and resilience: a case from Gujarat, India. J Hydrol: Regnal Stud 44(2):1–12

Anandhi A, Frei A, Pierson DC, Schneider EM (2011) Examination of change factor methodologies for climate change impact assessment. Water Resour Res 1–9

Arabameri A, Pal SC, Rezaie F (2021) Modeling groundwater potential using novel GIS-based machine-learning ensemble techniques. J Hydrol: Regnal Stud 36(2):100848. https://doi.org/10.1016/j.ejrh.2021.100848

Area WR (2019) Artificial groundwater recharge. U.S. Geological Survey, New York

Arnold JG, Srinivasan R, Muttiah RS, Williams JR (2007) Large area hydrologic modeling and assessment part I: model development. J Am Water Resour Assoc 34(1):73–89

Bahri D, Dasarathy AK (2024) Analysis of groundwater level fluctuations using AI & ML—a case study on Arkavathi Watershed, Karnataka, India. Int J Intell Syst Appl Eng 12(4):1169–1177

Bates B, Kundzewicz ZW, Wu S, Palutikof J (2008) Climate change and water. IPCC Secretariat, Geneva

Bhattacharya AK (2010) Artificial groundwater recharge with a special reference to India. Int J Recent Res Appl Stud 4(2):214–221

Candela L, Igel W, Elorza FJ, Aronica G (2009) Impact assessment of combined climate and management scenarios on groundwater resources and associated wetlands (Majorca, Spain). J Hydrol 376(3):327–604

Central Ground Water Board (2010) Guide on artificial recharge to ground water. New Delhi: ministry of water resources, Government of India

Chaussard E, Milillo P, Burgmann R (2017) Remote sensing of ground deformation for monitoring groundwater management practices: application to the Santa Clara Valley during the 2012–2015 California drought: groundwater monitoring with InSAR. J Geophys Res: Solid Earth 122(10):1–29

Chinnasamy P, Hubbart JA, Agoramoorthy G (2013) Using remote sensing data to improve groundwater supply estimations in Gujarat, India. Earth Interact 17(1):1–17

Dao PU, Heuzard, AG, Hoa Le TX, Zhao J (2024) The impacts of climate change on groundwater quality: a review. Sci Total Environ 912:169241. https://doi.org/10.1016/j.scitotenv.2023.169241

Davamani V, John JE, Poornachandhra C (2024) A critical review of climate change impacts on groundwater resources: a focus on the current status, future possibilities, and role of simulation models. Atmosphere 15(1). https://doi.org/10.3390/atmos15010122

Emmanuel MA, Jindal T, Hussain J, Swadhinata S (2018) Assessment of physico-chemical and biological parameters of Sanjay lake and old fort lake water in Delhi. J Eng Res Appl 8(6):44–48

Essamlali I, Nhaila H, El Khaili M (2024) Advances in machine learning and IoT for water quality monitoring: a comprehensive review. Heliyon 10(6):e27920

Hasan MF, Smith R, Vajedian S (2023) Global land subsidence mapping reveals widespread loss of aquifer storage capacity. Nat Commun 14:6180. https://doi.org/10.1038/s41467-023-41933-z

Hu K, Awange JL, Kuhn M (2023) Large-scale quantification of groundwater recharge threshold conditions using machine learning classifications: an attempt over the Australian continent. Groundw Sustain Dev 21(1):100941

Hughes A, Mansour M, Ward R (2021) The impact of climate change on groundwater recharge: national-scale assessment for the British mainland. J Hydrol 598:126336. https://doi.org/10.1016/j.jhydrol.2021.126336
—IGRAC—International Groundwater Resources Assessment Centre N (2021) Global groundwater information system (GGIS). Retrieved from global groundwater information system (GGIS). https://www.un-igrac.org/global-groundwater-information-system-ggis
Joshi P, Siddaiah NS, Dixit A (2020) Urban wetlands of Delhi, India: water quality and pollution status. Chem Ecol 104–131
Kabuya AK, Alowo R, Nkhonjera G (2021) Smart monitoring of groundwater. In: 2021 international conference on electrical, computer and energy technologies (ICECET). Capetown: University of Johannesburg, pp 1–8
Khan SA, Shekhar S, Kumar S (2021) Spatio-temporal dynamics of groundwater resources of national capital region, Delhi. Geostatistics and geospatial technologies for groundwater resources in India. Springer, New Delhi, pp 473–483
Lerner DN (1990) Groundwater recharge in urban areas. Atmos Environment Part B Urban Atmos 24(1):29–33
McCallum JL, Dogramaci S, Bai A (2020) Assessing temporal changes in groundwater recharge using spatial variations in groundwater ages. Water Resour Res 56(8):1–16. https://doi.org/10.1029/2020WR027240
Ministry of Housing and Urban Affairs (2018) NCR constituent areas. New Delhi
Ministry of Jal Shakti, Government of India (2018). Retrieved from India water resources information system: https://indiawris.gov.in/wris/
Mishra V, Dangar S, Tiwari VM, Lall U (2024) Summer monsoon drying accelerates India's groundwater depletion under climate change. Earth's Futur 12(8):1–20. https://doi.org/10.1029/2024EF004516
Motagh M, Indu J, Karanam V, Garg S (2022) Tracking hidden crisis in India's capital from space: implications of unsustainable groundwater use. Sci Rep 12:651–662
Mukherjee A, Scanlon BR, Aureli A (2021) Global groundwater: from scarcity to security through sustainability and solutions. In: Global groundwater-source, scarcity, sustainability, security, and solutions. Elsevier, pp 3–20
Nachson U, Silva CM, Sousa V (2022) A new modelling approach to optimize rainwater harvesting system for non-potable uses and groundwater recharge: a case study from Israel. Sustain Cities Soc 85(1):104097
Natarajan PM, Ponnavaiko M, Kalloikar S (2017) Sustainable water resources development in Tamil Nadu, India. In: XVI world water congress. Quintana Roo: International water resources association, pp 1–19
Nordin NF, Mohd NS, Koting S (2021) Groundwater quality forecasting modelling using artificial intelligence: a review. Groundw Sustain Dev 14(2):100643
Patel MP, Gami B, Patel A, Patel P (2020) Climatic and anthropogenic impact on groundwater quality of agriculture-dominated areas of southern and central Gujarat, India. Groundw Sustain Dev 10:100306
Pradeep P, Ramesh MV, Narendran S (2017) An internet of things (IoT) based sustainable water management. In: IEEE global humanitarian technology conference (GHTC). California, pp 1–7
Ramovha N, Chadyiwa M, Ntuli F, Sithole T (2024) The potential of stormwater management strategies and artificial intelligence modeling tools to improve water quality: a review. Water Resour Manag 38(2):3527–3560
Scanlon BR, Fakhreddine S, Rateb A, Graaf I (2023) Global water resources and the role of groundwater in a resilient water future. Nat Rev Earth Environ 4:87–101. https://doi.org/10.1038/s43017-022-00378-6
Shaikh M, Birajdar F (2024) Advancements in remote sensing and GIS for sustainable groundwater monitoring: applications, challenges, and future directions. Int J Res Eng Sci Manag 7(3):16–25
Shaikh M, Birajdar F (2024a) Advancements in remote sensing and GIS for sustainable groundwater monitoring: applications, challenges, and future directions. Int J Res Eng 7(3):16–23

Shaikh M, Birajdar F (2024b) Groundwater and ecosystems: understanding the critical interplay for sustainability and conservation. EPRA Int J Multidiscip Res 10(3):181–186. https://doi.org/10.36713/epra16111

Shishaye HA, Tait DR, Maher DT (2021) The legacy and drivers of groundwater nutrients and pesticides in an agriculturally impacted quaternary aquifer system. Sci Total Environ 753(1):142010. https://doi.org/10.1016/j.scitotenv.2020.142010

Tao H, Hameed MM, Marhoon HA (2022) Groundwater level prediction using machine learning models: a comprehensive review. Neurocomputing 489(2):271–308

Valhondo C, Martínez-Landa L, Carrera J (2020) Six artificial recharge pilot replicates to gain insight into water quality enhancement processes. Chemosphere 240(2):124826

Wang W, Chen Y, Chen Y (2022) Groundwater dynamic influenced by intense anthropogenic activities in a dried-up river oasis of Central Asia. Hydrol Res 53(4):532–546. https://doi.org/10.2166/nh.2022.049

Wu W-Y, Lo M-H, Wada Y (2020) Divergent effects of climate change on future groundwater availability in key mid-latitude aquifers. Nat Commun 3710. https://doi.org/10.1038/s41467-020-17581-y

Yu X, Luo L, Hu P (2022) Impacts of sea-level rise on groundwater inundation and river floods under changing climate. J Hydrol 614:128554. https://doi.org/10.1016/j.jhydrol.2022.128554. Webster PJ, Holland GJ, Curry JA, Chang HR (2005) Changes in tropical cyclone number, duration, and intensity in a warming environment. Science 309(5742):1844–1846

Abidin HZ, Andreas H, Gumilar I, Wibowo IRR (2015) On correlation between urban development land subsidence and flooding phenomena in Jakarta. In: Proceedings of the International Association of Hydrological Sciences 370 15–20.https://doi.org/10.5194/piahs-370-15-2015

Nyika J, Dinka MO (2023) Water challenges in rural and urban sub-Saharan Africa and their Management. Springer Nature Switzerland Cham

Assessment of Environmental Impacts of Arsenic and Fluoride Contamination and Remediation

Hamza Mohamed Flafel, Mohd Rafatullah, Japareng Lalung, Abduanaser AAli Ezhani, and Naseem Akhtar

Abstract Human health and general quality of life are greatly impacted by drinking water quality. With increasing reliance on groundwater due to declining surface water resources and rising population density, contamination by toxic elements such as arsenic and fluoride has become a critical global issue. This chapter explores the sources, toxicity levels, and regulatory limits of arsenic and fluoride in groundwater. Arsenic, found in over 200 mineral types, and fluoride, primarily from geogenic sources, pose severe health risks in elevated concentrations. Prolonged exposure to arsenic can lead to cardiovascular diseases and various cancers, while high fluoride levels cause dental and skeletal fluorosis. The chapter reviews Environmental Impact Assessment (EIA) methodologies to evaluate contamination levels, risk factors, and socioeconomic implications. It also examines remediation technologies for effectively removing these contaminants, aiming to mitigate their adverse effects on human health and the environment.

H. M. Flafel (✉) · J. Lalung · N. Akhtar
Environmental Technology Division, School of Industrial Technology, Universiti Sains Malaysia, Penang, Malaysia
e-mail: hamza121655@gmail.com

H. M. Flafel
Libyan Center for Studies and Research in Environmental Science and Technology, Brack Al-Shatti, Libya

Department of Environmental Science, Faculty of Environment & Natural Resources, Wadi Al-Shatti University, Brack Al-Shatti, Libya

H. M. Flafel · A. A. Ezhani · N. Akhtar
Libyan Centre for Research and Studies of Water, Soil and Combatting Desertification, Sabratha, Libya

M. Rafatullah
University Centre for Research and Development, Chandigarh University, Mohali, Punjab 140413, India

N. Akhtar
Nuclear Science Program, Faculty of Science and Technology, Universiti Kebangsaan Malaysia, Bangi, Selangor, Malaysia

A. Nath et al. (eds.), *Groundwater Depletion and Sustainability*, Advances in Geographic Information Science, https://doi.org/10.1007/978-3-032-09921-1_8

Keywords Environmental assessment · Arsenic · Fluoride · Remediation technologie

1 Induration

The quality of the water we drink significantly impacts our quality of life. Water is one of the most important resources for life to exist on Earth. There are 1,386 million km^3 of water on Earth, but only 30.1% of the 10.63 million km^3 of freshwater is found in groundwater. Increased demand for groundwater has resulted from a declining reliance on surface water resources and rising population density. (Jha and Tripathi 2021). According to World Health Organisation (WHO) estimates, 1.8 billion people consume tainted water. Different aquatic and terrestrial ecosystems are being contaminated with metals and non-metals due to various natural and human-caused activities. Every day, hazardous industrial chemicals, toxic metals, untreated sewage, agricultural pesticides, and herbicides are released into the environment from agricultural and industrial sectors. Fluoride and arsenic are two of the primary contributors to the global water crisis due to their toxicity, posing serious health risks when ingested. The increasing dependence on groundwater in many nations has exacerbated this issue. In several regions worldwide, groundwater contains elevated concentrations of fluoride and arsenic concentrations simultaneously. These contaminants can co-occur in groundwater due to either natural processes or human activities. The main pathway through which humans are exposed to high levels of fluoride and arsenic is by consuming water that has been naturally contaminated by geogenic sources (Bibi et al. 2017). Arsenic, a naturally occurring element in the environment, exhibits metallic and non-metallic characteristics in its chemical and physical makeup. The World Health Organization, the United States Environmental Protection Agency (US EPA), the European Environment Agency (EEA), and other international organizations have established that the maximum allowable limit for arsenic in drinking water, due to its toxicity even at low concentrations, is 10 μg/L. Furthermore, Long-term arsenic exposure has been connected to more severe health issues, including blood vessel and heart ailments, or cardiovascular diseases, and various malignancies, including skin, kidney, liver, and prostate cancers (Sandoval et al. 2021). On the other hand, fluoride is currently considered a low-priority pollutant by the U.S. EPA, with a maximum contaminant level of 4 mg/L under the National Primary Drinking Water Regulation. It's important to note that the Environmental Protection Agency's target for the optimal fluoride concentration ranges from 0.9 to 1.2 mg/L. The World Health Organisation states that human health requires fluoride, especially to strengthen tooth enamel [0.5 < F- < 1.5 mg L-1]. On the other hand, dental [1.5 < F- < 4.0 mg L-1] and skeletal [F- > 4.0 mg L-1] fluorosis can result from excessive fluoride levels, which results in tooth mottling and ligament calcification (Raj and Shaji 2017). In addition, skeletal fluorosis is a serious issue that requires attention due to its high level of accumulation over time, which prevents early symptoms and leads to significant disability. It primarily affects the major joints of the body, such as the neck, shoulders, knees, hips, and more, causing pain and stiffness in almost every age group. Chronic symptoms are persistent and irreversible, further deteriorating overall health (Srivastava and Flora 2020). Therefore, due to the problems caused by arsenic and fluoride, this chapter reviews various EIA methodologies used to evaluate contamination levels, risk factors, and the socio-economic implications of these pollutants. It also examines remediation technologies for the removal of arsenic and fluoride.

2 Sources of Arsenic and Fluoride in Groundwater Pollution

Since arsenic (As) is found in more than 200 different mineral types, it is believed that over 90% of As pollution is geogenic in nature. However, arsenic can also originate from anthropogenic, biogenic, and geogenic sources. Arsenopyrite is the most common form of As, and it is thought that around one-third of the As in the environment occurs naturally, mainly due to volcanic activity and low temperatures (Dhillon 2020). As is commonly found in the elements copper, silver, cadmium, zinc, tin, mercury, iron, uranium, gold, cobalt, selenium, phosphorus, molybdenum, lead, antimony, nickel, platinum, tellurium, sulfur, bismuth, and tungsten. Rock-water interactions in aquifer systems are the primary source of as pollution in groundwater. Aquifers promote as mobilization in physical and geochemical environments, particularly under reducing conditions. The main sources of anthropogenic As contamination are petroleum exploitation, coal mining, and mineral extraction. Globally, anthropogenic as pollution results from metal smelting, fossil fuel combustion, and the use of As in timber preservation. Agricultural activities, including using fertilizers, pesticides, and herbicides, also contribute to As pollution (Genchi et al. 2022). Additionally, As can originate from biogenic sources, primarily plants, aquatic microorganisms, and agricultural organisms. On the other hand, the weathering of fluoride-containing minerals such as fluorite, cryolite, sellaite, apatite, topaz, mica, and amphiboles is the main cause of fluoride (F) contamination, which is geogenic (Sahu and Debsarma 2023). Another large natural source is fluoride contamination from volcanic eruptions, with marine aerosols coming in third. Fluoride contamination is also caused by human activity; in developing countries like India and Pakistan, incorrect use of phosphatic fertilizers is a major contributing factor. Mining, smelting, coal combustion, cement and brick manufacturing, and landfill activities further contribute to groundwater fluoride contamination (Farooqi and Handboo 2015). Bundschuh et al. (2004) confirmed that As and F compounds were present in groundwaters and discovered a strong link between As and F. High As and F contents in shallow groundwaters in Robles County, Argentina, are linked to a layer of volcanic ash deposits in the pseudo-sedimentary sequence. The fact that these contaminants coexist in drinking water and present serious health hazards is alarming. Arsenic (As) and fluoride (F^-, the simplest fluorine anion) are found in the environment in

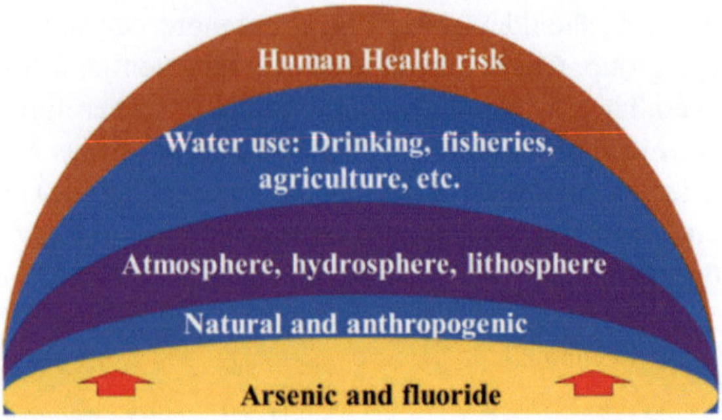

Fig. 1 Pathways for geogenic fluoride and arsenic mobilization through natural and man-made processes, as well as the health dangers involved (Maity et al. 2021)

organic and inorganic forms. Inorganic As and F can enter living things through food and water. The main source of arsenic and fluoride pollution in groundwater (GW) and surface water (SW) is geogenic processes, such as the dissolution of minerals that contain As and/or F that are present in rocks and soils (Fig. 1).

3 Toxicity Levels and Regulatory Limits for Arsenic and Fluoride

3.1 Arsenic

Arsenic and fluoride are naturally occurring elements in various environmental matrices such as soil, water, and air. While both elements are present in trace amounts in many regions, their toxicity at elevated levels has led to significant health concerns and the establishment of regulatory limits (Ghosh and Mukhopadhyay 2019). This station delves into the toxicity levels of arsenic and fluoride, their health impacts, and the regulatory frameworks established to control their concentrations in different mediums.

The best-described and characterized toxic symptoms of chronic exposure to arsenic, primarily in the adult population, are pigmentation alterations and hyperkeratosis. Arsenic mainly accumulates in the skin, causing cutaneous modification. Nonetheless, skin alterations were noted in kids aged 6 to 8. Genetic damage, decreased learning and memory, neuropathies, cardiovascular changes, endocrine dysfunctions, and immunosuppression are additional negative consequences that have been reported in youngsters (Limón-Pacheco et al. 2018). Arsenic contamination is often due to natural geological processes, industrial activities, and arsenic-containing pesticides and herbicides. It has been classified as a Group 1 carcinogen by the International Agency for Research on Cancer (IARC). Chronic exposure to arsenic can lead to various health issues, including skin lesions, cardiovascular diseases, and neurotoxicity. Several epidemiological studies conducted worldwide confirm that elevated levels of contaminants in drinking water increase the risk of lung, bladder, and skin cancers and other adverse health impacts (Ferreccio et al. 2000; Organization 2004). The toxicity of arsenic is dose-dependent, with higher doses leading to more severe health outcomes. The World Health Organization (WHO) and the United States Environmental Protection Agency (EPA) have established guidelines and regulatory limits to minimize arsenic exposure. The WHO has set a current drinking water arsenic guideline of 10 (μg/L) (Smith et al. 2002). The USEPA (United States Environmental Protection Agency, 2001) likewise embraced this suggestion, which is currently in effect (World Health Organisation 2011). This limit is tentative because of scientific concerns regarding the true dangers at low concentrations. Since no limit can guarantee innocuousness and there is no threshold value for As carcinogenic effects, the agreed limit value is a trade-off between the expenses of As decontamination and public health. Many nations presently follow the WHO's tentative recommendation. In certain countries, however, a cautious approach should be adopted, establishing a threshold below 10 μg As/L. For example, Denmark has maintained a criterion of 5 μg/L since 2003, as has New Jersey, USA, since 2006. The Netherlands' water sector implemented a voluntary limit of 1 μg/L in 2016, based on a societal cost–benefit analysis. To discourage using contaminated surface water over noncompliant well water which may pose more significant health risks due to pathogens some countries retain a threshold of 50 μg/L when lower thresholds are unfeasible because alternative water sources are unavailable. The current Consumer Affairs Authority (CAA) regulation sets a maximum limit of 200 μg/L for arsenic (As) in natural mineral waters and all products made with them, significantly higher than the acceptable limit for drinking water. Furthermore, there is no set standard for the information regarding the chemical makeup and As content of bottled waters. These rules have been in effect since 1994, when the population's use of bottled water was low (Código Alimentario Argentino, 2012). Even though bottled water usage has significantly increased in recent years, the laws have not been amended. Furthermore, some waters that have been artificially mineralized may have elevated As (Argentino 2012). The European Union has set maximum levels for arsenic in foodstuffs, such as rice and rice-based products, to reduce dietary exposure. Regulatory limits for arsenic in soil vary by country, depending on land use and the potential for human exposure. For instance, the United States has set a residential soil arsenic limit of 0.39 mg/kg.

3.2 Fluoride

An inorganic anion called fluoride (F −) is found naturally in minerals, especially fuorite (CaF_2). Fluoride salts in water in a wide range of concentrations and are very soluble. For instance, groundwater has been found to exhibit significantly wider concentration ranges (0.1 to 6 mg/L), and surface water levels typically remain below 0.5 mg/L (N. 2013). Although such high concentrations are rare, levels exceeding 10 mg/L have been documented, depending on the presence of specific minerals.

Fuoride is also present in seawater, although it is only in a minimal range (1.2–1.5 mg/L) (N. 2013).The rise in scientific publications suggests that fluoride's potential negative health effects have drawn more attention in recent years. Toxicological symptoms of fluoride include vomiting, diarrhea, salivation, heart depression, and respiratory arrest. Lethal dosages in humans have been documented to be between 40 and 80 mg/kg body weight (b.w.)(Ranjan and Ranjan 2015). Dental fluorosis is one of the most well-established long-term effects of fluoride in humans. Hypomineralization of the growing teeth results from excessive fuoride incorporation into dental enamel before tooth eruption. When enamel maturation is finished, which occurs at age 8, a person's susceptibility to dental fluorosis ends. The caries-preventive impact of fluoride should be considered when assessing the risk of dental fluorosis (Kondru 2014). For an adult weighing 70 kg, the median daily consumption of fluoride from water and water-based beverages in EU countries is 0.13 mg, or 1.86 μg/kg b.w./day. The intake of approximately 8.4 mg/day (120 μg/kg b.w./day for a person weighing 70 kg) is an extreme example, based on the maximum reported consumption (97.5th percentile) of tap water and a fluoride concentration of 3 mg/L (N 2013). According to the German drinking water legislation, tap water's fluoride content cannot exceed 1.5 mg/L. Fuoride levels in drinking water samples are typically below 0.3 mg/L and are rarely higher than this. To balance the benefits of reducing dental cavities against the danger of dental fluorosis, the EFSA recommended an AI of 0.05 mg fuoride/kg b.w. per day from all sources for children and adults, including pregnant and lactating women (N 2013). When drinking water is the only substantial source of fluoride, the amount consumed by adults does not surpass acceptable limits, as drinking water has a fluoride content of approximately 1 mg/L.

4 Environmental Impact Assessment (EIA) Methodologies for Arsenic and Fluoride

Environmental Impact Assessment (EIA) methodologies for assessing arsenic and fluoride involve systematic approaches to understanding the potential environmental and health risks these contaminants pose. Both arsenic and fluoride are naturally occurring elements that can enter water supplies and soils through natural processes, but they can also be introduced through human activities, such as industrial processes, agriculture, and waste disposal (Farooqi 2015). Proper EIA methodologies help identify, predict, and mitigate these contaminants' adverse effects on the environment and public health. The primary methodologies for conducting EIAs for arsenic and fluoride include risk assessment, site investigation, modelling and monitoring, stakeholder engagement, and impact mitigation strategies.

The Risk Assessment process is integral to Environmental Impact Assessment (EIA) for arsenic and fluoride contamination. This assessment evaluates toxicity, exposure pathways, and their impacts on human and ecological health (Trivedi et al. 2022). It consists of hazard identification, pinpointing sources and populations at risk;

dose–response assessment, analyzing exposure levels' effects on health considering demographics; exposure assessment, quantifying environmental concentrations in water, soil, and food; and risk characterization, which combines all assessment data to develop a comprehensive risk profile encompassing both non-carcinogenic risks (from fluoride) and carcinogenic risks (from arsenic). A thorough site investigation is critical to understand sources, contamination levels, and distribution. This involves water sampling, with regular surface and groundwater arsenic and fluoride concentrations analysis, accounting for seasonal variations and sources like natural deposits or industrial discharge. Soil and sediment analyses assess impacts on agriculture, food chains, and soil health (Ritter Paul Sibley et al. 2002). GIS mapping supports visualizing contamination patterns across regions, allowing targeted action in high-risk areas (Saha et al. 2022) (Yassin 2024). Modeling and Monitoring are essential in forecasting long-term contamination impacts and evaluating mitigation strategies. Hydrological modeling simulates groundwater flow and contaminant dispersion, which is valuable for predicting arsenic and fluoride behavior within diverse geological settings (Qiao et al. 2022). Ecotoxicological models analyze risks to ecosystems, assessing bioaccumulation and biomagnification potential, while long-term monitoring ensures ongoing observation of concentration changes from natural or human sources. Stakeholder engagement and social impact assessment are vital to effective EIA. Engaging with affected communities by sharing information about contamination, health risks, and proposed mitigations builds trust and collaboration. Conducting a Health Impact Assessment (HIA) alongside the EIA helps evaluate the health effects of long-term exposure, especially for vulnerable groups like children and the elderly. Stakeholder feedback enriches the process by incorporating cultural considerations influencing exposure and leveraging local knowledge in EIA methods. Finally, mitigation strategies are essential for managing arsenic and fluoride impacts. Effective water treatment technologies such as reverse osmosis, activated alumina filters, and coagulation/filtration systems are evaluated within the EIA for local suitability. Soil remediation techniques like phytoremediation, soil washing, and stabilization reduce soil contamination and are chosen based on soil properties, climate, and contamination severity. Additionally, robust policy and regulatory frameworks are crucial to ensure compliance across industries, agriculture, and municipalities, helping prevent further contamination.

5 Remediation Technologies for Arsenic and Fluoride

Remediation technologies aim to remove or reduce the concentrations of these hazardous elements to safe levels, helping communities access clean and safe drinking water and promoting sustainable land use. As shown in Fig. 2, some remediation technologies are used for arsenic and fluoride. Below is an explanation of some treatment methods:

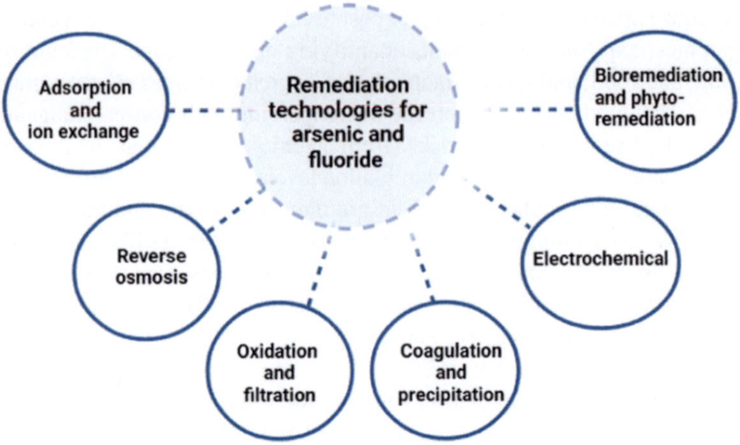

Fig. 2 Technologies for the removal of arsenic and fluoride

5.1 Adsorption

Adsorption is how fluoride and arsenic in a liquid or gaseous bulk phase stay on a solid surface. The substance employed for adsorption is called the adsorbent, and the pesticide molecules involved in this process are called adsorbates (Shah 2022). On occasion, liquids are employed as adsorbents as well. As a surface phenomenon, adsorption only occurs on the adsorbent's surface; the adsorbate shouldn't enter the adsorbent's internal structure. The adsorption process is shown in Fig. 3. Also, adsorption (AD) is a traditional technology frequently used to defluoridate water due to its affordability, durability, environmental friendliness, and effectiveness. Recent years have seen the development of new, cutting-edge materials for the efficient and affordable fluoride removal from drinking water. Numerous inexpensive adsorbents, such as alumina, red mud, clays, soils, activated carbon, calcite, brick powder, activated coconut shell, activated kaolinites, oxide ores, modified chitosan, bone char, and several other inexpensive materials, have also been used to remove fluoride (Jadhav et al. 2025). On the other hand, adsorption successfully removed arsenic from the water. The removal effectiveness of arsenic by adsorption using various element-based adsorbents varies. The kinds of adsorbents employed significantly impacted the removal effectiveness of As in the adsorption process. Natural materials were employed as adsorbents in past studies. However, their removal capabilities were subpar. When comparing the efficacy of natural and lab-synthesized adsorbents for the removal of As researchers discovered that the latter had a greater removal efficiency. Finding effective and affordable adsorbents has been the focus of numerous studies. It could be crucial to forecast the adsorption performance before choosing the elements for the As removal adsorbent, considering the elements' chemical properties (Uddin and Jeong 2020). Currently, a variety of adsorbents have been employed for the removal of arsenic from water, including activated carbon, biochar granular

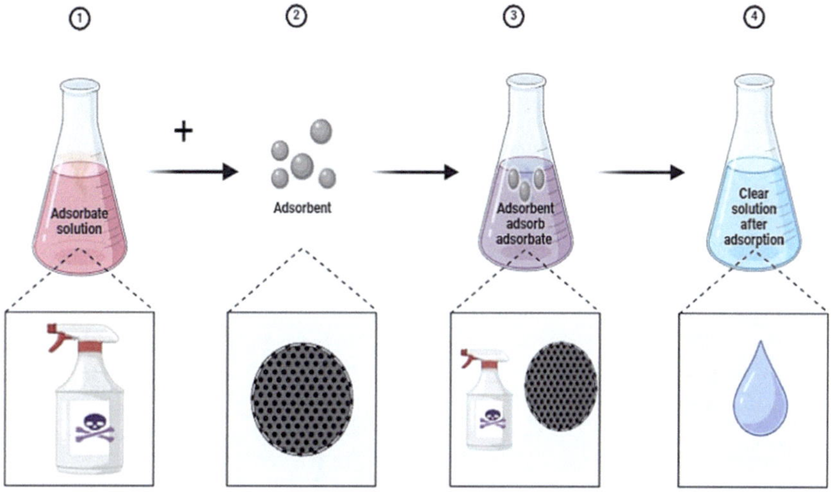

Fig. 3 Adsorption process

ferric hydroxide, iron oxide, magnetic graphene oxide, zeolite, nanocomposites, and ion exchange resins (Sun et al. 2022).

5.2 *Electrocoagulation*

An electrochemical process called electrocoagulation uses an electric current to remove contaminants from water. The process relies on the in-situ generation of coagulants by electrolytic oxidation of sacrificial anodes, typically made from iron or aluminum. These coagulants destabilize pollutants, forming flocs that can be removed by sedimentation or filtration. The electrochemical synthesis of destabilizing agents, which enables the elimination of polluting substances by charge neutralization and subsequent coagulation, is the foundation of electrochemical processes. Reduced sludge creation, no need for handling or storing chemicals, ease of use, and a low space requirement are just a few benefits of the EC process (Holt et al. 2002). Several studies show that the EC process is used to remove fluoride and arsenic from water. However, most of that research focused on eliminating a specific element. For example, a study compared the optimized coagulation/flocculation (CF) method with the electrocoagulation/flocculation (ECF) process for the removal of arsenic from the model and natural waters with low mineral concentrations. Without an oxidation phase to improve As (III) removal, the effect of organic matter on arsenic removal was investigated using iron electrodes and current densities ranging from 4 to 33 mA/cm^2. Arsenic concentrations, speciation, dissolved organic carbon (DOC), zeta potential, and turbidity were measured to evaluate the method. ECF showed some limitations

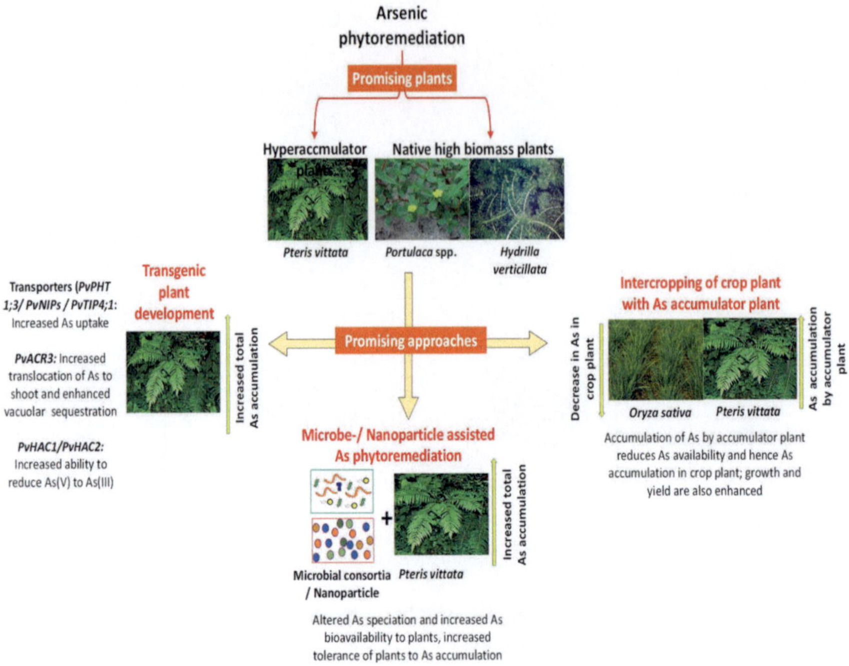

Fig. 4 lists several methods for arsenic phytoremediation, including the use of native high biomass and bioenergy plants or hyperaccumulator plants; intercropping arsenic accumulator plants with crop plants to reduce arsenic toxicity to crop plants; arsenic phytoremediation aided by microbes or nanoparticles; and genetic engineering techniques to increase plant phytoremediation potential (Srivastava et al. 2021)

compared to CF: lower As(V) removal efficiency due to ferrous iron release and negative zeta potential, higher DOC levels from the breakdown of large compounds, and increased turbidity from charge neutralization. However, ECF effectively oxidized As(III) to As(V), enabling significant As(III) removal without the need for additional oxidation steps that could create toxic byproducts (Pallier et al. 2011). On the other hand, another study used electrocoagulation with an aluminum electrode to investigate the simultaneous removal of fluoride and arsenic from synthetic groundwater. It looked at how beginning pH, current density, run time, inter-electrode distance, and NaCl content affected operating expenses and removal efficiency. Optimal conditions were an initial pH of 7, current density of 10 A/m^2, 95-min run time, 1 cm inter-electrode distance, and 0.71 g/L NaCl, achieving 98.51% arsenic removal (from 550 μg/L) and 88.33% fluoride removal (from 12 mg/L). Final concentrations were 8.19 μg/L for arsenic and 1.4 mg/L for fluoride, with an operating cost of 0.357 USD/m^3. Kinetic models were applied, and sludge analysis confirmed arsenic as As(III) and the presence of fluoride. This process meets WHO guidelines for maximum contaminant levels in drinking water (Thakur and Mondal 2017).

5.3 *Precipitation and Anion Exchange*

Precipitation is frequently used to clean contaminated groundwater, such as arsenate and fluoride, due to its ease of use and minimal equipment requirements (Huang et al. 2020). Following a pH adjustment, dissolved metal ions undergo a chemical reaction with a precipitant, such as lime, to form an insoluble solid phase. On the other hand, anion exchange is a widely used water treatment process that targets negatively charged ions in water. This technology relies on resins, which are synthetic polymers designed to attract and bind anions (Ahmer and Uddin 2024). These resins contain functional groups that can be exchanged with undesirable anions such as arsenate and fluoride, effectively removing them from water. In an important study on the removal of fluoride and arsenate from aqueous solutions using hydrocalumite through precipitation and anion exchange, hydrocalumite effectively reduced the levels of fluoride and arsenate in solutions to below 0.05 mmol/L and 0.005 mmol/L, respectively, when the initial concentrations were below 30 mmol/L for fluoride and 5 mmol/L for arsenate. However, as initial concentrations increased, the removal efficiency decreased. Arsenate was removed by forming minerals such as johnbaumite, while fluoride was primarily eliminated through the dissolution of hydrocalumite and the formation of precipitates like fluorite. The pH of the solution had minimal influence due to hydrocalumite's buffering properties, but temperature variations and competing anions significantly affected the removal process. Compared to other similar materials, hydrocalumite demonstrated superior adsorption capacities (719.1 mg/g for fluoride and 361.7 mg/g for arsenate), making it suitable for the treatment of industrial wastewater and groundwater with high fluoride or arsenic content (Guo and Tian 2013).

5.4 *Phytoremediation*

One efficient and inexpensive remediation method for clearing contaminants from soils is phytoremediation. According to Agarwal and Chauhan (Agarwal 2015), the leaves of the Hordeum vulgare variety RD 2052 exhibited the highest bioaccumulation of fluoride (F), measuring 9.948 mg kg^{-1}, in the presence of 18 mg kg^{-1} NaF, while its grains showed the lowest accumulation at 6.302 mg kg $^{-1}$. Polonskij and Polonskaya (Polonskij and Polonskaya 2013) reported that the tea bush (Camellia sinensis) can accumulate as much as 4000 mg/kg in its leaves without exhibiting any adverse effects. In the Gulf of Gabes (southeast Tunisia), Boukhris et al. (2015) studied the fluoride accumulation in the aboveground parts of native plants growing in soils near the coastal superphosphate industry. Their findings indicated accumulation levels between 37 and 360 mg/kg. For an in-situ phytoremediation on dry F^{-}polluted areas, they suggested three native promising perennials: *Rhanterium suaveolens, Atractylis serratuloides*, and *Erodium glaucophyllum.* When grown hydroponically in water containing 50 mg of F-L-1 for 10 days, *P. juliflora* can accumulate as much as

2222.83 mg g^{-1} (Mamta Baunthiyal and Vinay Sharma 2012). A study by Karmakar et al. (Karmakar et al. 2016) found that *Pistia stratiotes*, *Eichhornia crassipes*, and *Spirodela polyrhiza* could extract 12.71% to 19.87% of fluoride (F) from 3 mg F L^{-1} of contaminated water. For the repair of As contaminated sites, a variety of phytoremediation techniques can be applied carefully. Figure 4 summarizes several strategies, which are covered in more detail below. One study investigated the accumulation of arsenic (As) and the mechanisms of tolerance to As-induced damage in *Landoltia punctata*, assessing its potential for phytoremediation. The results showed that As absorption increased with higher As concentrations. Key antioxidant enzymes, such as superoxide dismutase, glutathione reductase, and anthocyanin levels, rose with increasing As levels. However, catalase and peroxidase activities increased only up to 1.0 mg L^{-1} of As before declining at higher concentrations. While lower As levels were associated with restrained reactive oxygen species, elevated As levels led to increased oxidative stress markers, including superoxide anion and lipid peroxidation. Biomass was reduced in As-exposed plants, and scanning electron microscopy revealed root structural damage at 3.0 mg L^{-1} As. Overall, *L. punctata* is identified as a hyperaccumulator species with potential for As phytoremediation at concentrations below 1.0 mg L^{-1}, significantly exceeding the safe level for drinking water (Canatto et al. 2021). Also, another study reported that As contaminated soils have been phytoremediation using Pteris vittata. The most common As species in aerobic soils is arsenate (AsV), which is absorbed by plants through phosphate transporters (Pht), including P. vittata, and is most likely what allows for effective As translocation in P. vittata. The findings imply that PvPht1;3 expression improves phytoremediation of As-contaminated soils by increasing As translocation and accumulation in plants (Cao et al. 2019).

6 Conclusion

The challenges posed by arsenic and fluoride contamination in water resources require urgent attention due to their profound environmental and public health impacts. Comprehensive Environmental Impact Assessments (EIA) enable evaluating and mitigating the risks associated with these pollutants. Effective EIA methodologies, including risk assessment, site investigation, modelling including risk assessment, site investigation, modelling, monitoring, and stakeholder engagement, facilitate a thorough understanding of contamination sources, exposure pathways, and health implications. Remediation technologies, such as adsorption, electrocoagulation, precipitation, phytoremediation, etc., are essential for mitigating contamination and ensuring safe water access. Furthermore, continuous advancements in water treatment and soil decontamination methods significantly reduce arsenic and fluoride levels, safeguarding both human health and ecosystems. An integrated approach that combines scientific rigor, policy enforcement, and community participation is crucial for the sustainable management and long-term prevention of arsenic and fluoride contamination.

7 Future Recommendations

Integrating advanced multi-disciplinary EIA methodologies is essential for effective mitigation of arsenic and fluoride contamination. Future research should focus on developing cost-efficient, eco-friendly remediation technologies like biochar adsorption and plant-based phytoremediation. Policymakers should emphasize global cooperation and community-driven projects for sustainable water resource management. Continuous monitoring and real-time data systems should be expanded for dynamic contamination tracking. Training programs for local stakeholders on health risks and treatment systems will help empower affected communities. Stronger enforcement of regulations is critical to ensure compliance in agricultural and industrial sectors, reducing contamination sources.

Acknowledgements The authors are grateful to all who worked together on this project. They also acknowledge each member of the Libyan Center for Studies and Research in Environmental Science, Faculty of Environment and Natural Resources, Wadi Alshatti University, who provided extensive personal, logistical, financial, and professional support.

References

Agarwal R (2015) Chauhan SS (2015) Bioaccumulation of sodium fluoride toxicity in plant parts at different phases and its impact on crop Hordeum vulgare (Barley) variety RD 2052. Int. J. Multidiscip. Res. Dev 2:16–21

Ahmer MF, Uddin MK (2024) Structure properties and industrial applications of anion exchange resins for the removal of electroactive nitrate ions from contaminated water. RSC Adv 14(45):33629–33648. https://doi.org/10.1039/D4RA03871A

Argentino CA (2012) Bebidas hídricas, aguas y aguas gasificadas. Capítulo XII

Bibi S, Kamran MA, Sultana J, Farooqi A (2017) Occurrence and methods to remove arsenic and fluoride contamination in water. Environ Chem Lett 15:125–149. https://doi.org/10.1007/s10311-016-0590-2

Boukhris A et al (2015) Screening biological traits and fluoride contents of native vegetations in arid environments to select efficiently fluoride-tolerant native plant species for in-situ phytoremediation. Chemosphere 119:217–223. https://doi.org/10.1016/j.chemosphere.2014.06.007

Bundschuh J et al (2004) Groundwater arsenic in the Chaco-Pampean plain, Argentina: case study from Robles county. Santiago Del Estero Province. Appl Geochemistry 19(2):231–243. https://doi.org/10.1016/j.apgeochem.2003.09.009

Canatto RA, de Oliveira JA, da-Silva CJ, Albino BÉS (2021) Tolerance of Landoltia punctata to arsenate: an evaluation of the potential use in phytoremediation programs. Int J Phytoremediation, 23(1), pp 102–110. https://doi.org/10.1080/15226514.2020.1797630

Cao Y et al (2019) Heterologous expression of Pteris vittata phosphate transporter PvPht1; 3 enhances arsenic translocation to and accumulation in tobacco shoots. Environ Sci Technol 53(18):10636–10644. https://doi.org/10.1021/acs.est.9b02082

Dhillon AK (2020). Arsenic Contamination of India's Groundwater: A Review and Critical Analysis. In: Fares A, Singh S (eds) Arsenic Water Resources Contamination. Advances in Water Security. Springer, Cham. https://doi.org/10.1007/978-3-030-21258-2_8

Farooqi, A (2015) Arsenic and Fluoride Pollution in Water and Soils. In: Arsenic and Fluoride Contamination. Springer, New Delhi. https://doi.org/10.1007/978-81-322-2298-9_1
Farooqi A. Handboo C (2015) Arsenic and fluoride contamination. A Pakistan Perspect
Ferreccio C, González C, Milosavjlevic V, Marshall G, Sancha AM, Smith AH (2000) Lung cancer and arsenic concentrations in drinking water in Chile. Epidemiology 11(6):673–679
Genchi G, Lauria G, Catalano A, Carocci A, Sinicropi MS (2022) Arsenic: a review on a great health issue worldwide. Appl Sci 12(12):6184. https://doi.org/10.3390/app12126184
Ghosh G, Mukhopadhyay DK (2019) Human Health Hazards Due to Arsenic and Fluoride Contamination in Drinking Water and Food Chain. In: Sikdar, P. (eds) Groundwater Development and Management. Springer, Cham. https://doi.org/10.1007/978-3-319-75115-3_15.
Guo Q, Tian J (2013) Removal of fluoride and arsenate from aqueous solution by hydrocalumite via precipitation and anion exchange". Chem Eng J 231:121–131. https://doi.org/10.1016/j.cej.2013.07.025
Holt PK, Barton GW, Wark M, Mitchell CA (2002) A quantitative comparison between chemical dosing and electrocoagulation. Colloids Surfaces A Physicochem. Eng. Asp. 211(2–3):233–248. https://doi.org/10.1016/S0927-7757(02)00285-6
Huang T, Ma B, Pang Z, Li Z, Li Z, Long Y (2020) How does precipitation recharge groundwater in loess aquifers? Evidence from multiple environmental tracers. J Hydrol 583:124532. https://doi.org/10.1016/j.jhydrol.2019.124532
Jadhav SV, Bringas E, Yadav GD, Rathod VK, Ortiz I, Marathe KV. Arsenic and fluoride contaminated groundwaters: A review of current technologies for contaminants removal. J. Environ. Manage., 162, pp https://doi.org/10.1016/j.jenvman.2015.07.020
Jha PK, Tripathi P (2021) Arsenic and fluoride contamination in groundwater: A review of global scenarios with special reference to India. Groundw. Sustain. Dev., 13, p 100576https://doi.org/10.1016/j.gsd.2021.100576
Karmakar S, Mukherjee J, Mukherjee S (2016) Removal of fluoride contamination in water by three aquatic plants. Int J Phytoremediation 18(3):222–227. https://doi.org/10.1080/15226514.2015.1073676
Kondru VLP (2014).Understanding Fluorosis: Implications for Dental and General Health. Trans. Latest Trends Heal. Sect., 6(6)
Limón-Pacheco JH, M. I., Cárdenas-González M, Retana IMS, Gonsebatt ME, Del Razo LM (2018) Potential co-exposure to arsenic and fluoride and biomonitoring equivalents for Mexican children. Ann. Glob. Heal., 84(2), p 257. https://doi.org/10.29024/aogh.913
Maity JP et al (2021) Seven 21st century challenges of arsenic-fluoride contamination and remediation. Groundw Sustain Dev 12:100538. https://doi.org/10.1016/j.gsd.2020.100538
Mamta Baunthiyal MB, Vinay Sharma VS (2012) Phytoremediation of fluoride contaminated water and soil: a search for fluoride hyperaccumulators. J Agric Technol, 8(6): 1965–1978. http://www.ijat-aatsea.com
N. EFSA Panel on Dietetic Products and Allergies (NDA), "Scientific opinion on dietary reference values for fluoride," EFSA J, 11(8), p. 3332
Organization WH (2004) Disability weights for diseases and conditions. Glob Burd Dis. 8. https://doi.org/10.3389/fped.2020.00359
Pallier V, Feuillade-Cathalifaud G, Serpaud B (2011) Influence of organic matter on arsenic removal by continuous flow electrocoagulation treatment of weakly mineralized waters. Chemosphere, 83(1), pp. 21–28.https://doi.org/10.1016/j.chemosphere.2011.01.038
Polonskij VI, Polonskaya DE (2013) Fluoride contamination of soil and phytoremediation. Agric Biol Plant Biol Sect
Qiao W et al (2022) Contrasting behaviors of groundwater arsenic and fluoride in the lower reaches of the Yellow River basin, China: Geochemical and modeling evidences. Sci Total Environ 851:158134. https://doi.org/10.1016/j.scitotenv.2022.158134
Raj D, Shaji E (2017) Fluoride contamination in groundwater resources of Alleppey, southern India. Geosci Front 8(1):117–124. https://doi.org/10.1016/j.gsf.2016.01.002
Ranjan R, Ranjan A (2015) Fluoride toxicity in animals. Springer

Ritter Paul Sibley KS, Ken Hall, Patricia Keen, Gevan Mattu, Beth Linton, Len (2002) Sources, pathways, and relative risks of contaminants in surface water and groundwater: a perspective prepared for the Walkerton inquiry. J Toxicol. Environ. Heal. Part A, 65(1), pp 1–142. https://doi.org/10.1080/152873902753338572

Saha A, Sen Gupta B, Patidar S, Martínez-Villegas N (2022) Spatial distribution and source identification of metal contaminants in the surface soil of Matehuala, Mexico based on positive matrix factorization model and GIS techniques. Front. Soil Sci., 2, p 1041377

Sahu P, Debsarma C (2023) Geochemical Sources, Aqueous Geochemistry, Human Health Risk of Fluoride-Enriched Groundwater, and Its Remedial Measures. Hydrogeochem. Aquat. Ecosyst., pp. 33–59. https://doi.org/10.1002/9781119870562.ch2

Sandoval MA, Fuentes R, Thiam A, Salazar R (2021) Arsenic and fluoride removal by electrocoagulation process: A general review. Sci. Total Environ., 753, p 142108.https://doi.org/10.1016/j.scitotenv.2020.142108

Shah DV (2022) Role of Absorption and Adsorption in the Removal of Waste in Emerging trends in environmental biotechnology, CRC Press, pp. 33–47

Smith AH, Lopipero PA, Bates MN, Steinmaus CM (2002) Arsenic epidemiology and drinking water standards. Sci 296(5576), pp 2145–2146.https://doi.org/10.1126/science.1072896

Srivastava S, Flora SJS (2020) Fluoride in drinking water and skeletal fluorosis: a review of the global impact. Curr Environ Heal Reports 7:140–146. https://doi.org/10.1007/s40572-020-00270-9

Srivastava S et al (2021) Arsenic remediation through sustainable phytoremediation approaches". Minerals 11(9):936. https://doi.org/10.3390/min11090936

Sun Y, Yu F, Han C, Houda C, Hao M, Wang Q (2022) Research progress on adsorption of arsenic from water by modified biochar and its mechanism: A review". Water 14(11):1691. https://doi.org/10.3390/w14111691

Thakur LS, Mondal P (2017) Simultaneous arsenic and fluoride removal from synthetic and real groundwater by electrocoagulation process: parametric and cost evaluation. J Environ Manage 190:102–112. https://doi.org/10.1016/j.jenvman.2016.12.053

Trivedi MH, Priyadarshi GV, Lalwani D, Eslamian S (2022) Risk Assessment Applications: Exposure, Safety, and Security. In: Furze JN, Eslamian S, Raafat SM, Swing K (eds) Earth Systems Protection and Sustainability. Springer, Cham. https://doi.org/10.1007/978-3-030-98584-4_9

Uddin MJ, Jeong YK (2020) Efficiently performing periodic elements with modern adsorption technologies for arsenic removal. Environ Sci Pollut Res 27(32):39888–39912. https://doi.org/10.1007/s11356-020-10323-z

Yassin MA et al (2024) Toward decontamination in coastal regions: groundwater quality, fluoride, nitrate, and human health risk assessments within multi-aquifer al-hassa, saudi arabia. Water, 16(10), p 1401. https://doi.org/10.3390/w16101401

Precipitation and Altitude Impact on Groundwater Arsenic and Fluoride Distribution in Chandrapur District, Central India

Rahul K. Kamble and Bhupen Barman

Abstract In the post-monsoon season, a stratified approach using grab sampling methodology was tried to sample from 36 locations in order to determine the impact of precipitation and altitude on the distribution of groundwater arsenic and fluoride in Chandrapur district, central India. The acid digestion method was used for arsenic concentration measurement with the help of Inductively Coupled Plasma-Optical Emission Spectroscopy and fluoride by SPANDS method. The generated primary data on the concentrations of these groundwater contaminants was divided into five altitudes and four precipitation ranges to ascertain their impact on their distribution. It is observed that a higher precipitation range (1801–1917 mm) leads to the maximum dissolution of arsenic (0.0338 mg/L) and fluoride (1.42 mg/L), which enter groundwater from the minerals and ores found in the Earth's crust. The lower precipitation ranges result in a lesser concentration of these contaminants. The maximum (0.0355 mg/L) average groundwater arsenic was observed at lower altitudes (152–179 m amsl) and minimum (0.0294 mg/L) at higher altitudes (261–287 m amsl). In the case of fluoride, contradictory to groundwater arsenic concentrations were observed. It was further observed that at various altitude levels, where groundwater arsenic was higher, fluoride concentration was lower, and vice versa, except in the altitude range of 207–233 m amsl, where both the contaminants concentrations were higher. The inhabitants residing in higher precipitation ranges (1801–1917 mm) and lower altitudes (152–179 m amsl) and medium altitudes (207–233 m amsl) are at higher health risk due to ingestion of groundwater contaminated with arsenic and fluoride. The precipitation has a positive impact on groundwater arsenic and fluoride distribution, whereas altitude has a negative impact on groundwater arsenic and a positive on groundwater fluoride distribution.

R. K. Kamble (✉)
Institution for Higher Learning, Research and Specialized Studies in Environmental Science, Sardar Patel College, Ganj Ward, Chandrapur, Maharashtra, India
e-mail: rahul@spm.ac.in

B. Barman
Department of Geography, Tufanganj Mahavidyalaya, Cooch Behar, West Bengal, India

A. Nath et al. (eds.), *Groundwater Depletion and Sustainability*, Advances in Geographic Information Science, https://doi.org/10.1007/978-3-032-09921-1_9

Keywords Altitude · Arsenic · Central India · Chandrapur · Fluoride · Precipitation

1 Introduction

Human existence is impractical without water. The 'quality of water' in a particular area governs the 'quality of life' enjoyed by the inhabitants of that area. Groundwater and surface water are the two main sources of water that is readily available. Since groundwater formations or aquifers are normally impermeable, they do not easily get contaminated. However, groundwater does undergo some natural filtering during penetration. Due to these characteristics, groundwater is now considerably more frequently used for drinking than surface water (UNEP 2002). More than 50% of the world's population rely on groundwater as their primary source of drinking water (Fry 2005). The groundwater is the main source of daily drinking water for around 2.5 billion (WWAP (United Nations World Water Assessment Programme) 2015). For some rural and small populations (Hani 1990), groundwater is the only source of drinking water (Hani 1990). McDonald et al. (2005) state that groundwater extraction is the only practical solution for addressing the distributed rural water demand because it is accessible from any location and requires less capital to maintain and grow (Bresline 2007; Habila 2005).

About 90% of Indians living in rural areas rely directly on groundwater for irrigation and drinking, according to (Narsimha et al. 2022). Drinking water pollution from different chemicals and heavy metals released from different anthropogenic sources has become a global problem, according to (Rapant and Krcmova 2007a). The ecology and human health are both significantly impacted by contamination of water resources (Emmanuel et al. 2009a; Muhammad et al. 2011a). According to the United Nations in 1997, 2.3 billion people worldwide have ailments that are related to water.

Elevated groundwater arsenic concentrations have been reported in Mexico (Bundschuh et al. 2012; Armienta and Segovia 2008), USA (Amasa et al. 1975; Haque and Johannesson 2006), China (Guo and Wang 2005; Yang et al. 2012), India (Kumar et al. 2010a, 2010; Shah 2012), Bangladesh (Halim et al. 2009; Kamal and Parkpian 2002), Vietnam (Berg et al. 2007; Winkel et al. 2011) and Pakistan (Farooqi et al. 2007a, 2007b; Muhammad et al. 2010). In the recent past, groundwater in Japan (Yoshizuka et al. 2010) and Korea (Ahn 2012), was also found to be contaminated with arsenic. However, China (Chaoke et al. 1997; Ali et al. 2016) and India (Arveti et al. 2011) are the worst-affected countries based on the data of the population exposed to the groundwater fluoride contaminant.

According to several studies (Cubadda et al. 2015; Karim 2000; Rossman et al. 2004; Tchounwou et al. 2004; Yoshida et al. 2004), prolonged exposure to arsenic-contaminated waters can lead to major health issues viz. skin lesions, cardiovascular problems, type II diabetes, as well as bladder, lung, and skin malignancies. Although fluorine at low concentrations (approximately 0.6 mg/L) protects teeth and bones, excessive consumption can lead to dental and skeletal disease (fluorosis) (Brindha and Elango 2011). These two elements commonly have concentrations in groundwater that are higher than the World Health Organization (WHO) guidelines for drinking water standards from 2017 (10.0 μg/L for arsenic and 1.5 mg/L for fluoride). Arsenic concentrations in drinking water that are higher than WHO guidelines pose a hazard to an estimated 200 million people worldwide (George et al. 2014). Fluorosis is a condition that affects over 200 million people in 25 different nations (Khatri and Tyagi 2015).

1.1 Study Area

The Chandrapur district (19° 25' to 20° 45' N and 78° 50' to 80° 10' E), which lies in the Vidarbha area of Maharashtra state in central India is 11,364 square kilometers in size and ranges in elevation from 106 to 589 m above mean sea level (amsl). The district is divided into 15 administrative blocks. Along with dense forests and wildlife, the region is rich in natural resources, such as coal, limestone, iron, copper, and other minerals. Because of the region's wealth in minerals and natural resources, it has seen the construction of numerous thermal power plants, cement factories, pulp and paper mills, and massive coal mines. The district also has Tadoba Andhari Tiger Reserve, which has one of the largest numbers of tigers in the world (CGWB 2009).

In addition to year-round general aridity, the district has a range of climatic extremes, such as a hot summer (May, with highs of 46 °C) and a cold winter (December, with lows of 7 °C). The climate of the area could be described as tropically hot. The humidity was recorded at 70% during the monsoon and 20% during the summer. Rainfall from the southwest monsoon occurred throughout the rainy season (June to September), with an average of 60 to 65 wet days per year and an annual rainfall range of 1200 to 1450 mm. Rainfall is distributed unevenly throughout the district. The Worora administrative block experiences relatively little rainfall, which gradually rises to a maximum close to the Bramhapuri administrative block (CGWB 2009). According to geology, the Chandrapur district is part of the Gondwana sedimentary basin. The lithology of the Chandrapur region is varied, ranging from the Archean to more recent alluvium and laterites.

A total of 10,73,946 women and 11,20,316 men make up the district's 21,94,262 inhabitants, according to the 2011 Census of India. According to reports, the population density was 192 people per square kilometer, the decadal growth rate from 2001 to 2011 was 6.0%, and 35.1% of people lived in urban areas. A detailed examination of census data reveals that 43.2% of the population's primary source of drinking water in the rural areas of the Chandrapur district comes from tube wells and hand pumps (36% and 7.2%, respectively). This statistical data indicates that groundwater is the primary source of drinking water for people living in the research area. (Census of India 2011).

2 Recent Developments

2.1 International Findings

Drinking water pollution from different chemicals and heavy metals released from different anthropogenic sources has become a global problem, according to Rapant and Krcmova (Rapant and Krcmova 2007b). The ecology and human health are significantly impacted by contamination of water resources (Emmanuel et al. 2009b; Muhammad et al. 2011b). The United Nations (United Nations Commission for Sustainable Development 1997) estimates that 2.3 billion people worldwide suffer from water-related illnesses. The global co-occurrence of arsenic and groundwater fluoride is shown in Table 1. The table illustrates how these two components are found in diverse parts of different nations worldwide.

Issanov et al. (2024) reported statistically significant correlations between arsenic exposure and the risk of lung cancer development or death. According to the meta-analysis, people who are exposed to low to moderate arsenic concentrations (less than 150 μg/L) have a higher chance of getting lung cancer or passing away from it. Furthermore, lung cancer risks may be significantly increased by arsenic exposure at the WHO provisional maximum acceptable concentration recommendation level of 10 μg/L. According to the findings, drinking water with 10 μg/L and 20 μg/L of arsenic is linked to an 11% and 32% higher risk of lung cancer, respectively.

Table 1 Global co-occurrence of fluoride and arsenic in groundwater

Region, Country	Groundwater arsenic concentration	Groundwater fluoride concentration	References
Brahmaputra Flood Plain, India	25.1 μg/L	14 mg/L	Das et al. (2018a)
Tripura, India	< 0.003–0.044 mg/L	< 0.005–4.8 mg/L	Bhattacharya et al. (2020a)
Bam, Iran	9.26–14.65 μg/L	0.91–1.12 mg/L	Toolabi et al. (2021a)
Zagros basin, Kurdistan region, Iraq	0.19–7.8 μg/L	0.01–2.1 mg/L	Mustafa et al. (2023a)
Latium, Central Italy	128.5 μg/L	6.1 mg/L	Parrone et al. (2020a)
San Luis Potosi, Mexico	2.50–30.0 μg/L	0.20–3.5 mg/L	Fernandez-Macias et al. (2020)
Inner Mongolia	0.31–47.0 μg/L	0.07–7.70 mg/L	Nakazawa et al. (2020)
Mexico	3.02–12.32 μg/L	361–1160 mg/L	Alarcon-Herrera et al. (2020a)
Punjab, Pakistan	32–1900 μg/L	2.47–21.1 mg/L	Farooqi et al. (2007c)

Moreover, individuals who are exposed to 20 μg/L of arsenic may have a 22% higher risk of developing lung cancer.

Xu et al. (2023) reported the presence of arsenic and fluoride in the inland plain in Huaihe River Basin, China. The findings from 62 groundwater samples indicate that the presence of increased arsenic and fluoride concentrations is a significant factor impacting quality. Since groundwater arsenic and fluoride are geogenic sources the co-contamination mechanisms are linked to flat terrain, high element abundance, alkaline groundwater, and reductive groundwater conditions. Mustafa et al. (2023b) reported, that fluoride and arsenic concentrations range from 0.01 to 2.1 mg/L and 0.19 to 7.8 g/L, respectively from the Kurdistan region of Iraq. The findings suggested a substantial risk of fluorosis in some places. The primary origins of arsenic appear to be geogenic in the research area.

Toolabi et al. (2021b) reported the concentration of arsenic, fluoride, and nitrate in groundwater samples of Bam, South-eastern Iran. In drinking water samples, the levels of arsenic, fluoride, and nitrate varied from 9.26–14.65 μg/L, 0.91–1.12 mg/L, and 8.50–10.85 mg/L, respectively. Health risk assessment findings show that the mean risk of arsenic oral hazard quotient (HQ) is 5.372 and dermal hazard quotient is 1.407. In the case of fluoride HQ was above 1 for children and HQ for adults was below 1. Among the three contaminants carcinogenic risk value for arsenic was more than 1. The findings indicated that for As and NO_3, the carcinogenicity ratio "children to adults" is 3.8 and 2.33, respectively. Children were more at risk for both cancer and non-cancer diseases than adults were.

Saha et al. (2021) reported arsenic contamination in the Brahmaputra, Ganges, and Meghna floodplain sediments is mild, but the delta plains are heavily contaminated (up to hundreds of mg/kg). The study indicates that the arsenic found in Bangladesh's groundwater and food is of inorganic origin and poses a greater risk to human health. In groundwater, the dominating species As (III) moves more frequently throughout a vast portion of Bangladesh.

Chen et al. (2021) reported regional variations of arsenic in groundwater, ranging from 0.5 to 5,000 parts per billion. Certain populations are particularly susceptible to the health concerns associated with arsenic exposure, such as developing children, pregnant women, and persons with underlying genetic risk factors. Compared to adults, children under the age of three are more likely to be exposed to arsenic. It has also been demonstrated that early childhood or prenatal arsenic exposure increases the risk of developing bladder and lung cancer in later life.

Podgroski et al. (2020) estimated that 94–220 million individuals may be exposed to high levels of arsenic in groundwater, with Asia accounting for the vast majority (94%) of these cases. According to Parrone et al. (2020b) in Latium, Central Italy, 55% and 29% of the groundwater samples had levels of arsenic (10.0 μg/L) and fluoride (1.5 mg/L) over the drinking water standards, respectively. There is a strong association between arsenic and fluoride (Pearson's $r = 0.61$, Spearman's $rs = 0.59$).

Jia et al. (2019) reported near the southern edge of the Chinese Loess Plateau increased fluoride levels in shallow groundwater (type I water) were found to offer a non-carcinogenic danger to human health for both adults and 60% of samples Hazard Index (HI) > 1 and children 94% of samples HI > 1 according to results from the human health risk assessment. All samples exhibited HI > 1 for adults, while 70% of the deep fracture groundwater (type IV water) samples had HI > 1 for children. HI > 1 is found in 73% of pediatric samples and 23% of adult samples pertaining to deep alluvial aquifers (type II water).

Alarcon-Herrera et al. (2020b) reported 3.05 million persons in Mexico are exposed to fluoride over the recommended daily intake of 0.06 mg/kg/day. Over the 10-μg/L limit, 8.81 million people are exposed to arsenic, and this exposure alone is predicted to cause an additional 13,070 lifetime instances of cancer. The dry states of north-central Mexico bear the brunt of this illness burden.

Numerous studies conducted in dry and semi-arid regions of Latin America have shown that the co-occurrence of fluoride and arsenic in drinking water is linked to alkaline pH, volcaniclastic particles in the loess or alluvium, and limited recharging, according to Alarcon-Herrera et al. (2012). Water–rock interactions cause arsenic and fluoride contamination, which can be increased by mining, geothermal energy, and over-exploitation of aquifers. Human epidemiological research conducted in Taiwan, Chile, Argentina, and Bangladesh has produced enough data to conclude that consuming arsenic in drinking water not only causes skin cancer but also increases the risk of lung and bladder cancer Centeno et al. (2007).

2.2 *National Findings*

According to Kamble (2024), the post-monsoon groundwater arsenic content in the Chandrapur district ranged from 0.015 to 0.041 mg/L, with an average of 0.031 mg/L. All the samples (n = 36) were above the acceptable limit but were within the permissible limits of the Indian Standard.

Kom et al. (2023) reported the results of the health risk assessment of fluoride in Coimbatore District. The oral pathway for fluoride was more severe than the dermal pathway. Non-carcinogenic health hazard evaluation revealed that the Total Hazard Index (THI) value for children ranges between 0.06–2.71 with an average of 1.09, for females 0.03–1.23 with an average of 0.49 and 0.02–0.97 for male with an average of 0.39. In the study area children were vulnerable to non-carcinogenic health hazards of fluoride around 48% of groundwater samples for children and 13% of females had THI > 1. Children are more susceptible to adverse health effects from drinking contaminated water due to smaller size and lower body weight.

In the Wardha sub-basin groundwater is the main source of drinking water, 83.87% of children, 62.37% of adult females, and 43.01% of adult males had $HQ_{(fluoride)}$ value > 1. For children and the female population high fluoride toxicity was observed in 1.1% samples (Nawale et al. 2021). Gudadhe and Manik (2020) reported groundwater fluoride concentration in dug well and bore well samples was in the range of 0.2–0.5 mg/L in Chandrapur city. All the samples were within the permissible limits.

Bhattacharya et al. (2020b) reported the co-occurrence of arsenic and fluoride in different areas of Dharmanagar town, North Tripura district. The average concentration of arsenic, fluoride, and iron was 0.012 mg/L, 1.8 mg/L, and 2.4 mg/L respectively. According to the data, 43% of the total examined samples had fluoride levels above the WHO drinking water norm of 1.5 mg/L, and 36% of the samples had arsenic concentrations above the WHO limit of 0.01 mg/L. Thirty-one out of the seventy-one samples had simultaneous violations of the corresponding WHO requirements. A noteworthy positive connection ($r = 0.6$) was discovered between the concentrations of arsenic (<0.003–0.044 mg/L) and fluoride (<0.005–4.8 mg/L) in groundwater. Children in the study area are at a high risk of acquiring dental fluorosis because their Estimated Daily Intake (EDI) of fluoride was already determined to be at the "Tolerable Upper Intake Level" (UL) value (0.1 mg/kg/day). The hazard indices for fluoride (HI Fluoride = 1.2–3.0) and arsenic (HI Arsenic = 1.5–5.0) were consistently determined to be harmful ($HI > 1$) in the population under study, regardless of the Central Tendency Exposure or Reasonable Maximum Exposure scenario.

Rajasthan and Gujarat (80–100%), Andhra Pradesh and Punjab (60–80%), Madhya Pradesh, Maharashtra, Kerala, Tamil Nadu, Jharkhand, West Bengal, and Sikkim (about 21–40%) are the states most impacted by fluoride contamination, according to Yadav et al. (2019). Ali et al. (2019) reported groundwater in rural regions has 1.85 times greater concentrations of fluoride than in urban areas. It was found that the concentration of fluoride in groundwater dramatically dropped ($p < 0.05$) with increases in longitude and rainfall. According to the risk assessment results, groundwater consumers in most Indian states are thought to be at non-carcinogenic risk for both adults and children, however, children are more susceptible to fluorosis because of their low body weight.

A severe issue in many nations worldwide, fluorosis is a health danger caused by fluoride that affects approximately 200 million individuals worldwide. More than 62 million people in 20 states in India are suffering from fluoride. Telangana (1,174 habitations), Rajasthan (7670 habitations), and Karnataka (1122 habitations) are the states most affected (Shrivas et al. 2018).

Bhattacharya (2017) observed arsenic concentration in groundwater was beyond the Bureau of Indian Standards maximum permissible limit in isolated pockets in 86 districts in 10 States. According to Das et al. (2018b), using energy dispersive spectroscopy, As, U, and Fe were found in the sediment matrix of the Brahmaputra floodplain (BFP) aquifer in India. Regression analysis showed that the co-occurrence of fluoride and arsenic was less correlated. When PO_4^{3-}, F^-, and HCO_3^- were present at higher pH levels, the upper BFP showed the greatest amount of arsenic desorption, indicating the highest likelihood for co-occurrence. Due to greater pH-induced desorption of adsorbed As and F^- from Fe (hydro)oxides, co-occurrence was primarily observed in isolated aquifers in the upper BFP. According to Hazard Indicators, children aged 3 to 8 are more likely to be at risk for health problems than adults.

Chakraborti et al. (2010) reported there are 411 million people (40% of India's total population) living in the 201 fluoride-endemic districts, and more than 66 million individuals, including 6 million children under the age of 14, are believed to have fluorosis. A total of 70.4 million people in six states along the Ganga–Brahmaputra Plain (GB-Plain) may be in danger of groundwater arsenic toxicity. According to Shankar et al. (2014) inorganic arsenic (III) and arsenic (V) are more toxic and prevalent than organic form.

Das (2013) reported fluoride concentration in Ballarpur city of Chandrapur district in the range of 0.0–0.5 mg/L, and all the samples were within the permissible limit. Chandekar et al. (2010) reported fluoride concentration in Rajura tehsil of Chandrapur district 10 sampling locations had a concentration within the Indian standard limit (1.0–1.5 mg/L), 6 sampling locations had a concentration above the permissible limit (>1.5 mg/L) and 9 sampling locations were below the acceptable limit (<1.0 mg/L).

Numerous studies on groundwater fluoride in the Chandrapur area were conducted, according to the literature review; however, none were conducted on groundwater arsenic, the co-occurrence of arsenic and fluoride, or their distribution based on altitude and precipitation. Thus, these are the identified gaps in the subject domain. To fill these gaps with the new knowledge this study was proposed to be carried out with the objectives to assess the distribution of groundwater arsenic and fluoride according to precipitation and altitude. The results of the investigation will provide fresh insight into the co-occurrence of fluoride and arsenic in the study area's groundwater.

3 Rationale

Hand pumps, covered and uncovered wells, tube wells/boreholes, and tap water from untreated sources are major drinking groundwater sources. Water being the universal solvent dissolves many contaminants in it like heavy metals, inorganic and organic contaminants. Arsenic is a metalloid and fluoride ionic form of the element fluorine are two geogenic contaminants that mix or dissolve in groundwater which is used by people for drinking purposes. Prolonged exposure to arsenic causes arsenicosis, cardiovascular diseases, cancer, and many other health issues. Exposure to fluoride above acceptable limits for a prolonged period causes dental and skeletal fluorosis, kidney and liver damage, and many health issues. Numerous studies have been conducted about the groundwater fluoride concentration in different regions in Chandrapur district, however, arsenic which is at present recognized as the world's greatest environmental risk is still disregarded. Arsenic and fluoride are two toxicants that have caused several health issues individually in humans, but their co-occurrence may have serious health effects on adults and children.

The Sustainable Development Goal (SDG) 6 is to 'ensure availability and sustainable management of water and sanitation for all' with the target to achieve 'universal and equitable access to safe and affordable drinking water for all' (Target 6.1) and 'sustainability increase water-use efficiency across all sectors ensure sustainable withdrawals and supply of freshwater to address water scarcity and substantially reduce the number of people suffering from water scarcity' (Target 6.4). Another SDG 3 is to 'ensure healthy lives and promote well-being for all at all ages' with the target to achieve 'substantially reduce the number of deaths and illnesses from hazardous chemicals and air, water and soil pollution and contamination' (Target 3.9). These two SDGs highlight the importance of studying the distribution of arsenic and fluoride spatially and temporally, and the variation in concentration according to precipitation and altitude.

Arsenic and fluoride are two geogenic contaminants. The International Agency for Research on Cancer listed arsenic and fluoride as Group I and Group III agents respectively. Several studies about groundwater fluoride have been carried out in the Chandrapur district. However, only one study was carried out for groundwater arsenic. According to the literature evaluation, no research has been done on the co-occurrence of fluoride and arsenic in groundwater in the Chandrapur district. In addition, Right to Information responses from government offices viz. Groundwater Survey and Development Agency, Chandrapur, and *Zilla Parishad*, Chandrapur revealed that groundwater arsenic is not being monitored by these agencies thus no information is available about groundwater arsenic of the Chandrapur district. Thus, this is the identified knowledge gap in the subject domain. To overcome this knowledge gap this study was proposed to be carried out with an objective to assess the spatial precipitation and altitude-wise distribution of groundwater arsenic and fluoride in the Chandrapur district of Central India. The expected outcome of this study will be utilized by the local government agency for the identification of hotspots of groundwater arsenic and fluoride and further computation of their health risk on adult males, females, and children. Moreover, the results will be utilized in policy formulation at the local level by concerned authorities.

4 Materials and Methods

There were four components to the strategy that was used to carry out this suggested study. These include (1) spatial distribution of groundwater arsenic, (2) spatial distribution of groundwater fluoride, (3) precipitation-wise distribution of co-occurrence of groundwater arsenic of fluoride, and (4) altitude-wise distribution of co-occurrence of groundwater arsenic of fluoride. The methodological flow diagram of the study is presented in Fig. 1.

4.1 Groundwater Arsenic and Fluoride Assessment

To determine the spatial distribution of groundwater arsenic and fluoride, 36 groundwater samples (Fig. 2) were collected by grab sampling method for general parameters and heavy metal into consideration (arsenic) in separate polyethylene containers of 1000 mL and 100 mL capacity each, respectively. The heavy metal groundwater sample was preserved by adding 2 mL of concentrated HNO3 (16 *N*, Merck) for 100 mL of the sample. Groundwater samples were analysed for different water

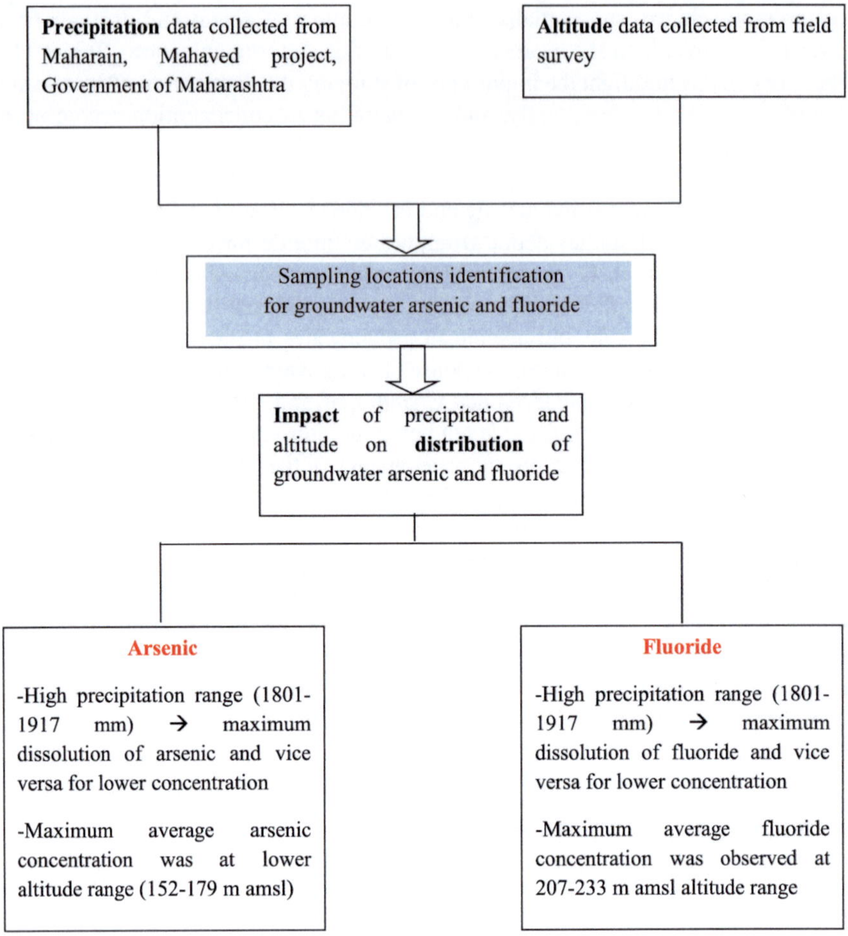

Fig. 1 Methodological flow diagram of the study

quality parameters and heavy metal into consideration (Table 2) according to the standard methods as described by the American Public Health Association (APHA 2023).

The concentration of arsenic in groundwater was measured using the acid digestion method and Inductively Coupled Plasma-Optical Emission Spectroscopy (PerkinElmer Dv 7000, Shelton, CT, USA). The 1,8–Dihydroxy–2–(4–sulfophenylazo)naphthalene–3,6–disulfonic acid trisodium salt (SPADNS) method, a colorimetric technique, was used to measure the fluoride content of groundwater. Fluoride reacts with zirconium reagent SPADNS solution in an acidic environment, bleaching the "lake" (color of SPADNS reagent) when ZrF6 is generated. This was the basis for the procedure. Because bleaching produces fluoride ions, there was a strong correlation between bleaching and fluoride concentration.

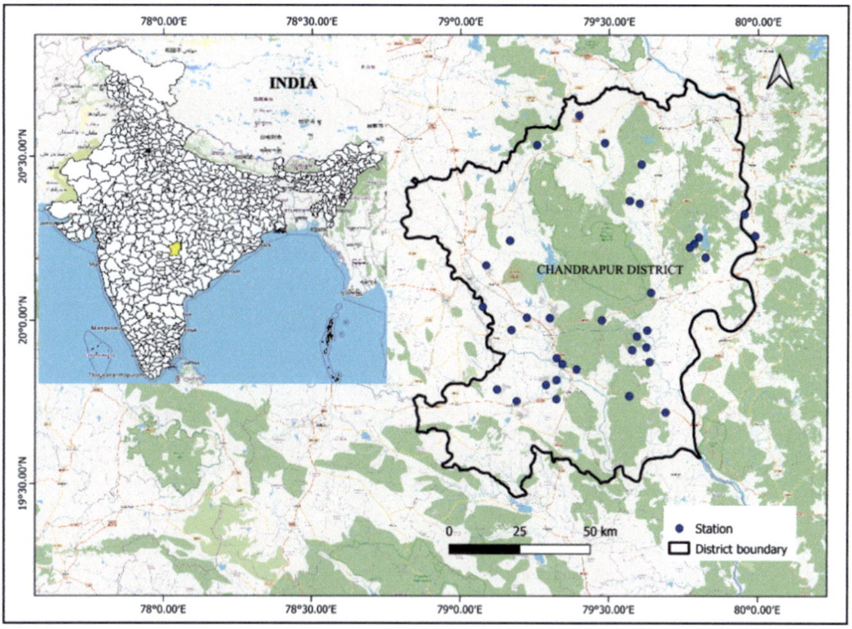

Fig. 2 Spatial distribution of groundwater sampling locations in the study area

Table 2 Standard methods for analysis of groundwater

Parameter	Standard method	APHA reference no
Temperature	Mercury thermometer	B of 2550
pH	Electrometric method	B of 4500-H^+
Conductivity	Conductivity meter	B of 2510
Total dissolved solids	Total dissolved solids dried at 180 °C	C of 2540
Alkalinity	Titration method	B of 2320
Hardness	EDTA titration method	C of 2340
Chlorides	Argentometric method	B of 4500-Cl
Sulphate	Barium sulphate	B of 4500- SO_4^{2-}
Sodium	Microprocessor flame photometer	B of 3500-Na^+
Potassium	Microprocessor flame photometer	B of 3500-K^+
Arsenic	Inductively coupled Plasma-OES	C of 3500-As
Fluoride	SPANDS method	D of 4500-F^-

4.2 Precipitation

The influence of precipitation on the distribution of groundwater arsenic and fluoride was carried out taking rainfall data from the 15 administrative blocks of the

Table 3 Rainfall in different administrative blocks of Chandrapur district

Administrative block	Total for the rainy season (mm)
Chandrapur	1661
Mul	1708
Gondpipari	1721
Warora	1833
Bhadravati	1788
Chimur	1660
Bramhapuri	1912
Nagbhid	1440
Sindhewahi	1396
Rajura	1726
Korpana	1786
Sawali	1454
Ballarpur	1693
Pombhurna	1916
Jivati	1680

study area. Rainfall data during the rainy season (June–September) was collected from www.maharain.gov.in (Table 3). Rainfall averaged 1692 mm, with a range of 1396 mm (Sindhewahi) to 1916 mm (Pombhurna). Five classes were identified from this rainfall range: 1395–1500 mm, 1501–1600 mm, 1601–1700 mm, 1701–1800 mm, and 1801–1917 mm (Fig. 3). The study area's administrative blocks and additional sampling sites were categorized according to these classes. Five (33.33%) of the administrative blocks had rainfall between 1701 and 1800 mm, four (26.66%) had rainfall between 1601 and 1700 mm, and three (20%) each had rainfall between 1396–1500 mm and 1801–1917 mm. Rainfall in the 1501–1600 mm range did not fall on any administrative unit. The administrative blocks with sampling locations were allocated based on these rainfall classes.

4.3 Altitude

Any location's altitude is measured in meters above mean sea level (m amsl). The topography of a place determines its altitude. The physical topography of the Earth in the region is the focus of topographic factors. The surface of the Earth is not uniform; certain areas may exhibit a number of anomalies. These topographic characteristics may have an impact on the minerals and ores found in the Earth's crust in addition to biotic factors. Temperature, pressure, wind speed, humidity, solar radiation intensity, and other factors change as one's height above mean sea level rises. The distribution of elements and heavy metals may also be caused by variations in altitude, which

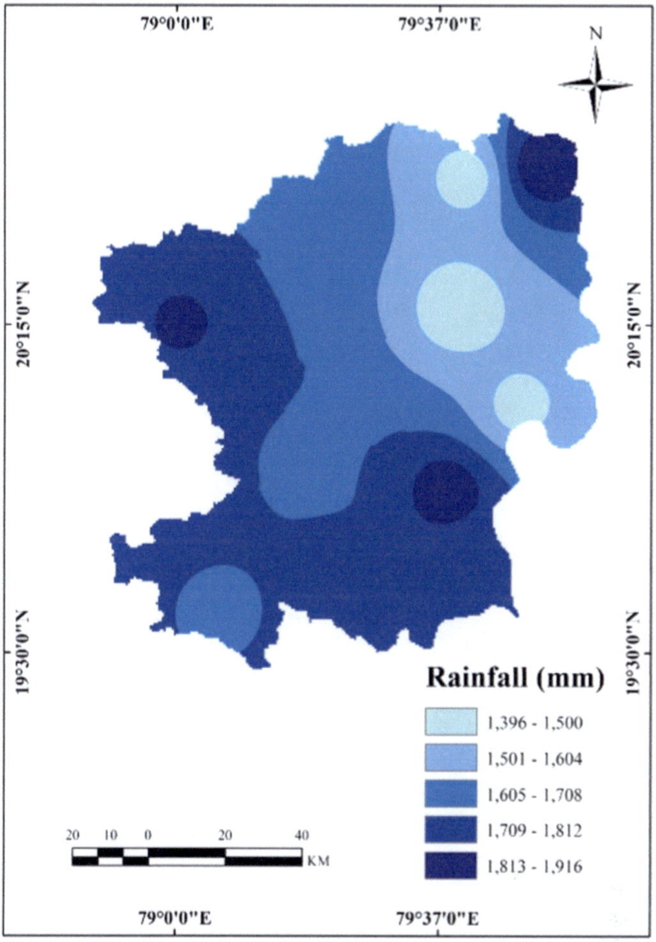

Fig. 3 Spatial distribution of precipitation in the study area

can affect the Earth's crust's geology and, eventually, the flow of ores and minerals into it.

The distribution of fluoride and arsenic at the height of sampling sites was attempted to take these factors into account. The sampling sites ranged in elevation from 152 m above sea level (Visapur, HP) to 287 m above sea level (Bhisi, HP) (Fig. 4). The mean elevation was 211 m above sea level. Five ranges of altitude were identified: 152–179 m amsl, 180–206 m amsl, 207–233 m amsl, 234–250 m amsl, and 261–287 m amsl. Five (13.88%) in 234–260 m amsl, two (5.55%) in 261–287 m amsl, three (8.33%) in 152–179 m amsl, eleven (30.55%) in 180–206 m amsl, and fifteen (41.66%) in 207–233 m amsl were the sampling locations with altitude in these classes. In these altitude classes, the average concentrations of fluoride and arsenic in groundwater were dispersed appropriately.

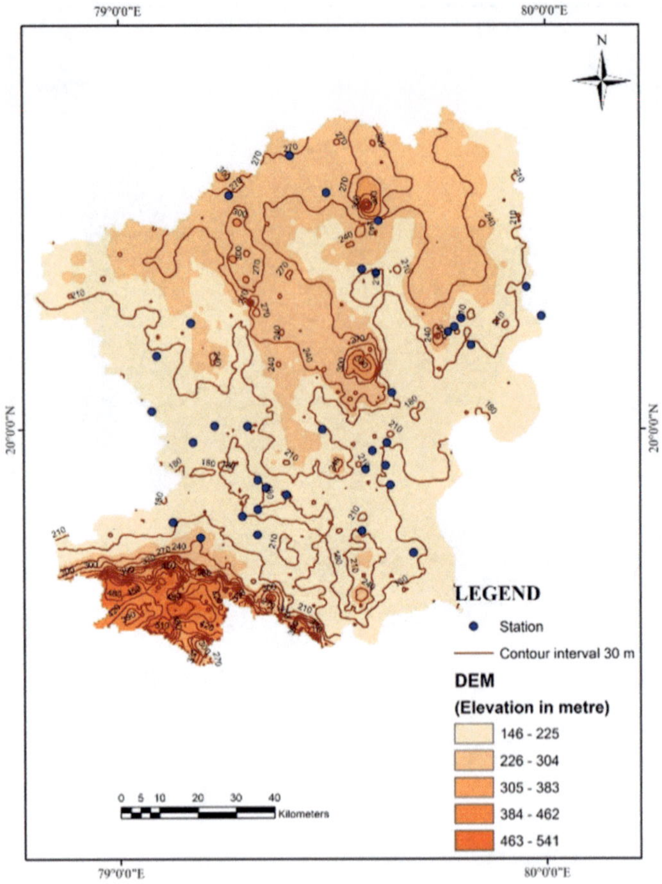

Fig. 4 Spatial distribution of elevation in the study area

5 Results and Discussion

The groundwater arsenic and fluoride distribution in various precipitation ranges is depicted in Fig. 5a whereas their distribution for altitude is presented in Fig. 5b.

The average groundwater arsenic concentration at various precipitation ranges was in the order of 0.0338 mg/L (1801–1917 mm) > 0.0326 mg/L (1601–1700 mm) > 0.0322 mg/L (1395–1500 mm) > 0.0296 mg/L (1701–1800 mm) (Table 4). In the case of average fluoride concentration, it was 1.42 mg/L (1801–1917 mm) > 1.28 mg/L (1701–1800 mm) > 1.12 mg/L (1395–1500 mm) > 0.99 mg/L (1601–1700 mm) (Table 4). It can be arrived at higher precipitation range (1801–1917 mm) results in the maximum dissolution of arsenic and fluoride present in the ores and minerals present in the Earth's crust into groundwater. The lower precipitation ranges result in lesser concentration of these contaminates. These findings are inconsistent with

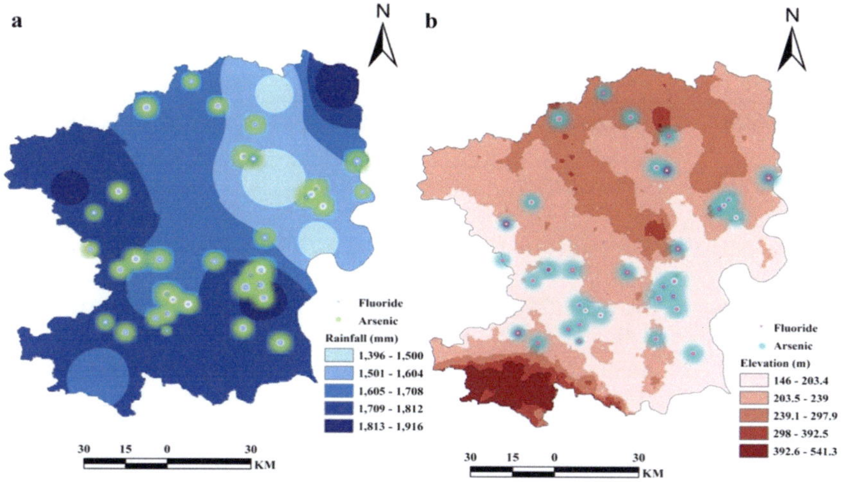

Fig. 5 **a** Precipitation ranges with the distribution of arsenic and fluoride, **b** Altitude ranges with the distribution of arsenic and fluoride

Alarcon-Herrera et al. (2020b). However, the findings are in agreement with Ali et al. (2019) that the fluoride in groundwater dramatically dropped with an increase in rainfall.

At various altitude ranges of groundwater sampling locations the average arsenic concentrations were found to be 0.0355 mg/L (152–179 m amsl) > 0.323 mg/L (207–233 m amsl) > 0.0310 mg/L (234–260 m amsl) > 0.0308 mg/L (180–206 m amsl) >0.0294 mg/L (261–287 m amsl) (Table 5). For average groundwater fluoride, it was 1.30 mg/L (207–233 m amsl) > 1.23 mg/L (261–287 m amsl) >1.20 mg/L (180–206 m amsl) > 0.96 mg/L (152–179 m amsl) > 0.85 mg/L (234–260 m amsl) (Table 5). The maximum (0.0355 mg/L) average groundwater arsenic concentration was observed at lower altitudes (152–179 m amsl) and minimum (0.0294 mg/L) at higher altitudes (261–287 m amsl). Thus, residents at lower altitudes from the study area are comparatively more exposed to arsenic than at higher altitudes. In the case of fluoride, contradictory groundwater arsenic concentrations were observed. It was further observed that at various altitude levels, where groundwater arsenic concentration was higher fluoride concentration was lower, and vice versa except in the altitude range of 207–233 m amsl where both the average contaminants concentrations were higher. The inhabitants residing in higher precipitation ranges (1801–1917 mm) and lower altitudes (152–179 m amsl) and medium altitudes (207–233 m amsl) are at higher risk to groundwater arsenic and fluoride exposure. The limitation of the study is that it is carried out at 36 sampling locations in the study area. To have a comprehensive understanding of the study topic it is suggested to increase the number of sampling locations in future studies.

Table 4 Groundwater arsenic and fluoride distribution w.r.t. precipitation ranges

1395–1500 mm n = 3, 20%	1601–1700 mm n = 4, 26.66%	1701–1800 mm n = 5, 33.33%	1801–1917 mm n = 3, 20%
Govindpur (0.0295, 1.37)	Sonegaon (0.0305, 1.05)	Dabgaon (0.0314, 1.1)	Mowada (0.0230, 1.1)
Chikmara (0.0334, 1.3)	Lohara (0.0339, 0.5)	Naleshwar (0.0414, 0.97)	Dongargaon (0.0345, 2.16)
Gunjewahi (0.0352, 0.74)	Chichpalli (0.0312, 1.5)	Karwan (0.0302, 1.76)	Pombhurna (0.0358, 1.1)
Mangali Chak (0.0334, 0.74)	Durgapur (0.0355, 1.48)	Ganpur (0.0330, 1.1)	Jam (Tukum) (0.0365, 1.3)
Pathri (0.0317, 0.63)	Morwa (0.0367, 0.93)	Gondpipari (0.0336, 1.15)	Dongar Haldi (0.0396, 1.44)
Ratnapur (0.0337, 0.78)	Pethbhansouli (0.0339, 1.05)	Telwasa (0.0268, 1.1)	(n = 5, 13.88%) Average As = 0.0338 Average F = 1.42
Antargaon (0.0290, 2.32)	Bhisi (0.0251, 1.3)	Belora (0.0263, 2.0)	
(n = 7, 19.44%) Average As = 0.0322 Average F = 1.12	Pimpalgaon (0.0291, 0.9)	Sagra (0.0312, 0.8)	
	Visapur (0.0366, 0.93)	Sasti (0.0295, 1.1)	
	Ballarpur (0.0322, 0.63)	Gowari (0.0290, 1.3)	
	Kem (Tukum) (0.0335, 0.65)	Arvi (0.0158, 1.15)	
	(n = 11, 30.55%) Average As = 0.0326 Average F = 0.99	Awarpur (0.0265, 1.82)	
		Lakhmapur (0.0308, 1.3) (n = 13, 36.11%) Average As = 0.0296 Average F = 1.28	

Values in parenthesis are of groundwater arsenic and fluoride in mg/L

Table 5 Groundwater arsenic and fluoride distribution w.r.t. altitude

152–179 m amsl	180–206 m amsl	207–233 m amsl	234–260 m amsl	261–287 m amsl
Visapur (152, 0.0336, 0.93)	Mowada (198, 0.023, 1.1)	Sonegaon (215, 0.0305, 1.05)	Sagra (240, 0.0312, 0.8)	Bhise (287, 0.0251, 1.3)
Kem Tukum (178, 0.0335, 0.65)	Lohara (202, 0.0339, 0.5)	Telwasa (207, 0.0268, 1.1)	Pimpalgaon (246, 0.0291, 0.9)	Govindpur (271, 0.0336, 1.15)
Jam Tukum (174, 0.0365, 1.3)	Karwan (205, 0.0302, 1.76)	Belora (210, 0.0263, 2.0)	Pathari (240, 0.0317, 0.63)	(n = 2, 5.55%) Average As = 0.0294 Average F = 1.23
(n = 3, 8.33%) Average As = 0.0355 Average F = 0.96	Sasti (198, 0.0295, 1.1)	Pethbhansouli (209, 0.0339, 1.05)	Ballarpur (243, 0.0322, 0.63)	
	Gowari (198, 0.029, 1.3)	Dongargaon (222, 0.0345, 2.16)	Lakhmapur (243, 0.0308, 1.3)	
	Arvi (202, 0.0158, 1.15)	Chichpalli (226, 0.0312, 1.5)	(n = 5, 13.88%) Average As = 0.0310 Average F = 0.85	
	Ganpur (199, 0.033, 1.1)	Dabgaon Tukum (215, 0.0314, 1.1)		
	Gondpipari (195, 0.0336, 1.15)	Naleshwar (215, 0.0414, 0.97)		
	Pombhurna (189, 0.0358, 1.1)	Chikmara (214, 0.0334, 1.3)		
	Dongar Haldi (187, 0.0396, 1.44)	Gunjewahi (230, 0.0352, 0.74)		
	Durgapur (201, 0.0355, 1.48)	Mangali Chak (224, 0.0334, 0.74)		
	(n = 11, 30.55%) Average As = 0.0308 Average F = 1.20	Ratnapur (225, 0.0337, 0.78)		
		Antargaon (230, 0.029, 2.32)		

(continued)

Table 5 (continued)

152–179 m amsl	180–206 m amsl	207–233 m amsl	234–260 m amsl	261–287 m amsl
		Awarpur (216, 0.0265, 1.82)		
		Morwa (218, 0.0367, 0.93) (n = 15, 41.66%) Average As = 0.0323 Average F = 1.30		

Values in parenthesis (altitude in m amsl, average arsenic, and fluoride concentration in mg/L)

6 Conclusion

From the findings of the study area, it can be concluded that precipitation and altitude have a pronounced impact on groundwater arsenic and fluoride distribution in the study area. This impact is due to the dilution of the ores and minerals present in the Earth's crust by precipitation and further their presence in various quantities at various altitudes. For future extraction of groundwater, the precipitation and altitude at a particular location should be taken into consideration. The health risk assessment on inhabitants for the co-occurrence of arsenic and fluoride in various precipitation and altitudes needs to be carried out to assess the vulnerability and disease burden on them. Those areas that pose a greater risk to human health be well communicated to the inhabitants of the area. The local authorities should initiate appropriate measures by providing safe potable water, and studies on the removal of these contaminants together should be attempted.

AI Disclosure The principal author (Corresponding author) utilized QuillBot and Grammerly to check spelling and punctuation and enhance writing style when preparing this book chapter. The writers took full responsibility for the publication's content after utilizing these tools and software, reviewing and editing it as necessary.

Disclose Conflict of Interest The writers attest that they are not associated with or involved with any organization or institution that has a financial or non-financial interest in the topics or resources covered in this work.

Reference Data The article displays all of the processed data utilized in the investigation. Upon request, the corresponding author may provide raw data.

References

Ahn JS (2012) Geochemical occurrences of arsenic and fluoride in bedrock groundwater: a case study in Geumsan County, Korea. Environ Geochem Health 34:43–54. https://doi.org/10.1007/s10653-011-9411-5

Alarcon-Herrera MT, Bundschuh J, Nath B, Nicolli HB, Gutierrez M, Reyes-Gomez VM, Nunez D, Martin-Dominguez IR, Sracek O (2012) Co-occurrence of arsenic and fluoride in groundwater of semi-arid regions in Latin America: Genesis, mobility and remediation. J Hazard Mater 262(15):960–969

Alarcon-Herrera MT, Martin-Alarcon DA, Gutierrez M, Reynoso-Cuevas L, Martin-Dominguez A, Olmos-Marquez MA, Bundschuh J (2020a) Co-occurrence, possible origin, and health-risk assessment of arsenic and fluoride in drinking water sources in Mexico: Geographical data visualization. Sci Total Environ 698:134168. https://doi.org/10.1016/j.scitotenv.2019.134168

Alarcon-Herrera MT, Martin-Alarcon DA, Gutierrez M, Reynoso-Cuevas L, Martin-Dominguez A, Olmos-Marquez MA, Bundschuh J (2020b) Co-occurrence, possible origin, and health-risk assessment of arsenic and fluoride in drinking water sources in Mexico: Geographical data visualization. Sci Total Environ 698:13416

Ali S, Thakur SK, Sarkar A, Shekhar S (2016) Worldwide contamination of water by fluoride. Environ Chem Lett 14:291–315. https://doi.org/10.1007/s10311-016-0563-5

Ali S, Fakhri Y, Golbini M, Thakur S, Alinejad A, Parseh I, Shekhar S, Bhattacharya P (2019) Concentration of fluoride in groundwater of India: A systematic review, meta-analysis and risk assessment. Groundw Sustain Dev 9:1–10

Amasa SK, Nyarko BJB, Adotey DK, Dampare SK, Adomako D (1975) Arsenic pollution at Obuasi goldmine, town and surrounding countryside. RN-Environ Health Perspect 12:131–135

APHA (American Public Health Association) (2023) Standard methods for the examination of water and wastewater (24th ed.). Washington D.C.: APHA, AWWA, WPCF

Armienta M, Segovia N (2008) Arsenic and fluoride in the groundwater of Mexico. Environ. Geochem Health, 30, 345e353. https://doi.org/10.1007/s10653-008-9167-8

Arveti N, Sarma MRS, Aitkenhead-Peterson J, Sunil K (2011) Fluoride incidence in groundwater: a case study from Talupula, Andhra Pradesh, India. Environ Monit Assess 172:427–443. https://doi.org/10.1007/s10661-010-1345-3

Berg M, Stengel C, Pham TKT, Pham HV, Sampson ML, Leng M, Samreth S, Fredericks D (2007) Magnitude of arsenic pollution in the Mekong and Redriver deltas-Cambodia and Vietnam. Sci Total Environ 372:413–425. https://doi.org/10.1016/j.scitotenv.2006.09.010

Bhattacharya AK (2017) Arsenic Contamination in Indian groundwater. J Inst Public Health Eng India 40(2):18–36

Bhattacharya P, Adhikari S, Samal AC, Das R, Dey D, Deb A, Ahmed S, Hussein J, De A, Das A, Jordar M, Panigrahi AK, Roychowdhury T, Santra SC (2020a) Health risk assessment of co-occurrence of toxic fluoride and arsenic in groundwater of Dharmanagar region, North Tripura (India). Groundw Sustain Dev 11:100430. https://doi.org/10.1016/j.gsd.2020.100430

Bhattacharya P, Adhikari S, Samal AC, Das R, Dey D, Deb A, Ahmed S, Hussein J, De A, Das A, Jordar M, Panigrahi AK, Roychowdhury T, Santra SC (2020b) Health risk assessment of co-occurrence of toxic fluoride and arsenic in groundwater of Dharmanagar region, North Tripura (India). Groundw Sustain Dev 11:10043

Bresline E (2007) Sustainable water supply in developing countries. Geol Soc Am. Paper No. 194–1

Brindha K, Elango L (2011) Fluoride in groundwater: causes, implications and mitigation measures. In: Monroy SD (Ed), Fluoride Prop, Appl Environ Manag, 111–136. Available from: https://www.researchgate.net/publication/220000345_Fluoride_in_Groundwater_Causes_Implications_and_Mitigation_Measures [Accessed September 28, 2022]

Bundschuh J, Litter MI, Parvez F, Roman-Ross G, Nicolli HB, Jean JS, Liu CW, Lopez D, Armienta M, Guilherme LRG, Cuevas AG, Cornejo L, Cumbal L, Toujaguez R (2012) One century of arsenic exposure in Latin America: a review of history and occurrence from 14 countries. Sci Total Environ 429:2–35. https://doi.org/10.1016/j.scitotenv.2011.06.024

Census of India (2011) Chandrapur district profile. directorate of census operation, Maharashtra. Minist Home Aff, Gov Ment India, Mumbai 2011:1–8

Cento JA, Tseng C, Van der Voet GB, Finkelman BR (2007) Global impact of geogenic arsenic: A medical geology research case. A J Hum Environ, 36(1):78–81

CGWB (Central Ground Water Board) (2009) Ground water information chandrapur district, Maharashtra. Ministry of Water Resources. Government of India, Central Ground Water Board, Central Region, Nagpur. pp 1–10

Chakraborti D, Rahman MM, Das B, Murrill M, Dey S, Chandra Mukherjee S, Dhar RK, Biswas BK, Chowdhury UK, Roy S, Sorif S, Selim M, Quamruzzaman Q (2010) Status of groundwater arsenic contamination in Bangladesh: a 14-year study report. Water Res 44(19):5789–5802. https://doi.org/10.1016/j.watres.2010.06.051

Chandekar AV, Kamble R (2010) Groundwater fluoride in Rajura Tehsil in Chandrapur district. Maharashtra. Indian J Environ Prot 30(10):823–831

Chaoke L, Rongdi J, Shouren C (1997) Epidemiological analysis of endemic fluorosis in China. J Environ Sci Health-Part C Environ Carcinog & Ecotoxicol Rev 15:123–138. https://doi.org/10.1080/10590509709373493

Chen QY, Costa M (2021) Arsenic: A global environmental challenge environmental challenge. 2021. Annu Rev Pharmacol Toxicol 61:47–63

Cubadda F, D'Amato M, Mancini FR, Aureli F, Raggi A, Busani L, Mantovani A (2015) Assessing human exposure to inorganic arsenic in high-arsenic areas of Latium: a biomonitoring study integrated with indicators of dietary intake. Ann Ig 27:39–51. https://doi.org/10.7416/ai.2015.2021

Das NC (2013) Physico-chemical characteristics of selected groundwater samples of Ballarpur city of Chandrapur district, Maharashtra, India. Int Res J Environ Sci 2(11):96–100

Das N, Das A, Sarma KP, Kumar M (2018a) Provenance, prevalence and health perspective of co-occurrences of arsenic, fluoride and uranium in the aquifers of the Brahmaputra River floodplain. Chemosphere 194:755–772. https://doi.org/10.1016/j.chemosphere.2017.12.021

Das N, Das A, Sarma KP, Kumar M (2018b) Provenance, prevalence and healthperspective of co-occurrences of arsenic, fluoride and uranium in the aquifers of the Brahmaputra Riverfloodplain. Chemosphere 194:755–772. https://doi.org/10.1016/j.chemosphere.2017.12.021

Emmanuel E, Pierre MG, Perrodin Y (2009a) Groundwater contamination by microbiological and chemical substances released from hospital wastewater and health risk assessment for drinking water consumers. Environ Int 35:718–726

Emmanuel E, Pierre MG, Perrodin Y (2009b) Groundwater contamination by microbiological and chemical substances released from hospital wastewater and health risk assessment for drinking water consumers. Environ Int 35(4):718–726. https://doi.org/10.1016/j.envint.2009.01.011

Farooqi ABF, Asuda HAM, Usakabe MIK, Aseem MUN, Irdous NOF (2007a) Distribution of highly arsenic and fluoride contaminated groundwater from East Punjab, Pakistan, and the controlling role of anthropogenic pollutants in the natural hydrological cycle. Geochem J 41(4):213–234

Farooqi A, Masuda H, Firdous N (2007b) Toxic fluoride and arsenic contaminated groundwater in the Lahore and Kasur districts, Punjab, Pakistan and possible contaminant sources. Environ Pollut 145(3):839–849. https://doi.org/10.1016/j.envpol.2006.05.007

Farooqi ABF, Asuda HAM, Usakabe MIK, Aseem MUN, Irdous NOF (2007c) Distribution of highly arsenic and fluoride contaminated groundwater from East Punjab, Pakistan, and the controlling role of anthropogenic pollutants in the natural hydrological cycle. Geochem J 41(4):213–234. https://doi.org/10.2343/geochemj.41.213

Fernandez-Macias JC, Ochao-Garcia ST, Varela-Silva JA, Perez-Maldonado IN (2020) Probabilistic human health risk assessment associated with fluoride and arsenic co-occurrence in drinking water from the metropolitan area of San Luis Potosí. Mexico. Environ Monit Assess 192:712. https://doi.org/10.1007/s10661-020-08675-7

Fry A (2005) Water facts and trends. World Bus Counc Sustain Dev

George CM, Sima L, Arias MHJ, Mihalic J, Cabrera LZ, Danz D, Checkley W, Gilman RH (2014) Arsenic exposure in drinking water: an unrecognized health threat in Peru. Bull World Health Organ 92:565–572. https://doi.org/10.2471/blt.13.128496

Gudadhe SK, Manik VS (2020) Evaluation of groundwater quality o Chandrapur city, Vidarbha region, Central India. Vidyabharti Int Interdiscip Res J 11(1):223–227

Guo H, Wang Y (2005) Geochemical characteristics of shallow groundwater in Datong basin, northwestern China. J Geochem Explor 87:109–120. https://doi.org/10.1016/j.gexplo.2005.08.002

Habila O (2005) Groundwater and the millennium development goals. In: Proceedings groundwater and poverty reduction in Africa. International association of hydrogeologist, London

Halim MA, Majumder RK, Nessa SA, Hiroshiro Y, Uddin MJ, Shimada J, Jinno K (2009) Hydrogeochemistry and arsenic contamination of groundwater in the Ganges delta plain. Bangladesh. J Hazard Mater 164(2–3):1335–1345

Hani H (1990) The analysis of inorganic and organic pollutants in soil with special regard to their bioavailability. Int J Environ Anal Chem 39:197–208

Haque S, Johannesson KH (2006) Arsenic concentrations and speciation along a groundwater flow path: the Carrizo Sand aquifer, Texas, USA. Chem Geol 228(1–3):57–71

Issanov A, Betty A, Saint-Jacques N, Dummer TJB (2024) Arsenic in drinking water and lung cancer: A systematic review of 35 years of evidence. Toxicol Appl Pharmacol 483:1–33

Jia H, Qian H, Qu W, Zheng L, Feng W, Ren W (2019) Fluoride occurrence and human risk in drinking water wells from southern edge of Chinese Loess Plateau. Int J Environ Res Public Health, 16

Kamal AS, Parkpian P (2002) Arsenic contamination in Hizla, Bangladesh: Sources, Effects and Remedies. Sci Asia 28:181–189

Kamble R (2024) Groundwater arsenic in Chandrapur District Central India, Sustain, Agri, Food Environ Res, 12 (X)1–23. http://dx.doi.org//https://doi.org/10.7770/safer

Karim MM (2000) Arsenic in groundwater and health problems in Bangladesh. Water Res 34:304–310

Khatri N, Tyagi S (2015) Influences of natural and anthropogenic factors on surface and groundwater quality in rural and urban areas. Front Life Sci, 8 (1): 25–39. https://doi.org/10.1080/21553769.2014.933716

Kom KP, Gurugnanam B, Bairavi S, Chidambaram S (2023) Sources and geochemistry of high fluoride groundwater in hard rock aquifer of the semi-arid region. A special focus on human health risk assessment. Total Environ Res Themes, 5

Kumar M, Furumai H, Kurisu F, Kasuga I (2010) A comparative evaluation of mobile heavy metal pool in the soakaway sediment, road dust and soil through sequential extraction and isotopic dilution techniques. Water Sci Technol 62(4):920–928

Kumar M, Kumar P, Ramanathan AL, Bhattacharya P, Thunvik R, Singh UK, Tsujimura M, Sracek O (2010a) Arsenic enrichment in groundwater in the middle Gangetic plain of Ghazipur district in Uttar Pradesh, India. J Geochem Explor, 105, 83–94. https://doi.org/10.1016/j.gexplo.2010.04.008

MacDonald A, Davies J, Calow R, Chilton J (2005) Developing groundwater: A guide to rural water supply. Rugby: Practical Action Publishing Ltd. pp. 358. https://doi.org/10.3362/9781780441290 (Accessed July 12, 2015)

Muhammad S, Shah MT, Khan S (2010) Arsenic health risk assessment in drinking water and source apportionment using multivariate statistical techniques in Kohistan region, northern Pakistan. Food Chem Toxicol 48(10):2855–2864

Muhammad S, Shah MT, Khan S (2011a) Health risk assessment of heavy metals and their source apportionment in drinking water of Kohistan region, northern Pakistan. Microchem J 98:334–343

Muhammad S, Shah MT, Khan S (2011b) Health risk assessment of heavy metals and their source apportionment in drinking water of Kohistan region, northern Pakistan. Microchem J 98(2):334–343. https://doi.org/10.1016/j.microc.2011.03.003

Mustafa O, Mohmmud R, Sracek O, Seeyan S (2023a) Geogenic sources of arsenic and fluoride in groundwater: Examples from the Zagros basin, the Kurdistan region of Iraq. Water 15(11):1981. https://doi.org/10.3390/w15111981

Mustafa O, Mohmmud R, Sracek O, Seeyan S (2023b) Geogenic Sources of Arsenic and Fluoride in Groundwater: Examples from the Zagros Basin, the Kurdistan Region of Iraq. Water 15(11):1981

Nakazawa K, Nagafuchi O, Otede U, Chen J, Kanefuji K, Shinozuka K (2020) Risk assessment of fluoride and arsenic in groundwater and a scenario analysis for reducing exposure in Inner Mongolia. RSC Advance 10:18296–18304. https://doi.org/doi:10.1039/D0RA00435A

Narsimha A, Manne R, Zhang Y-H, Xu P, Qian H (2022) Evaluation of groundwater quality and its suitability for drinking purposes in semi-arid region of Southern India: An application of GIS. Geocarto Int 37(25):1–13. https://doi.org/10.1080/10106049.2022.2040603

Nawale VP, Malpe DB, Marghade D, Yenki R (2021) Non-carcinogenic health risk assessment with source identification of nitrate and fluoride polluted groundwater of Wardha sub-basin, central India. Ecotoxicol Environ Saf 208:1–11

Parrone D, Ghergo S, Frollini E, Rossi D, Preziosi E (2020a) Arsenic-fluoride co-contamination in groundwater: Background and anomalies in a volcanic-sedimentary aquifer in central Italy. J Geochem Explor 217:106590. https://doi.org/10.1016/j.gexplo.2020.106590

Parrone D, Ghergo S, Frollini E, Rossi D, Preziosi E (2020) Arsenic-fluoride co-contamination in groundwater: Background and anomalies in a volcanic-sedimentary aquifer in central Italy. J Geochem Explor, 217, 106590

Podgroski J, Berg M (2020) Global threat of arsenic in groundwater. Sci 368:845–850

Rapant S, Krcmova K (2007a) Health risk assessment maps for arsenic groundwater content, application of national geochemical databases. Environ Geochem Health 29:131–141

Rapant S, Krcmova K (2007b) Health risk assessment maps for arsenic groundwater content, application of national geochemical databases. Environ Geochem Health 29(2):131–141. https://doi.org/10.1007/s10653-006-9072-y

Rossman TG, Uddin AN, Burns FJ (2004) Evidence that arsenite acts as a cocarcinogen in skin cancer. Toxicol Appl Pharmacol 198:394–404

Saha S, Selim Reza AHM, Roy MK (2021) Arsenic geochemistry of the sediments of the shallow aquifer and its correlation with groundwater, Rangpur. Applied Water Science, Bangladesh, p 11

Shah BA (2012) Role of quaternary stratigraphy on arsenic-contaminated groundwater from parts of Barak valley, Assam, north-east India. Environ Earth Sci 66:2491–2501. https://doi.org/10.1007/s12665-011-1472-3

Shankar S, Shankar U, Shikha (2014) Arsenic contamination of groundwater: A review of source, prevalence, health risk, and strategies for mitigation. Sci World J, Article ID: 304524

Shrivas SK, Bhatia RK, Verma S (2018) Review of fluoride contamination and health risk assessment of groundwater. Int J Innov Res Technol 5(3):128–133

Tchounwou PB, Centeno JA, Patlolla AK (2004) Arsenic toxicity, mutagenesis, and carcinogenesis—a health risk assessment and management approach. Mol Cell Biochem 255:47–55

Toolabi A, Bonyadi Z, Paydar M, Najafpoor AA, Ramavandi B (2021a) Spatial distribution, occurrence, and health risk assessment of nitrate, fluoride, and arsenic in Bam groundwater resource. Iran. Groundw Sustain Dev 12:100543. https://doi.org/10.1016/j.gsd.2020.100543

Toolabi A, Bonyadi Z, Paydar M, Najafpoor AA, Ramavandi B (2021b) Spatialdistribution, occurrence, and health risk assessment of nitrate, fluoride, and arsenic in Bam groundwater resource. Iran. Groundw Sustain Dev 12:10054

UNEP (2002) Vital Water Graphics (Nairobi, Kenya)

United Nations Commission for Sustainable Development, Comprehensive assessment of freshwater resources of the world, USA, (1997)

Winkel LHE, Trang PTK, Lan VM, Stengel C, Amini M, Ha NT, Viet PH, Berg M (2011) Arsenic pollution of groundwater in Vietnam exacerbated by deep aquifer exploitation for more than a century. Proceedings of the National Academy of Sciences United States of America 108:1246–1251. https://doi.org/10.1073/pnas.1011915108

WWAP (United Nations World Water Assessment Programme) (2015) The United Nations World Water Development Report 2015: Water for a Sustainable World. UNESCO, Paris

Xu N, Liu L, Mengjiao T, Tao X, Li Liang, Wang H, Gong J (2023) Co-occurrence, sources and co-enrichment mechanism of arsenic fluoride in groundwater from Huaihe River Basin, China. Emerg Contam, 1–23

Yadav KK, Kumar S, Pham QB, Gupta N, Rezania S, Kamyab H, Ydav S, Vymazal J, Kumar V, Tri DQ, Talaiekhozani A, Prasad S, Reece LM, Singh N, Cho MJ (2019) Fluoride contamination, health problems and remediation methods in Asian groundwater: A comprehensive review. Ecotoxicol Environ Saf, 182

Yang X, Hou Q, Yang Z, Zhang X, Hou Y (2012) Solid-solution partitioning of arsenic (As) in the paddy soil profiles in Chengdu Plain, Southwest China. Geosci Front 3:901–909. https://doi.org/10.1016/j.gsf.2012.03.006

Yoshida T, Yamauchi H, Fan Sun G (2004) Chronic health effects in people exposed to arsenic via the drinking water: dose-response relationships in review. Toxicol Appl Pharmacol, 198 (3) (2004), 243–252

Yoshizuka K, Nishihama S, Sato H (2010) Analytical survey of arsenic in geothermal waters from sites in Kyushu, Japan, and a method for removing arsenic using magnetite. Environ Geochem Health 32:297–302. https://doi.org/10.1007/s10653-010-9300-3

Environmental Impact Evaluation of Arsenic and Fluoride Contamination and Remediation Strategies

Karthika Arumugam and Mary Arpana

Abstract Geogenic contaminants Arsenic and Fluoride threaten the wellness of groundwater worldwide. These pollutants enter groundwater in several ways, either from natural or anthropogenic sources. Though groundwater is a crucial resource for over 2.5 billion people worldwide, human society's main concern is providing good-quality drinking water. A high human health risk could result from the high percentage of As and F contaminants in groundwater. In the last decade, there have been a huge number of reports on groundwater contamination by arsenic and fluoride from Southeast Asian countries (Bangladesh, China, India, etc.) as well as a few recently impacted areas in other countries. This chapter examines the current situation about arsenic and fluoride contaminations of groundwater in different parts of the world, as well as the environmental impact assessment (occurrence and co-occurrence) and recommends appropriate remedies for various spatial concentration zones.

Keywords Arsenic · Fluoride · Groundwater · Environmental impact · Remediation

1 Introduction

Nature's precious gift to mankind is water. It's an essential resource for maintaining life on Earth. The Earth contains 1,386 million cubic kilometers of water. Of the whole, 10.63 million km^3 are freshwater, with just 30.1% present as groundwater. Groundwater is experiencing increased stress as a result of population growth and reduced reliance on surface water resources. (Mukherjee 2018). Groundwater is a vital water resource for humanity, providing drinking water entirely or partially for

K. Arumugam (✉)
Department of Chemistry, Jyoti Nivas College Autonomous, Bengaluru, India
e-mail: drkarthikaarumugam@gmail.com

M. Arpana
Department of Biochemistry, Jyoti Nivas College Autonomous, Bengaluru, India

A. Nath et al. (eds.), *Groundwater Depletion and Sustainability*, Advances in Geographic Information Science, https://doi.org/10.1007/978-3-032-09921-1_10

around 50% of the global population and accounting for 43% of all water used for cultivation.

Around the world, approximately 2.5 billion people depend only on groundwater supplies to fulfil their basic daily water needs. (The Ground Water Project, Ontario, Canada). Overexploitation of several major acquifiers have resulted due to unequal distribution of human populations as well as differences in ground water availability and accessibility. The deterioration in groundwater quality is a global concern. In many parts of the world, arsenic and fluoride are major groundwater contaminants. According to research, arsenic pollution has spread to more than 70 countries.

High levels of arsenic are reported in countries of East and Southeast Asia. The Gangetic lowlands of West Bengal, the Ganga plains of Uttar Pradesh, and Bihar are among the regions in India where ground water is tainted with arsenic.

The principal mechanisms that release arsenic into groundwater are mining of iron-containing minerals, oxidation of sulphide ore, desorption of alkali and the influence of microorganisms on organic matter. Arsenic is usually found adsorbed onto iron oxides and the reductive process of iron release arsenic into the environment which is a predominant process in the Indo-Gangetic belt. Ingestion of arsenic-contaminated water can result in several health issues, including cancer, keratosis, and arsenicosis, Presently, 24% of global aquifers are severely overexploited, primarily due to the need for groundwater irrigated agriculture aimed at sustaining human and cattle populations (Gleeson et al. 2012). Groundwater has the potential to solubilize a variety of chemical substances and hence contains high amounts of fluoride, arsenic, iron, manganese, and radionucleotide, to name a few, which have adverse effects on the environment (Brands et al. 2017).

High levels of arsenic in groundwater are a global issue. Arsenic concentrations in drinking water that are higher than WHO guidelines concern an estimated 200 million people worldwide (George et al. 2014). The primary mechanisms contributing to elevated groundwater arsenic levels encompass alkali desorption, oxidation of sulphide rocks, geothermal activity, disintegration of rock bearing minerals, ion exchange with surface water, exchange with surface water, and increasing anthropogenic influences like agricultural runoff, industrial discharge, and over-extraction of groundwater (Gonzalez-Rodriguez and Fernandez-Marcos 2021; Erbs et al. 2010; Blowes et al. 2013; Bundschuh and Maity 2015; Raychowdhury et al. 2014; Mukherjee et al. 2011). Activities, such as ore smelting and pesticide application, release arsenic into the environment (Bhattacharya et al. 1997a, 1997b). The highest levels of arsenic contamination in Asia have been deducted in India and Bangladesh (Sracek et al. 2004; Mukherjee et al. 2009). It causes a number of neurological, reproductive, cardiovascular, immunological, and endocrine disorders, skin cancer, and lung cancer (Ameer et al. 2015; Kapaj et al. 2006; Ahmad and Bhattacharya 2019; Centeno et al. 2006; Ferreccio et al. 2000). A rare hepatic manifestation may also be brought by chronic arsenic exposure (Sarkar 2002). Various remediation strategies have been explored to address arsenic contamination in water, including membrane-based technologies (such as reverse osmosis and nanofiltration), precipitation and adsorption techniques, ion exchange, bioremediation, and the use of alternative water sources with naturally low arsenic concentrations. (Ahmad and Bhattacharya 2019;

Sen et al. 2010; Harper and Kingham 1992; Baskan and MAjiPala 2010; Maji et al. 2011; Greenleaf et al. 2006; Ahmad and Ganjo 2020; Kao et al. 2013).

Fluoride is one of the vital elements for humans, although elevated concentrations can cause fluorosis in man. Natural processes such as eroding of mineral rocks like calcium fluorophosphate ($Ca_5(PO_4)_3F$) and fluorites (CaF_2) result in the infiltration of these minerals in groundwater. Fluoride is advantageous for oral health in low concentrations, but it can be hazardous at elevated levels, impacting teeth, bones, thyroid function, endocrine dysfunction, and brain development.

High levels of fluoride in drinking water (more than 1.5 mg/L) lead to dental fluorosis, and more than 3.0 mg/L causes skeletal fluorosis (DenBesten and Li 2011; Dar et al. 2011). It reduces cognitive function in children, chronic kidney illnesses, osteoporosis and osteosclerosis, disturbed intracellular homeostasis, oxidative stress and lipid peroxidation, decrease in crop yield and soil fertility, and bioaccumulation in soil and food crops (Choi et al. 2014; Dharmaratne 2015; Joshi et al. 2011a; Bhattacharya and Samal 2018; Yadav et al. 2015; Kundu and Mandal 2009; Ali et al. 2016).

Countries across the globe have all observed high amounts of fluoride in groundwater. (Ali et al. 2016). Numerous conventional methods for removing fluoride from groundwater exist, including coagulation, reverse osmosis, adsorption, nanofiltration, ion-exchange, precipitation, and electrochemical approaches (Maity et al. 2018). The membrane and ion exchange procedures are not widely favored due to their significant establishment and operational costs (Waghmare 2015). Various remediation procedures were implemented to facilitate the removal of fluoride from groundwater, electro-coagulation, and adsorption techniques were assessed.

Given the challenges in directly observing and tracking groundwater, monitoring, mapping, and modeling are essential to comprehending its storage, distribution, and flow patterns within the subsurface. The health effects of arsenic and fluoride poisoning in groundwater are significant. Alleviating these effects necessitates synthesizing technological solutions, including filtration systems, source management, public awareness, and governmental regulation. It must be meticulously managed and safeguarded against overexploitation and contamination to guarantee its enduring survival. Enhanced water conservation methods, improved monitoring, and rules that promote responsible water usage are essential to satisfy the needs of both present and future generations.

2 Materials and Methods

The literature search was done for research articles, reports, case studies, reviews, and book chapters on arsenic, fluoride, and co-existence of arsenic-fluoride. The articles were retrieved digitally from accessible databases such as Science Direct, American Chemical Society, Royal Society of Chemistry, Springer, and PubMed. The study included full-text, English language articles, various study designs. The study analyses over 100 articles on arsenic levels, poisoning of groundwater sources by arsenic in Southeast Asia, fluoride levels, fluoride contamination, analysing its effects on humanity, and Remediation techniques.

3 Results and Discussion

3.1 Arsenic in Water

Origin of Arsenic Contamination

Arsenic contamination is primarily caused by the natural sources include geologic formations, geothermal, coal, and volcanic activities. Geothermal water contaminates surface water and groundwater with inorganic arsenic (Welch et al. 2000), and the average level of arsenic concentration is reported to be 1.5–5.0 mg/kg. Sedimentary, hydrothermal, and magmatic rocks are found with significant amounts of arsenic, widely available in sulfide ores of metals. Secondly, anthropogenic has been considered another source of arsenic contamination. Around 54 nations have reported arsenic contamination in groundwater, primarily due to mining, coal, petroleum extraction, and human activity. In Asia, Europe, and Africa, sedimentary deposits and mining are the largest contributors, surpassing fossil fuels, coal, and volcanic materials (Shaji et al. 2021). In America, excavation, volcanic rocks, and sedimentary deposits all make equal contributions, and every sector in Australia makes an equal contribution. There are various types of anthropogenic arsenic contamination, including coal-related, mining-related, and coal-burning. The extraction and smelting of gold from ores pose environmental risks due to the possibility of arsenic leakage into surface and groundwater from extraction basins, waste dumps, slag pits, and mines. Almost 74 nations worldwide are experiencing arsenic contamination from mining-related activities (coal mining). Arsenic from petroleum has impacted 17 nations worldwide. Arsenic groundwater contamination in India mostly originates from riverine sediments, which are primarily produced by the erosion of Himalayan strata. Kumar et al. (Kumar et al. 2016a, 2016b) reported that Fe^{3+}-oxyhydroxides undergo reductive dissolution in a reducing environment, which mobilizes arsenic. Figure 1 shows the extent to which arsenic impacted many countries worldwide, as indicated by the various sizes of plots. According to Ali et al., (Ali et al. 2019) report, South America and South Asia are the most impacted.

Arsenic Contaminant in Groundwater

Numerous factors, including pH, organic content in sediments, fluctuations in the water table, microbial activity, water saturation of sediments, low sulfur supply, the flow direction of groundwater, geography, age of groundwater, and oceanic transgression, can contribute to the release of arsenic into groundwater (Hinkle and

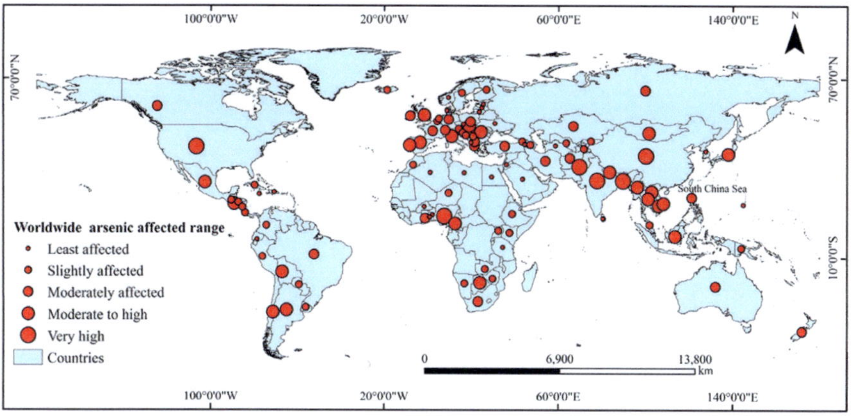

Fig. 1 World countries affected by arsenic (*Source* Ali et al. (2019) and Shaji et al. (2021))

Polette 1999; Rodrıguez et al. 2004; Matisoff et al. 1982; Chapelle and Lovley 1992; Lovley 1997; Fendorf et al. 2010; Trafford et al. 1996; Berg et al. 2001). Groundwater is exposed to arsenic through the following mechanisms: (i) the oxidation and dissolution of As and Fe-containing minerals (Kumar et al. 2016a; McArthur et al. 2001; Smedley and Kinniburgh 2002) (ii) erosion and degradation of primary and secondary minerals that contain arsenic, especially when natural organic matter is present (Berg et al. 2001) (iii) the deterioration of arsenic-containing ferric oxides and oxyhydroxides via oxidative and reductive processes (McArthur et al. 2001; Nickson et al. 1998; Kinniburgh and Smedley 2001) (iv) The competitive exchange of As with phosphate, bicarbonate, and nitrate (Chowdhury et al. 1999; Nickson et al. 2000). Quaternary deposits of Holocene-aged alluvial sediments are associated with arsenic-rich minerals (Mukherjee et al. 2009; Shah et al. 2010). Rivers from the Himalayas transmit arsenic-rich sediments, which are deposited in deltaic regions and downstream basins, causing arsenic contamination. Because arsenic-bearing minerals dissolve reductively (Postma et al. 2016). The organic components trapped with the sediment are absorbed by microbes for metabolic processes. The microbes reduce ferric to ferrous through the utilization of arsenic-containing ferroxy hydroxides, thereby allowing arsenic enter the water (Drahota et al. 2013; Verma et al. 2016). As mentioned above, these kinds of arsenic contamination can occur in high pH environments, reducing aquifer conditions, oxidative weathering of sulfide minerals, and geothermal activity (Nickson et al. 2005). Hoque et al. (2009) reported that soil texture is vital in creating the right conditions for releasing arsenic into groundwater. Shaji et al., (Shaji et al. 2021) reported that pH has a direct correlation with arsenic level indicates that alkaline water promotes the discharge of arsenic into the groundwater.

According to Fig. 2, a detailed examination of the global distribution of As-enriched areas shows that the arsenic-enriched aquifers are connected to sedimentary basins adjacent to significant orogenic belts.

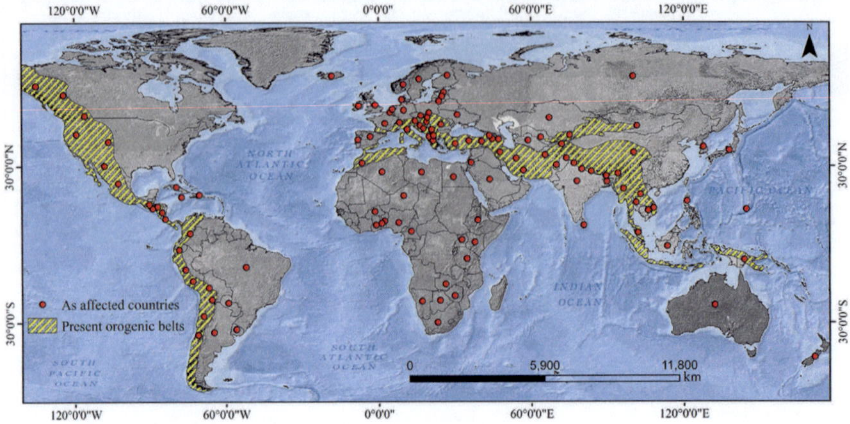

Fig. 2 World tectonic map with arsenic-affected regions superimposed (*Source* Shaji et al. (2021))

Groundwaters throughout the Himalayan regions and the foothills of recent mountain belts such as the Cordilleran-Andean and Alpine-Himalayan-Tibet belts are significantly contaminated with arsenic, as shown in Fig. 2. High arsenic concentration is typically seen in groundwaters found in Holocene aquifers, which are made up of sediments from the Himalayas that were deposited under deltaic circumstances by the Asian rivers. The Ganga–Brahmaputra River system is one of the world's greatest contemporary deltas and is considered the main contributor to the Bengal fan. According to Guillot and Charlet (2007), one of the primary sources of arsenic may be the arc-related rocks of the Indus-Tsangpo suture zone and arsenic-rich serpentinites. They further claim that the remobilization, transportation, and disposal of arsenic may have been caused by the active seismic deformation in the Himalayan frontal zone during the Holocene, which was accompanied by high rainfall conditions.

Toxicity and health impact of arsenic

The primary way that humans are exposed to arsenic is through water (Zhao and Babatunde 2011). Mainly, consuming seafood and arsenic-containing water exposes large populations to arsenic toxicity. Arsenic enters the bloodstream after being absorbed by the body and attaches to the hem protein. It is redistributed in the gastrointestinal system, liver, kidney, and lung (Khan and Ho 2011). Animal and human arsenic toxicity is significantly influenced by metabolism. The mechanism depends on the type of arsenic involved. For example, As^{5+} can be replaced through an ionic substitution process, thereby impacting the health of organisms. But As^{3+} inhibits the activity of enzymes by interacting with the thiol group present in proteins and enzymes, which lowers the production of acetyl-CoA and also decreases the biosynthesis of glucose from non-carbohydrates (neo-glucogenesis) and β-oxidation (Hughes et al. 2011), which will lead to gene amplification, interruption in cellular respiration, and create problems in oxidative phosphorylation. As a result, more reactive species (such as peroxide, hydroxyl and peroxyl radicals, and nitrogen species, etc.) may be produced, which could lead to genetic toxicity, problem in cell signalling, suppression of DNA repair, cell division, and the inhibition of antioxidant production (Chouhan and Flora 2010; Kitchin and Conolly 2010; Wu et al. 2016). Inorganic arsenic that has been consumed is eliminated through urine, and As^{3+} ion is more harmful than As^{5+} ion. The "King of Poisons" arsenic is known to be a teratogen, mutagen, and carcinogen (Sanjrani et al. 2019). Notably, genetic changes caused by arsenic can result in cancer (Hubaux et al. 2013; Speer et al. 2023). In the modern world, it is a significant concern due to its capability to outbreak infectious disease in large populations. Numerous studies mentioned conditions like melanoma, necrosis, leucomelanosis, keratoderma, dermal lesions, melasma, hepatomegalia, arsenicosis, tumor, black foot, and more that were linked to arsenic toxicity (Chouhan and Flora 2010; Kitchin and Conolly 2010; Wu et al. 2016; Sanjrani et al. 2019; Hubaux et al. 2013; Speer et al. 2023).

3.2 *Fluoride in Water*

Origin of Fluoride Contamination

Fluoride contamination falls into two basic categories: geogenic and anthropogenic. Figure 3 depicts the sources and distribution of fluoride in the water system and its physical environment. It also symbolizes the man-made sources of fluoride in water as well as the rocks and minerals that contain fluoride.

Geogenic sources include volcanic eruptions, mineral weathering, hydrothermal exertion, and oceanic aerosols are the main causes of fluoride contamination in drinking water (Chowdhury et al. 2019; Edmunds et al. 2005; Vithanage and Bhattacharya 2015). Rocks that contain fluorides, such as granites, mafic igneous rocks, and alkali rhyolites, as well as minerals like fluorspar, cryolite, apatite, mica, and others, are the primary mineral sources of fluoride (Kinnunen et al. 2003; Jha and Tripathi 2021a). Metamorphic rocks typically have higher levels of fluoride, specifically the cryptozoic metamorphic basements are located in Indian ocean region. Fluoride minerals can be released into phreatic and fractured aquifers under favorable conditions, such as rock-water interaction, arid environment, high alkalinity, and calcium ion (Ca^{2+}) depletion.

Fluoride levels as high as 200 mg/kg is also present in sedimentary rocks, mostly in limestone, in dry and semi-arid environments (Subba Rao 2003). High fluoride levels range from 1 to 10 mg/L were reported in various geothermal hotspots like Iceland, New Zealand, France, arid and semi-arid regions of Algeria, the active continental rift zone in East Africa the American west (Nordstrom and Jenne 1977; Ellis and Mahon 1967; Ayenew 2008). Smedley et al. (Smedley et al. 2002) noticed an adverse relationship between fluoride, pH, and HCO_3 concentrations in the sedimentary aquifer region. Figure 4 shows the world climate map with fluoride-affected

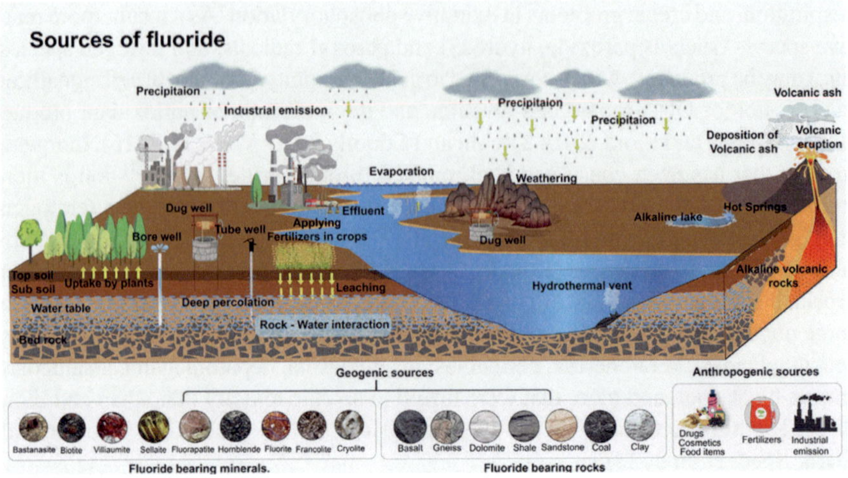

Fig. 3 Representation of fluoride contamination in various water systems by geogenic and anthropogenic sources (*Source* Edmunds and Smedley (2012); Mukherjee and Singh (2022), Shaji et al. (2024))

regions. Groundwater resources in the edges of tropical deserts and subtropical desert regions are significantly impacted by fluoride.

Industries are the primary anthropogenic factor. Globally, the amount of fluoride used for industrial processes is rising and has surpassed 15 million metric tonnes. It has grown from 0.5 to 1.5 million metric tonnes in the USA, a fourfold rise (Singh et al. 2011). Overusing phosphate-based fertilizer, coal incineration, aluminum and cement processing, would contribute to the release of fluoride into the ecosystem.

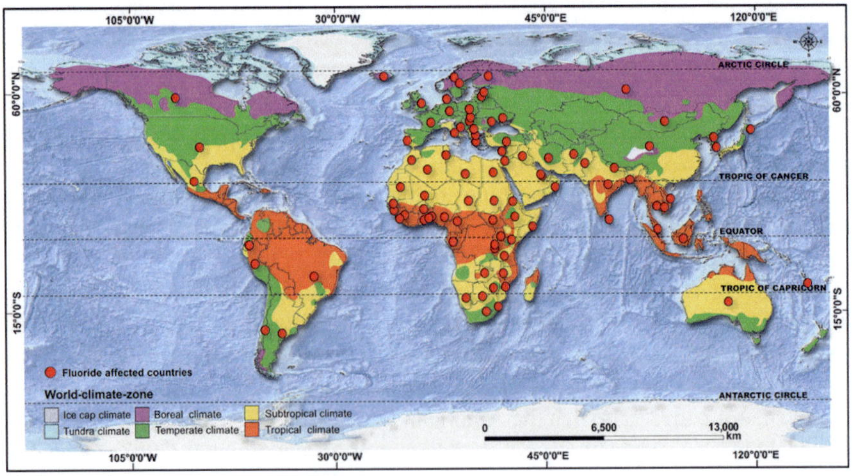

Fig. 4 A map of the global climate with fluoride-affected regions shown (*Source* Jha et al. (2021b))

The manufacture of charcoal, glass and ceramic materials, and electronic devices, as well as the processing of steel and the use of insecticides, fertilizer, and electroplating processes, are additional anthropogenic sources. Aluminum electrolysis is aided by cryolite, and fluoride contamination around aluminum factories has been extensively documented in numerous nations. Drinking water is the primary way that most people are exposed to fluoride (Jagtap et al. 2012; Abouleish 2016). Sometime, fluoride can enter the body through aerosols, food products (including tea, dairy products, fatty substances, beverages, soupy drinks, meat, poultry, fish, and some vegetables), beauty products, and more (Dabeka and Mckenzie 1995; Schamschula et al. 1988; Bergmann and Bergmann 1995; Whitford et al. 1997; Chen et al. 1996; Cao et al. 2004).

Fluoride Contaminant in Groundwater

When fluoride concentrations in water exceed 5 ppm, it is believed to be oversaturated with fluorite, which increases its solubility. The presence of aluminum and pH substantially impact the amount of fluoride in groundwater, because aluminum and fluoride can make strong complexes. The concentration of free fluoride drastically drops in drinking water when fluoride and aluminum form the AlF_2 strong complex at $pH < 5$. Highly developed magma produces rock with a plagioclase composition that is rich in fluoride minerals and is loaded with more sodium rather than calcium. This makes fluoride and sodium-rich soft groundwater (Ozsvath 2006). High fluoride in the groundwater is sometimes directly correlated with high nitrate concentrations (Reddy et al. 2010). Due to their distinct leaching and protracted water–mineral interactions, arid and humid tropical environments, and high atmospheric temperature are susceptible to high fluoride concentration (Edmunds and Smedley 2013). Generally, the groundwater's ecological conditions (physical, chemical, geological, etc.), characteristic weather, penetrability, depth, rock acidity, and degree of weathering, all affect the fluoride content of groundwater. Aquifer replenishment, area, interaction time, ionic concentrations, and the disintegration of fluoride-rich minerals by pH change are some of the mechanisms that facilitate the release of fluoride into groundwater. On the whole, the soil and groundwater are contaminated only through natural and anthropogenic sources, mentioned above. Figure 5 illustrates the process by which fluoride dissolves and cycles in groundwater.

Toxicity and Health Impact of Fluoride

Many health issues in people of all ages, including enamel hypomineralization, bone weakening, renal impairment, osseous distortion, sexual organs inactivation, nerve and muscle deterioration, skeletal fluorosis, and more, are brought on by consuming more fluoride than is permitted (Yadav et al. 2019; Singh et al. 2013). The human body needs fluoride at low concentrations (0.5–1.0 ppm) because it helps prevent

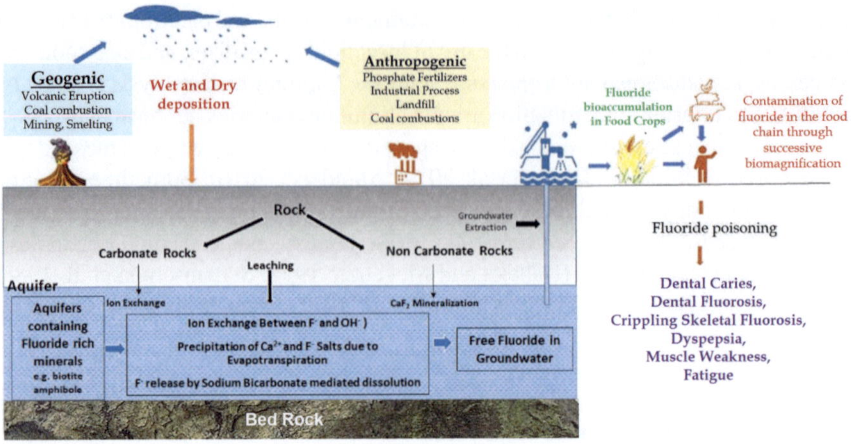

Fig. 5 Suitable process for the dissolution and cycling of fluoride in groundwater (*Source* Jha et al. (2021b))

dental cavities and promotes bone growth, mineralization, and tooth enamel development. The high level of fluoride content develops a solid layer on the tooth enamel. Dental cavities, enamel formation, weakening of bones, and arthritis are all consequences of fluorosis (Edmunds and Smedley 2013). The primary component of tooth enamel, hydroxyapatite, is formed as fluorapatite by replacing hydroxide ions. Dental fluorosis, which is caused by prolonged exposure, causes teeth to become brittle and hard, and constant and prolonged exposure to high level of fluoride (>4 ppm) can calcify bones, leading to osteoporosis and osteosclerosis (Viswanathan et al. 2010; Joshi et al. 2011b). Fluoride has been identified by Barbier et al. as the cause of several malfunctions in humans, such as disrupted intracellular redox homeostasis, oxidative imbalance, cell death, protein oxidation marker, oxidative lipid damage, and altered expression of genes related to the stress response, intercellular signalling, biocatalysts, cellular replication cycle, and molecular signaling (Barbier et al. 2010).

3.3 *Arsenic and Fluoride Co-Occurrence*

Arsenic and fluoride are found in groundwater samples, indicating a common source or pathway of both contaminants, which is corroborated by reports of a positive correlation (Rasool et al. 2015). The most common sources of fluoride and arsenic are found in nature. For example, arsenic mobilization is favored by redox conditions, and the Fe release is favored by reducing conditions, which helps add arsenic to groundwater. Geogenic fluoride release is mainly caused by the interaction between rocks and water (Bibi et al. 2017). However, the rapid rate of recharge and outflow limits the interaction between the groundwater and aquifer minerals. (Kumar et al.

2016a). According to experts, the primary source of arsenic and fluoride contamination is water–rock interaction, which can be exacerbated by mining, geothermal, and over-aquifer use. Growing concern are being raised about the enhancement of multi-nutrients in groundwater systems. When they coexist, various synergistic-antagonistic interactions can result in an improved effect. The coexistence of arsenic and fluoride can culminate to an amplified effect through various complex synergistic-antagonistic interactions (Patel et al. 2019). Compared to their individual effects, the coexistence will give serious health problems due to coupled and complicated interactions with the groundwater, according to Kumar et al. (Kumar et al. 2016a). Fe hydroxides have a strong attraction for both arsenic and fluoride oxyanions, and several research studies indicate that they are the primary source of both (Bundschuh et al. 2017). The river largely controls the concentration of arsenic and fluoride, which also influences the recharge conditions and the deposition of new sediments (Das et al. 2016). There are very few reports on the coexistence of fluoride and arsenic in groundwater. According to Kumar et al., the primary cause of fluoride and arsenic coexistence is through anthropogenic activities such as excavation and extraction, recharge, and agricultural chemical products (Kumar and Singh 2020). In the semi-arid regions of Latin America, Alarcon-Herrera et al. (2001) evaluated the co-existence of arsenic and fluoride, while Rasool et al. (Rasool et al. 2015) examined the prevalence of arsenic and fluoride in Punjab (Pakistan). The geochemical characteristics of Mexican aquifers were examined by Reyes-Gòmez et al. (2013) to study the co-existance of arsenic and fluoride. They discovered that the rocks, sediments, and surface water were of geogenic origin. Kumar et al. (Kumar et al. 2016a) found that in the assam valley floodplains, isolated aquifers may exhibit increased fluoride contents, indicating that oxidizing conditions may produce co-occurrence. Chen et al. (2017a, 2017b) examined the source of the coexistence of arsenic and fluoride in groundwater around Dawukou District, China. Their findings indicated that the precipitation and disintegration of arsenic and fluoride rich ores in weak basic conditions greatly increases the concentration of these contaminants in groundwater. Arsenic and fluoride concentrations were comparatively more significant in the subsurface water. The co-occurrence was documented in Sri Lanka by Chandrajith et al. (2020), who found that over eighty percent of the people drank groundwater and experienced chronic fluoride poisoning. However, arsenic was deposited on stratified aquifers and was not a major problem. Iran's groundwater was examined by Toolabi et al. (2020), who discovered that both arsenic and fluoride were over allowable limits.

3.4 Arsenic-Fluoride Toxicity and Health Impact

Researchers carefully researched the effects of arsenic-fluoride coexistence, indicating that dentin and calcified tissues do not pile up substantial quantities of arsenic and fluoride. Nonetheless, neurological harm, including impaired learning capacity and memory, is inclined due to concurrent intake to arsenic and fluoride. Mishra et al.

(2010) observed that simultaneous exposure to arsenic and fluoride resulted in aberrant cerebral tissue structure analysis, where arsenic exhibit an adverse effect than fluoride. The genotoxic effects of combined exposure to fluoride and arsenic are more pronounced than their individual effects. Through synergistic mechanisms, arsenic may elevate fluoride's toxicity (Rao and Tiwari 2006). Additionally, several investigations on experimental animals have found that simultaneous exposure to fluoride and arsenic harms the rat brain by increasing lipid peroxidation and lowering the activity of antioxidant enzymes (Birkner et al. 2008). Sariñana-Ruiz et al. (2017) found that the intestinal phase of fluoride and arsenic had a higher bioavailability of fluoride, while the gastric phase had a higher bioavailability of arsenic. This highlights the need for ongoing ecological and biometric monitoring in Laguna Region, Mexico. According to Das et al. (2018), who investigated the effect of arsenic and fluoride coexistence along with uranium in and around Brahmaputra Valley aquifer system and identified that the young children are more vulnerable to related diseases than adults because of their reduced physiologies and higher water intake to body weight ratio. In the Dharmanagar region of Tripura, an elevated Cumulative Noncancerous Lifetime Hazard Index value for fluoride and arsenic, fluoride suggests a high risk of chronic toxic syndrome, according to Bhattacharya et al. (2020). Large populations in China, India, and Pakistan are suffering from fluoride and arsenic co-occurrence since arsenic causes arsenicosis and hydrofluoric acid degrades bones (Kumar and Singh 2020). Maity et al. (2021) documented fluorosis and cancers in South Asian countries according to their study of the worldwide effects of arsenic and fluoride coexistence.

3.5 Remediation of Arsenic-Fluoride

As both these elements co-exist, the remediation strategies for their removal were also studied. Sandoval et al. (2019) employed a photoflow reactor made of aluminum to electro-coagulate fluoride and arsenic from surface water. In the study investigated by Saika et al. (2017) to remediate the coexistence from an aqueous solution using a surface technique, they discovered that activated biochar was a good, economical, and environmentally friendly adsorbent because of its many active adsorption sites and greater specific surface area. Using Aluminum-Iron hydroxides, Qiao et al. (2014) simultaneously eliminated fluoride and arsenate from water. They discovered that arsenate adsorption on Aluminum-Iron hydroxides was highly pH dependent and more significant adsorption observed at increased pH values. Jing et al. (2012) also investigated the simultaneous removal of arsenic (V) and fluoride. They created an adsorbent made up of titanium (Ti)and lanthanum (La) oxides on activated carbon (TLAC) and discovered it was more efficient than other adsorbents. In order to simultaneous removal of arsenic and fluoride from ground water samples, Thakur and Mondal (2017) employed electrocoagulation using an aluminum electrode, and the ideal experimental conditions remove roughly 98% arsenic and 88.33% fluoride and it was a successful procedure for the simultaneous removal. In the study of arsenic and fluoride contamination in Assam, Patel et al. (2019) found that Fe hydroxides are useful for remediation since they are crucial for the adsorption of arsenic and fluoride from water. For purifying water samples, Haldar et al. also employed a metal–organic framework to trap fluoride and arsenic (Haldar et al. 2020).

4 Perspectives

Absolutely, arsenic and fluoride contamination in water sources is a pressing global concern. The United Nations is working to guarantee that everyone can access pure drinking water and sanitary facilities by implementing its Sustainable Development Goals (SDG 6). Globally, the United Nations Development Programs (UNDP) addresses new issues pertaining to water resources and implements water management strategies that support all of the Sustainable Development Goals. To address the medical threats, mitigation, and pollution of arsenic and fluoride, plenty of research is being conducted every moment. Due to their reliance on untreated arsenic- and fluoride-rich water, rural communities are especially vulnerable to the severe and pervasive health issue of arsenic and fluoride poisoning. Several factors, the majority of which are unknown, affect a concurrent release of arsenic and fluoride in groundwater. Adequate knowledge of the hydrogeochemistry of pollutants is lacking, especially in regions where both are present in high concentrations. In Bangladesh, the geogenic contribution is significantly greater than the anthropogenic contribution. Higher levels of arsenic were found in Sri Lankan aquifers. Future research should examine the potential interactions between the co-occurring contaminants and the other pollutants. The groundwater contaminated with arsenic and fluoride needs to be treated. Prospective investigations should examine the potential interplay between co-occurring contaminants and other pollutants. It was discovered that the probabilistic technique revealed the health risk deeper than the deterministic models, so future research should use probabilistic models as much as possible to evaluate the health risk. Governmental and non-governmental groups must collaborate with the researcher to properly mitigate arsenic and fluoride while considering the framework for safe drinking water and the Sustainable Development Goals (SDG). It is essential to educate and empower the local populace to facilitate the appropriate and affordable remedial technology.

5 Conclusions

Groundwater quality is deteriorating globally due to arsenic and fluoride contaminants. Arsenic contamination has spread across 107 countries, including most affected Asian countries. The Gangetic plains, Uttar Pradesh, and Bihar are the worst-affected areas in India. The release of arsenic occurs through the various disintegrations of minerals, microorganisms and organic entities. Consumption of arsenic-contaminated water can lead to various chronic physical ailments. Fluoride contamination is mainly from geogenic sources and has been reported in East, South and Southeast Asian countries. Remediation techniques for arsenic and fluoride include membrane technology, flocculation, molecular adhesion, cation/anion exchange, and electrochemical coagulation. Factors like mineral dissolution, aquifer recharge, and carbonate mobilization influence groundwater fluoride. It can cause chronic fluoride toxicity, cavities, and reduced bone density. The coexistence of arsenic and fluoride can synergistically affect health. Addressing this problem requires increased awareness, strict regulations, proper mapping of contaminated wells, and research on affordable restoration approaches. The UNEP and WHO should prioritize global arsenic poisoning, launch surveillance drives, raise awareness, and work on practical solutions. Financial and logistic assistance are crucial, with innovative interdisciplinary research that should focus on comprehending the presence, source, distribution pattern, and elimination of arsenic and fluoride contaminants in groundwater.

Acknowledgements The authors thank The Principal, Jyoti Nivas College Autonomous, Bengaluru for her constant support and encouragement.

AI Disclosure During the preparation of this chapter, the author(s) did not use AI tools.

Disclose Conflict of Interest The authors declare no conflict of interest.

References

Abouleish MYZ (2016) Evaluation of fluoride levels in bottled water and their contribution to health and teeth problems in the United Arab Emirates. Saudi Dent J 28(4):194–202

Ahmad A, Bhattacharya P (2019) Arsenic in drinking water: is 10 μ g/L a safe limit? Curr Pollut Rep. 5(1):1–3

Ahmad TA, Ganjo DG (2020) Fungal isolates and their bioremediation for pH, chloride, tph and some toxic heavy metals. Eur. Asian J BioSciences 14(1):149–160

Alarcon-Herrera MT, MartIn-Dominguez IR, Trejo-V´azquez R et al. (2001) Well water fluoride, dental fluorosis, and bone fractures in the Guadiana Valley of Mexico. Fluoride 34 (2), 139–149

Ali S, Thakur SK, Sarkar A, Shekhar S (2016) Worldwide contamination of water by fluoride. Environ Chem Lett 14:291–315

Ali W, Rasool A, Junaid M, Zhang H (2019) A comprehensive review on current status, mechanism, and possible sources of arsenic contamination in groundwater: a global perspective with prominence of Pakistan scenario. Environ Geochem Health 41(2):737–760

Ameer SS, Engstr¨ om K, Harari F, Concha G, Vahter M, Broberg K (2015) The effects of arsenic exposure on blood pressure and early risk markers of cardiovascular disease: evidence for population differences. Environ. Res. 140, 32–36

Ayenew T (2008) The distribution and hydrogeological controls of fluoride in the groundwater of central Ethiopian rift and adjacent highlands. Environ Geol 54:1313–1324

Barbier O, Arreola-Mendoza L, Del Razo LM (2010) Molecular mechanisms of fluoride toxicity. Chem Biol Interact. https://doi.org/10.1016/j.cbi.2010.07.011

Baskan MB, MAjiPala A (2010) A statistical experiment design approach for arsenic removal by coagulation process using aluminum sulfate. Desalination, 254(1–3), 42–48

Berg M, Tran HC, Nguyen TC, Pham HV, Schertenleib R, Giger W (2001) Arsenic contamination of groundwater and drinking water in Vietnam: a human health threat. Environ Sci Technol 35(13):2621–2626

Bergmann KE, Bergmann RL (1995) Salt fluoridation and general health. Adv Dent Res 9(2):138–143

Bhattacharya P, Samal AC (2018) Fluoride contamination in groundwater, soil and cultivated foodstuffs of India and its associated health risks: A review Fluoride contamination in groundwater, soil and cultivated foodstuffs of India and its associated health risks: A review. Res J Recent Sci 7(4):36–47

Bhattacharya P, Chatterjee D, Jacks G (1997a) Occurrence of arsenic-contaminated groundwater in alluvial aquifers from delta plains, eastern India: options for safe drinking water supply. Int J Water Resour Dev 13:79–92

Bhattacharya P, Chatterjee D, Jacks G (1997b) Occurrence of arsenic-contaminated groundwater in alluvial aquifers from delta plains, eastern India: options for safe drinking water supply. Int J Water Resour Dev 13(1):79–92

Bhattacharya P, Adhikari S, Samal AC, Das R, Dey D, Deb A, Ahmed S, Hussein J, De A, Das A, Joardar M, Panigrahi AK, Roychowdhury T, Santra SC (2020) Health risk assessment of co-occurrence of toxic fluoride and arsenic in groundwater of Dharmanagar region, North Tripura (India). Groundw Sustain Dev 11:100430

Bibi S, Kamran MA, Sultana J, Farooqi A (2017) Occurrence and methods to remove arsenic and fluoride contamination in water. Environ Chem Lett 15:125–149

Birkner E, Grucka-Mamczar E, Kasperczyk S, Zalejska-Fiolka J, Kasperczyk A, Chlubek D, Filipowska B, Birknera B (2008) The influence of sodium fluoride on the concentration of malondialdehyde and 7-ketocholesterol and the activity of superoxide dismutase in blood plasma of rabbits with experimental hypercholesterolemia. Fluoride 41:199–205

Blowes DW, Ptacek,CJ, Jambor JL, Weisener CG, Paktunc D, Gould WD, Johnson DB (2013) The geochemistry of acid mine drainage. In: Treatise on Geochemistry, second ed

Brands E, Rajagopal R, Eleswarapu U, Li P (2017) International encyclopedia of geography: people, the earth, environment and technology: groundwater. John Wiley & Sons, Ltd

Bundschuh J, Maity JP (2015) Geothermal arsenic: occurrence, mobility and environmental implications. Renew Sustain Energy Rev 42:1214–1222

Bundschuh J, Maity JP, Mushtaq S, Vithanage M, Seneweera S, Schneider J, Bhattacharya P, Khan NI, Hamawand I, Guilherme LRG, Reardon-Smith K, Parvez F, Morales-Simfors N, Ghaze S, Pudmenzky C, Kouadio L, Chen CY (2017) Medical geology in the framework of the sustainable development goals. Sci Total Environ 581–582:87–104

Cao J, Luo SF, Liu JW, Li Y (2004) Safety evaluation on fluoride content in black tea. Food Chem 88(2):233–236

Centeno JA, Tchounwou PB, Patlolla AK, Mullick FG, Murakata L, Meza E, Gibb H, Longfellow D, Yedjou CG (2006) Environmental pathology and health effects of arsenic poisoning: a critical review. In: Managing Arsenic in the Environment: from Soil to Human Health, pp 311–327

Chandrajith R, Diyabalanage S, Dissanayake CB (2020) Geogenic Fluoride and Arsenic in Groundwater of Sri Lanka and its Implications to Community Health. Groundwater for Sustainable Development, p. 100359

Chapelle FH, Lovley DR (1992) Competitive exclusion of sulfate reduction by Fe (lll)-re ducing bacteria: a mechanism for producing discrete zones of high-iron ground water. Groundwater 30(1):29–36

Chen YX, Lin MQ, He ZL, Xiao YD, Chen C, Min D, Yu MH (1996) Relationship between total fluoride intake and dental fluorosis in areas polluted by airborne fluoride. Fluoride 29(1):7–12

Chen J, Qian H, Wu H, Gao Y, Li X (2017a) Assessment of arsenic and fluoride pollution in groundwater in Dawukou area, Northwest China, and the associated health risk for inhabitants. Environ Earth Sci 76:314

Chen J, Wu H, Qian H, Gao Y (2017b) Assessing nitrate and fluoride contaminants in drinking water and their health risk of rural residents living in a semiarid region of northwest China. Expo Heal 9:183–195

Choi H-H, Merriënboer JJGV, Paas F (2014) Effects of the physical environment on cognitive load and learning: Towards a new model of cognitive load. Educ Psychol Rev 26:225–244

Chouhan S, Flora SJS (2010) Arsenic and fluoride: two major ground water pollutants. Indian J Exp Biol 48:666–678

Chowdhury TR, Basu GK, Mandal BK, Biswas BK, Samanta G, Chowdhury UK, Chanda CR, Lodh D, Roy SL, Saha KC, Roy S (1999) Arsenic poisoning in the Gan ges delta. Nature 401(6753):545–546

Chowdhury A, Adak MK, Mukherjee A, Dhak P, Khatun J, Dhak D (2019) A critical review on geochemical and geological aspects of fluoride belts, fluorosis and natural materials and other sources for alternatives to fluoride exposure. J Hydrol 574:333–359

Dabeka RW, Mckenzie AD (1995) Survey of lead, cadmium, fluoride, nickel, and cobalt in food composites and estimation of dietary intakes of these elements by Canadians in 1986–1988. J AOAC Int 78(4):897–909

Dar MA, Sankar K, Dar IA (2011) Fluorine contamination in groundwater: a major challenge. Environ Monit Assess 173:955–968

Das N, Deka JP, Shim J, Patel AK, Kumar A, Sarma KP, Kumar M (2016) Effect of river proximity on the arsenic and fluoride distribution in the aquifers of the Brahmaputra Floodplains, Assam, Northeast India. Groundw. Sustain. Dev. 2– 3, 130–142. 2–3, 130–142

Das N, Das A, Sarma KP, Kumar M (2018) Provenance, prevalence and health perspective of co-occurrences of arsenic, fluoride and uranium in the aquifers of the Brahmaputra River floodplain. Chemosphere 194, 755–772. https://doi.org/10.1016/j.chemosphere.2017.12.021

DenBesten P, Li W (2011) Chronic fluoride toxicity: dental fluorosis. Monogr Oral Sci 2011(22):81–96

Dharmaratne RW (2015) Fluoride in drinking water and diet: the causative factor of chronic kidney diseases in the North Central Province of Sri Lanka. Environ Health Prev Med 20:237–242

Drahota P, Falteisek L, Redlich A, Rohovec J, Matoušek T, Čepička I (2013) Microbial effects on the release and attenuation of arsenicinthe shallow subsurface of a natural geochemical anomaly. Environ Pollut 180:84–91

Edmunds WM, Smedley PL (2012) Fluoride in natural waters. In Essentials of medical geology: Revised Edition, 311–336, Dordrecht: Springer Netherlands

Edmunds WM, Smedley PL (2013) Fluoride in natural waters. In: EssTials Med Geol: Revis Ed. pp 311–336

Edmunds WM, Smedley PL (2005) Fluoride in natural waters. In: Selinus O (ed) Essentials of medical geology. Springer, Dordrecht, pp 311–336

Ellis AJ, Mahon WAJ (1967) Natural hydrothermal systems and experimental hot water/rock interactions (Part II). Geochim Cosmochim Acta 31(4):519–538

Erbs JJ, Berquó TS, Reinsch BC, Lowry GV, Banerjee SK, Penn RL (2010) Reductive dissolution of arsenic-bearing ferrihydrite. Geochim Cosmochim Acta 74(12):3382–3395

Fendorf S, Michael HA, van Geen A (2010) Spatial and temporal variations of ground water arsenic in South and Southeast Asia. Sci 328(5982):1123–1127

Ferreccio C, Gonz´alez C, Milosavjlevic V, Marshall G, Sancha AM, Smith AH (2000) Lung cancer and arsenic concentrations in drinking water in Chile. Epidemiology 11, 673–679

George CM, Sima L, Arias M, Mihalic J, Cabrera LZ, Danz D, Gilman RH (2014) Arsenic exposure in drinking water: an unrecognized health threat in Peru. Bull World Health Organ 92:565–572

Gleeson T, Wada Y, Bierkens MF, Van Beek LP (2012) Water balance of global aquifers revealed by groundwater footprint. Nature 488(7410):197–200

Gonzalez-Rodriguez S, Fernandez-Marcos ML (2021) Sorption and desorption of vanadate, arsenate and chromate by two volcanic soils of Equatorial Africa. Soil Systems 5(2):22
Greenleaf JE, Lin JC, Sengupta AK (2006) Two novel applications of ion exchange fibers: arsenic removal and chemical-free softening of hard water. Environ Prog 325(4):300–311
Guillot S, Charlet L (2007) Bengal arsenic, an archive of Himalaya orogeny and paleohydrology. J Environ Sci Health A 42(12):1785–1794
Haldar D, Duarah P, Purkait MK (2020) MOFs for the treatment of arsenic, fluoride and iron contaminated drinking water: a review. Chemosphere 251:126388
Harper TR, Kingham NW (1992) Removal of arsenic from wastewater using chemical precipitation methods. Water Environ Res 64(3):200–203
Hinkle SR, Polette DJ (1999) Arsenic in Ground Water of the Willamette Basin. US. De partment of the Interior, US Geological Survey 98, Oregon, p 4205
Hoque MA, Khan AA, Shamsudduha M, Hossain MS, Islam T, Chowdhury SH (2009) Near surface lithology and spatial variation of arsenic in the shallow groundwater: southeastern Bangladesh. Environ Geol 56(8):1687–1695
Hubaux R, Becker-Santos DD, Enfield KSS, Rowbotham D, Lam S, Lam WL, Martinez VD (2013) Molecular features in arsenic-induced lung tumors. Mol Canc 12:20
Hughes MF, Beck BD, Chen Y, Lewis AS, Thomas DJ (2011) Arsenic exposure and toxicology: a historical perspective. Toxicol Sci 123(2):305–332
Jagtap S, Yenkie MK, Labhsetwar N, Rayalu S (2012) Fluoride in drinking water and defluoridation of water. Chem Rev 112(4):2454–2466
Jha PK, Tripathi P (2021a) Arsenic and fluoride contamination in groundwater: a review of global scenarios with special reference to India. Groundw Sustain Dev 13:100576
Jing C, Cui J, Huang Y, Li A (2012) Fabrication, characterization, and application of a composite adsorbent for simultaneous removal of arsenic and fluoride. ACS Appl Mater Interfaces 4:714–720
Joshi S, Hlaing T, Whitford GM, Compston JE (2011b) Skeletal fluorosis due to excessive tea and toothpaste consumption. Osteoporos Int 22(9):2557–2560
Joshi A, Bhangoo RS, Kumar K (2011) Quality of nutrition related information on the internet for osteoporosis patients: a critical review. Technology and Health Care 19.6: 391–400
Kao AC, Chu YJ, Hsu FL, Liao VHC (2013) Removal of arsenic from groundwater by using a native isolated arsenite-oxidizing bacterium. J Contam Hydrol 155:1–8
Kapaj S, Peterson H, Liber K, Bhattacharya P (2006) Human health effects from chronic arsenic poisoning—a review. J. Environ. Sci. Health—Part A Toxic/Hazard. Substan Environ Eng 41:2399–2428
Khan MA, Ho YS (2011) Arsenic in drinking water: a review on toxicological effects, mechanism of accumulation and remediation. Asian J Chem
Kinniburgh DG, Smedley P (2001) Arsenic contamination of groundwater in bangladesh british geological survey, 630pp (WC/00/019) (unpublished)
Kinnunen H, Holopainen T, Räisänen ML, Kärenlampi L (2003) Fluoride in birch leaves, ground vegetation, litter and humus in the surroundings of a fertilizer plant and apatite mine in Siilinjärvi, eastern Finland. Boreal Environ Res 8(2):185
Kitchin KT, Conolly R (2010) Arsenic-induced carcinogenesis-oxidative stress as a possible mode of action and future research needs for more biologically based risk assessment. Chem Res Toxicol 23(2):327–335
Kumar A, Singh CK (2020) Arsenic enrichment in groundwater and associated health risk in Bari doab region of Indus basin, Punjab. India. Environ Pollut. 256:113324
Kumar M, Das A, Das N, Goswami R, Singh UK (2016a) Co-occurrence perspective of arsenic and fluoride in the groundwater of Diphu, Assam, Northeastern India. Chemosphere 150:227–238
Kumar M, Rahman MM, Ramanathan AL, Naidu R (2016b) Arsenic and other elements in drinking water and dietary components from the middle Gangetic plain of Bihar, India: health risk index. Sci Total Environ 539:125–134

Kundu CM, Mandal B (2009) Assessment of potential hazards of fluoride contamination in drinking groundwater of an intensively cultivated district in West Bengal, India. Environ Monit Assess 152:97–103

Lovley DR (1997) Microbial Fe (III) reduction in subsurface environments. FEMS Microbiol Rev 20(3–4):305–313

Maity JP, Hsu CM, Lin TJ, Lee WC, Bhattacharya P, Bundschuh J, Chen CY (2018) Removal of fluoride from water through bacterial-surfact in mediated novel hydroxyapatite nanoparticle and its efficiency assessment: adsorption isotherm, adsorption kinetic and adsorption Thermodynamics. Environ. Nanotechnol. Monit. Manag. 9:18–28

Maity JP, Vithanage M, Kumar M, Ghosh A, Mohan D, Ahmad A, Bhattacharya P (2021) Seven 21st century challenges of arsenic-fluoride contamination and remediation, 100538

Maji SK, Kao YH, Liu CW (2011) Arsenic removal from real arsenic-bearing groundwater by adsorption on iron-oxide-coated natural rock (IOCNR). Desalination 280(1–3):72–79

Matisoff G, Khourey CJ, Hall JF, Varnes AW, Strain WH (1982) The nature and source of arsenic in Northeastern Ohio ground water. Groundw 20(4):446–456

McArthur JM, Ravenscroft P, Safiulla S, Thirlwall MF (2001) Arsenic in groundwater: testing pollution mechanisms for sedimentary aquifers in Bangladesh. Water Resour Res 37(1):109–117

Mishra SP, Das M, Dash UN (2010) Review on adverse effects of water contaminants like arsenic, fluoride and phosphate and their remediation. J Sci Ind Res (India)

Mukherjee I, Singh UK (2022) Exploring a variance decomposition approach integrated with the Monte Carlo method to evaluate groundwater fluoride exposure on the residents of a typical fluorosis endemic semi-arid tract of India. Environ Res 203:111697

Mukherjee A, Bhattacharya P, Shi F, Fryar AE, Mukherjee AB, Xie ZM, Jacks G, Bundschuh J (2009) Chemical evolution in the high arsenic groundwater of the Huhhot basin (Inner Mongolia, PR China) and its difference from the western Bengal basin (India). Appl Geochem 24(10):1835–1851

Mukherjee A, Zimmerman AR, Harris W (2011) Surface chemistry variations among a series of laboratory-produced biochars. Geoderma 163(3–4):247–255

Mukherjee A (2018) Overview of the groundwater of South Asia. https://doi.org/10.1007/978-981-10-3889-1_1.

Nickson R, McArthur J, Burgess W, Ahmed KM, Ravenscroft P, Rahman M (1998) Ar senic poisoning of Bangladesh groundwater. Nature 395:338

Nickson RT, McArthur JM, Ravenscroft P, Burgess WG, Ahmed KM (2000) Mecha nism of arsenic release to ground water. Bangladesh and West Bengal. Appl Geochem. 15(4):403–413

Nickson RT, McArthur JM, Shrestha B, Kyaw-Myint TO, Lowry D (2005) Arsenic and other drinking water quality issues, Muzaffargarh District. Pakistan Appl Geochem. 20(1):55–68

Nordstrom DK, Jenne EA (1977) Fluorite solubility equilibria in selected geothermal waters. Geochim Cosmochim Acta 41(2):175–188

Ozsvath DL (2006) Fluoride concentrations in a crystalline bedrock aquifer Marathon County. Wisconsin. Environ Geol. 50:132–138

Patel AK, Das N, Goswami R, Kumar M (2019) Arsenic mobility and potential co- leaching of fluoride from the sediments of three tributaries of the Upper Brahmaputra floodplain, Lakhimpur, Assam. India. J Geochem Explor. 203:45–58

Postma D, Pham TKT, Sø HU, Vi ML, Nguyen TT, Larsen F, Pham HV, Jakobsen R (2016) A modelfor the evolution in water chemistry of an arsenic contaminated aquifer over the last 6000 years, Red River floodplain. Vietnam Geochim Cosmochim Acta 195:277–292

Qiao J, Cui Z, Sun Y, Hu Q, Guan X (2014) Simultaneous removal of arsenate and f luoride from water by Al-Fe (hydr)oxides. Front Environ Sci Eng 8:169–179

Rao MV, Tiwari H (2006) Amelioration by melatonin of chromosomal anomalies induced by arsenic and/or fluoride in human blood lymphocyte cultures. Fluoride 39(4):255–260

Rasool A, Xiao T, Baig ZT, Masood S, Mostofa KMG, Iqbal M (2015) Co- occurrence of arsenic and fluoride in the groundwater of Punjab, Pakistan: source discrimination and health risk assessment. Environ Sci Pollut Res 188(2):319–333

Raychowdhury N, Mukherjee A, Bhattacharya P, Johannesson K, Bundschuh J, Sifuentes GB, Nordberg E, Martin RA, Storniolo A del R (2014) Provenance and fate of arsenic and other solutes in the Chaco-Pampean Plain of the Andean foreland, Argentina: from perspectives of hydrogeochemical modeling and regional tectonic setting. J Hydrol. 518, 300–316

Reddy AGS, Reddy DV, Rao PN, Prasad KM (2010) Hydrogeochemical characterization of fluoride rich groundwater of Wailpalli watershed, Nalgonda District, Andhra Pradesh, India. Environ. Monit. Assess. 171, 561–577. https://doi.org/10.1007/s10661-009-1300-3

Reyes-Gòmez VM, Alarcon-Herrera MT, Gutierrez M, L´opez DN (2013) Fluoride and arsenic in an alluvial aquifer system in Chihuahua, Mexico: contaminant levels, potential sources, and co-occurrence. Water Air Soil Pollut. 224, 1433

Rodrıguez R, Ramos JA, Armienta A (2004) Groundwater arsenic variations: the role of local geology and rainfall. Appl Geochem 19(2):245–250

Saikia R, Goswami R, Bordoloi N, Senapati KK, Pant KK, Kumar M, Kataki R (2017) Removal of arsenic and fluoride from aqueous solution by biomass based activated biochar: optimization through response surface methodology. J Environ Chem Eng 5(6):5528–5539

Sandoval MA, Fuentes R, Nava JL, Core˜no O, Li Y, Hern´ andez JH (2019) Simultaneous removal of fluoride and arsenic from groundwater by electrocoagulation using a filter-press flow reactor with a three-cell stack. Separ. Purif. Technol. 208, 208–216

Sanjrani MA, Zhou B, Zhao H, Bhutto SA, Muneer AS, Xia SB (2019) Arsenic contaminated groundwater in China and its treatment options, a review. Appl Ecol Environ Res 17(2):1655–1683

Sariñana-Ruiz YA, Vazquez-Arenas J, Sosa-Rodríguez FS, Labastida I, Armienta MA, Aragon-Pi˜ na A, Escobedo-Bretado MA, Gonzalez-Valdez LS, Ponce-Pe˜na P, Ramírez-Aldaba H, Lara RH (2017) Assessment of arsenic and fluorine in surface soil to determine environmental and health risk factors in the Comarca Lagunera, Mexico. Chemosphere. 178, 391–401

Sarkar B (2002) Heavy Metals in the Environment. CRC Press

Schamschula RG, Un PS, Sugar E, Duppenthaler JL, Toth K, Barmes DE (1988) The fluoride content of selected foods in relation to the fluoride concentration of water. Acta Physiol Hung 72(2):217–227

Sen M, Manna A, Pal P (2010) Removal of arsenic from contaminated groundwater by membrane-integrated hybrid treatment system. J Membr Sci 354:108–113

Shah AQ, Kazi TG, Baig JA, Arain MB, Afridi HI, Kandhro GA, Wadhwa SK, Kolachi NF (2010) Determination of inorganic arsenic species (As^{3+} and As^{5+}) in muscle tissues of fish species by electrothermal atomic absorption spectrometry (ETAAS). Food Chem 119(2):840–844

Shaji E, Santosh M, Sarath KV, Prakash P, Deepchand V, Divya BV (2021) Arsenic contamination of groundwater: A global synopsis with focus on the Indian Peninsula. Geosci Front 12:101079

Shaji E, Sarath KV, Santosh M, Krishnaprasad PK, Arya BK, Babu MS (2024) Fluoride contamination in groundwater: A global review of the status, processes, challenges, and remedial measures. Geosci Front 15:101734

Singh CK, Rina K, Singh RP, Shashtri S, Kamal V, Mukherjee S (2011) Geochemical modeling of high fluoride concentration in groundwater of Pokhran area of Rajasthan. India. Bull Environ Contam Toxicol. 86:152–158

Singh CK, Kumari R, Singh N, Mallick J, Mukherjee S (2013) Fluoride enrichment in aquifers of the Thar desert: controlling factors and its geochemical modelling. Hydrol Process 27:2462–2474

Smedley PL, Nicolli HB, Macdonald DMJ, Barros AJ, Tullio JO (2002) Hydrogeochemistry of arsenic and other inorganic constituents in groundwaters from La Pampa. Argentina. Applied Geochemistry 17 (3), 259 284

Smedley PL, Kinniburgh DG (2002) A review of the source, behaviour and distribution of arsenic in natural waters. Appl Geochem 17(5):517–568

Speer RM, Zhou X, Volk LB, Liu KJ, Hudson LG (2023) Arsenic and cancer: evidence and mechanisms. Adv Pharmacol 96:151–202

Sracek O, Bhattacharya P, Jacks G, Gustafsson JP, Von Brömssen M (2004) Behavior of arsenic and geochemical modeling of arsenic enrichment in aqueous environments. Appl Geochem 19(2):169–180

Subba Rao N (2003) Groundwater quality: focus on fluoride concentration in rural parts of Guntur district. Andhra Pradesh. India. Hydrol Sci J 48(5):835–847

Thakur LS, Mondal P (2017) Simultaneous arsenic and fluoride removal from synthetic and real groundwater by electrocoagulation process: parametric and cost evaluation. J Environ Manag 190:102–112. https://doi.org/10.1016/j.jenvman.2016.12.053

Toolabi A, Bonyadi Z, Paydar M, Najafpoor AA, Ramavandi B (2020) Spatial distribution, occurrence, and health risk assessment of nitrate, fluoride, and arsenic in bam groundwater resource. Groundw Sustain Dev, Iran, p 100543

Trafford JM, Lawrence AR, Macdonald DMJ, Nguyen VD, Tran DN, Nguyen TH (1996) The effect of urbanisation on the groundwater quality beneath the city of Hanoi, Vietnam. BGS technical report WC/96/ 22. British Geological Survey, Keyworth, UK

Verma PK, Verma S, Pande V, Mallick S, Deo Tripathi R, Dhankher OP, Chakrabarty D (2016) Overexpression of rice glutaredoxin OsGrx_C7 and OsGrx_C2. 1 reduces in tracellular arsenic accumulation and increases tolerance in Arabidopsis thaliana. Front. Plant Sci. 7, 740

Viswanathan G, Gopalakrishnan S, Siva Ilango S (2010) Assessment of water contribution on total fluoride intake of various age groups of people in fluoride endemic and Non-endemic Areas of Dindigul District, Tamil Nadu. South India. Water Res. 44(20):6186–6200

Vithanage M, Bhattacharya P (2015) Fluoride in the environment: sources, distribution and defluoridation. Environ Chem Lett 13:131–147

Waghmare (2015) Fluoride removal from water by various techniques - review. Int. J. Innov Sci Eng Technol. 2, 560–571

Welch AH, Westjohn DB, Helsel DR, Wanty RB (2000) Arsenic in ground water of the United States: occurrence and geochemistry. Groundwater 38(4):589–604

Whitford GM, Pashley DH, Garman RH (1997) Effects of fluoride on structure and function of canine gastric mucosa. Dig Dis Sci 42:2146–2155

Wu X, Cobbina SJ, Mao G, Xu H, Zhang Z, Yang L (2016) A review of toxicity and mechanisms of individual and mixtures of heavy metals in the environment. Environ Sci Pollut Res 23:8244–8259

Yadav IC, Devi NL, Syed JH, Cheng Z, Li J, Zhang G, Jones KC (2015) Current status of persistent organic pesticides residues in air, water, and soil, and their possible effect on neighbouring countries: A comprehensive review of India. Sci Total Environ 511:123–137

Yadav KK, Kumar V, Gupta N, Kumar S, Rezania S, Singh N (2019) Human health risk assessment: study of a population exposed to fluoride through groundwater of Agra city. India. Regul Toxicol Pharmacol. 106:68–80

Zhao YQ, Babatunde AO (2011) Journal of environmental science and health - Part A toxic/ hazardous substances and environmental engineering: foreword. J Environ Sci Health—Part A Toxic/Hazard. Substan Environ Eng. 46 (7), 677–679

Sustainable Groundwater Systems: Recharge, Depletion, and Emerging Pollution Control Technologies

Assessment of Aquifers Recharge Potential Zones Using an Integrated Approach for Sustainable Water Supply Management

Dheeraj Sonkar, Dhiraj Kumar, and G. V. Prajapati

Abstract The Replenishment of underground water is essential for combating water scarcity as global water sources face growing pressure. Arid and semi-arid regions worldwide face severe water scarcity due to low rainfall and increasing population, resulting in higher demand for water resources. Groundwater resources, often secondary sources, are in a vulnerable phase due to the actions of both natural variability and anthropogenic interventions. The main aim of this chapter is to devise a method for accurately determining the rate at which underground water is replenished in dry regions experiencing rapid groundwater depletion. This chapter covers (1) how to assess the zone where replenishment is needed and find water sources using geological, climatic, and remote sensing data; (2) how to investigate an aquifer property through the analysis of groundwater isotopes; and (3) evaluation of water resources sustainability by quantifying replenishment volume using distinct hydrological/hydraulic models. This chapter, it is also discussing how changes in local water extraction practices, shifts in rainfall patterns, and the expansion of irrigated areas influence water storage, depletion, and replenishment dynamics over time and highlights the valuable insights gained from understanding the complex interactions that affect the underground water behaves, essential for managing groundwater sustainably in arid and semi-arid regions.

Keywords Groundwater · Aquifer · Remote sensing · Replenishment · Water resources

D. Sonkar (✉)
Ph.D. Scholar, Department of Hydrology, Indian Institute of Technology, Roorkee, Uttarakhand, India
e-mail: dhirajsonkar.nihr@gmail.com

D. Kumar
Ph.D. Scholar, Department of Soil and Water Conservation Engineering, Junagadh Agricultural University, Gujarat, India

G. V. Prajapati
Prof. & Head, Department of Irrigation and Drainage Engineering, Junagadh Agricultural University, Junagadh, Gujarat, India

A. Nath et al. (eds.), *Groundwater Depletion and Sustainability*, Advances in Geographic Information Science, https://doi.org/10.1007/978-3-032-09921-1_11

1 Introduction

Groundwater underpins global water resources. It supplies almost half of the world's drinking water and irrigates around 40% of total cropland (Doll and Siebert 2002). While groundwater is protected from evaporation and, to some extent, from pollution, unlike surface water, it is a more reliable resource, especially in water-scarce areas. Its importance grows during dry years when it acts as a reserve against deficits in surface water. Moreover, groundwater also supports wetlands, rivers, and other ecosystems by sustaining base flows and is thus critical for maintaining biodiversity. Groundwater is a reliable substitute for accessing domestic and agricultural supplies in developing nations where surface water storage infrastructure is usually lacking (Tyagi 1987). Nonetheless, there must be a balance between withdrawal and recharge to maintain sustainability. On the flip side, insufficient recharge results in the over-extraction of aquifers, eventually leading to declining groundwater levels, deterioration in quality, and total depletion. Thus, groundwater recharge is a hydrological need and a socio-economic need for sustainable development (Nandi and Swain 2024). Groundwater is of critical importance but faces many global challenges. Over-extraction in response to the growing demand for water in agriculture, urbanization, and industrial activities is one of the essential problems (McDonald et al. 2014). This over-extraction is usually more significant than the natural recharge, resulting in a gradual decline in groundwater levels. These aquifers are being depleted at alarming rates, for example, in highly populated areas like the Indo-Gangetic Plains (Ambast et al. 2006). Pollution is another big problem. Agricultural runoff laced with fertilizers and pesticides, industrial effluents, and untreated sewage seep into aquifers, polluting their water. In several places, polluted aquifers become unusable, worsening water scarcity. Urbanization presents further challenges by diminishing natural permeability and creating impervious surfaces (e.g., roads and buildings). This changes natural recharge processes, making rainfall go into stormwater systems rather than allowing it to seep into the earth. Climate change further compounds these problems by changing precipitation patterns, higher evapotranspiration rates, and more severe droughts (Lesk et al. 2022). As these factors impair the quantity and timing of recharge, traditional water management practices are not cutting it. Identifying and managing recharge potential zones requires innovative approaches and a combination of technological advancements with community-driven initiatives.

This chapter investigates groundwater recharge potential zones as a fundamental sustainable water resource management component. An integrated approach that utilizes geospatial technologies, hydrogeological assessment, and decision-making frameworks to delineate high-potential recharge zones is emphasized. The methodologies include Geographic Information Systems (GIS) for spatial analysis, remote sensing for data acquisition, and multi-criteria decision analysis (MCDA) for evaluating recharge suitability. The chapter additionally investigates interactions of (natural and anthropogenic) factors affecting groundwater recharge, including soil type, land use, slope, and rainfall. Incorporating these parameters into advanced analytical models offers an integrated approach to identifying recharge zones.

Through case studies in different geographical contexts, the practical application of these techniques is demonstrated along with what has been achieved and what has been learned in implementation. Moreover, the chapter explores the consequences of identifying recharge zones for decision-making, infrastructure building, and citizen participation. It ends with recommendations for adopting sustainable water management practices, improved groundwater governance, and building resilience against climate change impacts. Research and development of groundwater recharge for future generations This chapter provides guidelines for researchers, water managers, and policymakers in developing and implementing groundwater recharge technologies to ensure sustainable water supplies.

2 Overview of Aquifer Recharge

2.1 Aquifers: Types and Characteristics

Aquifers are underground water-bearing rock or sediment levels that store and diffuse groundwater (Idris et al. 2018). Their capability to supply water depends on their permeability and porosity, which vary notably across different classes. Aquifers are broadly classified into the following (Table 1).

Table 1 Aquifer types, characteristics and recharge potential

Type of aquifer	Characteristics	Recharge potential
Unconfined aquifer	Located near the surface, directly recharged by precipitation. Water levels fluctuate significantly based on seasonal variations. Highly vulnerable to pollution from surface activities	High, as recharge occurs directly through the surface
Confined aquifer	Separated from the surface by an impermeable layer. Water stored under pressure is accessible through artesian wells. It is challenging to recharge artificially	Low due to limited direct recharge opportunity
Perched aquifers	Found above the regional water table, resting on low-permeability material. Provides localized groundwater storage	Moderate, limited by small extent and localized conditions
Fractured or Karst aquifers	It is found in rocks like limestone with fractures or solution cavities. Highly variable in storage and flow characteristics, making recharge assessment complex	Variable depends on geological features and fracture networks

2.2 Concept of Recharge Potential Zones and Factors Influencing Recharge Potential

Recharge potential zones are regions where surface water infiltration into aquifers is favored, thus augmenting groundwater storage. These zones are significant for groundwater level stabilization and supporting aquifer systems. Analysis of various physical and hydrological characteristics that govern water infiltration, storage, and flow is used to identify the recharge potential zones. The potential groundwater recharge is affected by geological, topographical, climatic, land use, hydrogeological, and anthropogenic factors. Soil texture and rock type are geological factors that play an essential role in deciding the ease and storage of infiltration. The porous sandy and gravelly types of soil favor the processes of infiltration and recharge. In contrast, the impervious clayey soil is not conducive to infiltration and recharge, as it has low permeability. While karst or fractured rocks are helpful for recharge through natural openings and cavities, compact rocks such as shale restrict water movement. Gentle slopes promote water percolation, giving recharge time, so topography also affects recharge. On steep slopes, surface runoff is favored over infiltration and recharge possibility, emphasizing the importance of contouring or vegetation cover in steep areas to achieve runoff control. Rainfall and temperature are key climatic drivers of recharge dynamics. Moderate rain allows sufficient percolation into the soil, while heavy rain leads to surface runoff and low efficiency of groundwater recharge. The evapotranspiration also increases at high temperatures, making less water available for percolation, another adverse factor. Optimizing the recharge potential is a climate balance. Recharge processes are influenced by land use and land cover. Vegetated zones such as forests facilitate infiltration by decreasing runoff and breaking into soil through root systems. In contrast, impervious surfaces (such as roads and buildings) in cities inhibit infiltration, which results in increased runoff and lessened potential for recharge. Incorporating green infrastructure and sustainable land-use planning to alleviate the burdens of urbanization (Table 2).

The knowledge of the regions with appropriate hydrogeological conditions is necessary for focused groundwater management. Recharge potential is also mediated by human activities, including unsustainable practices (e.g., deforestation, intensive agriculture) that disrupt natural recharge processes. Deforestation leads to more runoff and soil erosion, and agricultural machines compact soil, blocking its permeability. On the other hand, artificial recharge structures introduce check dams, percolation tanks, and recharge wells to improve groundwater honeypot levels in regions that do not get adequate natural recharge. To ensure the sustainability of these physical structures over time, careful design and investment could lead to improved groundwater availability and resilience, helping to secure an urban water future in the face of both urbanization and climate variability.

Table 2 Recharge zone category

Recharge zone category	Characteristics	Key factors influencing recharge	Importance
High-potential zones	Due to permeable soils, gentle slopes, and sufficient rainfall, high infiltration rates. Minimal surface runoff and effective natural recharge processes	Sandy or gravelly soils, gentle slopes, and good vegetation cover	Criteria for sustaining aquifers and replenishing groundwater reserves. Supports sustainable water supply and mitigates over-extraction risks
Moderate-potential zones	Intermediate recharge condition has limitations, such as soil compaction or reduced permeability. Slope and rainfall are moderately favorable for infiltration	Silty soils with moderate permeability, mildly sloped terrain, and intermittent rainfall	Contributes to groundwater recharge but may require artificial enhancement for management strategies. Provides a secondary recharge source for maintaining aquifer balance
Low-potential zones	Minimal recharge due to impermeable surfaces (e.g., urban areas), steep slopes, or clay-rich soils: high surface runoff and low infiltration capacity	Urbanization with impervious surfaces (concrete, asphalt), steep slopes, or highly compacted soils	Requires intervention through artificial recharge structures like recharge wells or check dams. Protecting such areas is crucial to prevent further groundwater depletion or contamination from pollutants

3 Integrated Approach to Assess Recharge Potential Models

3.1 Use of Remote Sensing and GIS in Hydrogeology

Remote sensing and Geographic Information Systems (GIS) play a critical role in identifying the potential zones of aquifer recharge (Agarwal and Garg 2016). Information on land use, vegetation cover, soil types, and topography required to understand recharge drivers can be obtained through remote sensing that affords spatial data. Standard practices such as satellite imagery analysis (e.g., Landsat, Sentinel, MODIS) and Digital Elevation Models (DEMs) are employed to determine slope, drainage patterns, and surface water availability. GIS combines those spatial datasets to create maps showing areas with recharge potential. It also allows for overlay analysis, integrating thematic layers like rainfall, soil permeability, and land use through weighted overlay or multi-criteria decision analysis (MCDA). GIS became a cornerstone part of recharge zone assessments by processing large datasets with accuracy and precision. Hydrological models depict water flow through the hydrological cycle, such as groundwater recharge, surface runoff, and infiltration. These scenarios, including changes in rainfall patterns or land use, provide quantitative estimates of frost-susceptible weather types and recharge rates through modeling (Mensah et al. 2022). Popular hydrological models include SWAT (Soil and Water Assessment Tool), a popular choice for simulating watershed-scale hydrology. This comprehensive model integrates various processes, including the rainfall-runoff response, soil moisture dynamics, and landscape use influences to estimate recharge and water balance.

3.2 Applications for SWAT (Soil and Water Assessment Tool) in Groundwater Recharge

SWAT is an integrated, watershed-scale, process-based hydrological model developed to predict the effects of land use, climate, and management practices on water resources. QGIS has a robust framework that makes it appropriate for groundwater recharge applications (Zhao et al. 2024). Below are its key applications (Table 3).

3.3 Applications for MIKE SHE in Groundwater Recharge

MIKE SHE is an integrated, physically based hydrological modeling system simulating the entire hydrological cycle. GDMs are commonly used to interpret and manage groundwater recharge dynamics. It is significantly advanced in coupling surface and subsurface hydrology, making it particularly valuable for recharge studies (Graham et al. 2005). Below are key applications of MIKE SHE in groundwater recharge (Table 4).

3.4 Applications for MODFLOW in Groundwater Recharge

MODFLOW is a popular software for numerical modeling of groundwater flow developed by the U.S. Geological Survey. These characteristics make it a valuable tool for analyzing groundwater recharge processes and aquifer dynamics (Harbaugh 2005). Below are its key applications in groundwater recharge (Table 5).

Table 3 Application of SWAT Model in groundwater recharge

Application area	Key functions
Estimating recharge potential	Simulates infiltration and percolation to aquifers. Calculates net recharge, accounting for runoff, evapotranspiration, and soil storage. Evaluate spatial/seasonal variability
Assessing land use changes	Analyze the impact of urbanization, deforestation, and agricultural practices on recharge. Considers effects on soil permeability and water retention
Evaluating climate change impacts	Incorporates climate variables to assess future recharge trends. Uses GCM/RCM data to predict changes in recharge from altered rainfall and rising temperatures
Artificial recharge integration	Simulates impacts of recharge interventions (e.g., wells, basins). Estimates contribution of structures to aquifer replenishment. Reduces runoff and erosion
Watershed management	Produces spatial/temporal recharge and runoff maps. Identifies priority areas for conservation and recharge interventions. Guides sustainable aquifer management
Recharge under irrigation practices	Evaluate aquifer recharge from irrigation practices—Assesses impacts of irrigation techniques (e.g., drip vs. flood)
Multi-scenario analysis	Combines land use, climate, and management scenarios. Models best-case, worst-case, and policy-driven scenarios (e.g., afforestation or urban expansion)
MIKE SHE integration	Simulates integrated surface–subsurface interactions. Addresses groundwater recharge influenced by surface processes and subsurface flow
Simulation of recharge processes	Model's interaction between precipitation, infiltration, evapotranspiration, and groundwater flow. Calculates recharge rates based on soil permeability, precipitation, and land cover
Assessment of land use and land cover impact	Incorporates land use and land cover changes. Simulates reduced infiltration due to urbanization and enhanced recharge in reforested areas
Integrated surface–groundwater interaction	Model's dynamic interaction between surface water and groundwater. Accounts for runoff, streamflow, and river-aquifer exchanges
Evaluation of artificial recharge systems	Simulates performance of recharge interventions like wells and ponds. Evaluate managed aquifer recharge techniques

(continued)

Table 3 (continued)

Application area	Key functions
Climate change impact analysis	Integrates climate models to assess effects on recharge potential. Simulates future recharge patterns and the impact of increased evapotranspiration
Hydrological management and policy support	Provides outputs for sustainable water resource management. Identifies high-priority recharge areas for aquifer protection
Evaluation of agricultural practices	Evaluates percolation under different irrigation practices. Assesses recharge potential from agricultural return flows
Contamination risk assessment	Simulates water movement through the vadose zone. Assesses transport of contaminants to groundwater during recharge

Table 4 Application of MIKE SHE Model in groundwater recharge

Application area	Key functions
Simulation of recharge processes	Model's interaction between precipitation, infiltration, evapotranspiration, and groundwater flow. Calculates recharge rates based on soil permeability, precipitation, and land cover
Assessment of land use and land cover impact	Incorporates land use and land cover changes. Simulates reduced infiltration due to urbanization and enhanced recharge in reforested areas
Integrated surface–groundwater interaction	Model's dynamic interaction between surface water and groundwater. Accounts for runoff, streamflow, and river-aquifer exchanges
Evaluation of artificial recharge systems	Simulates performance of recharge interventions like wells and ponds. Evaluate managed aquifer recharge techniques
Climate change impact analysis	Integrates climate models to assess effects on recharge potential. Simulates future recharge patterns and the impact of increased evapotranspiration
Hydrological management and policy support	Provides outputs for sustainable water resource management. Identifies high-priority recharge areas for aquifer protection
Evaluation of agricultural practices	Evaluates percolation under different irrigation practices. Assesses recharge potential from agricultural return flows
Contamination risk assessment	Simulates water movement through the vadose zone. Assesses transport of contaminants to groundwater during recharge

Table 5 Application of SWAT Model in groundwater recharge

Application area	Key functions	Examples
Groundwater flow simulation	Simulates groundwater flow under natural and artificial recharge conditions	Analyzing aquifer response to seasonal recharge in semi-arid regions
Recharge zone delineation	Identifies and maps zones with high recharge potential based on hydrological data	Mapping high recharge areas to guide sustainable development plans
Artificial recharge assessment	Evaluates the impact of artificial recharge structures like recharge wells and basins on aquifer levels	Assessing recharge basin efficiency in increasing water table levels
Climate change impact analysis	Assesses changes in recharge patterns and groundwater levels due to varying precipitation and temperature scenarios	Modeling groundwater sustainability under projected climate change in tropical regions
Integrated water resource management	Integrates recharge estimation with other water resource components for sustainable management	Balancing recharge and withdrawal rates in water-scarce regions
Contamination transport modeling	Simulates contaminant movement in groundwater during recharge processes to assess pollution risks	Evaluating nitrate transport during agricultural recharge in rural areas

3.5 Applications for HEC-HMS in Recharge Water

The Hydrologic Engineering Center-Hydrologic Modeling System (HEC-HMS) Most of the latest open-source software, areas such as Hydrologic Engineering Center-Hydrologic Modeling System (HEC-HMS), a software modeling package produced by the United States Army Corps of Engineers. The code was initially designed to simulate the hydrological processes occurring within a watershed, making it a valuable tool for evaluating recharge potential and water resources management strategies and interventions (Feldman 2000). Below are its key applications in groundwater recharge (Table 6).

Combined with geospatial data, these models outline the potential for recharge across a watershed and predict impacts based on future climate and land undulation scenarios.

Table 6 Application of SWAT Model in groundwater recharge

Application area	Key functions
Estimating recharge potential	Simulates infiltration and percolation to aquifers. Calculates net recharge, accounting for runoff, evapotranspiration, and soil storage. Evaluate spatial/seasonal variability
Assessing land use changes	Analyze the impact of urbanization, deforestation, and agricultural practices on recharge. Considers effects on soil permeability and water retention
Evaluating climate change impacts	Incorporates climate variables to assess future recharge trends. Uses GCM/RCM data to predict changes in recharge from altered rainfall and rising temperatures
Artificial recharge integration	Simulates impacts of recharge interventions (e.g., wells, basins). Estimates contribution of structures to aquifer replenishment. Reduces runoff and erosion
Watershed management	Produces spatial/temporal recharge and runoff maps. Identifies priority areas for conservation and recharge interventions. Guides sustainable aquifer management
Recharge under irrigation practices	Evaluate aquifer recharge from irrigation practices—Assesses impacts of irrigation techniques (e.g., drip vs. flood)
Multi-scenario analysis	Combines land use, climate, and management scenarios. Models best-case, worst-case, and policy-driven scenarios (e.g., afforestation or urban expansion)
MIKE SHE integration	Simulates integrated surface–subsurface interactions. Addresses groundwater recharge influenced by surface processes and subsurface flow
Simulation of recharge processes	Model's interaction between precipitation, infiltration, evapotranspiration, and groundwater flow. Calculates recharge rates based on soil permeability, precipitation, and land cover
Assessment of land use and land cover impact	Incorporates land use and land cover changes. Simulates reduced infiltration due to urbanization and enhanced recharge in reforested areas
Integrated surface–groundwater interaction	Model's dynamic interaction between surface water and groundwater. Accounts for runoff, streamflow, and river-aquifer exchanges
Evaluation of artificial recharge systems	Simulates performance of recharge interventions like wells and ponds. Evaluate managed aquifer recharge techniques

(continued)

Table 6 (continued)

Application area	Key functions
Climate change impact analysis	Integrates climate models to assess effects on recharge potential. Simulates future recharge patterns and the impact of increased evapotranspiration
Hydrological management and policy support	Provides outputs for sustainable water resource management. Identifies high-priority recharge areas for aquifer protection
Evaluation of agricultural practices	Evaluates percolation under different irrigation practices. Assesses recharge potential from agricultural return flows
Contamination risk assessment	Simulates water movement through the vadose zone. Assesses transport of contaminants to groundwater during recharge

3.6 Combining Ground Data with Geospatial Analysis

Field data fills a gap in remote sensing and GIS: ground truth for spatial analyses. Geospatial models can augment such data through field surveys, boreholes, and piezometers. Key steps include.

Ground Data Collection and Geospatial Analysis Integration

The initial step in determining recharge potential involves the gathering of terrestrial data. Observation data of well and piezometer monitoring detail temporal and spatial recharge trends and, thus, aquifer responses to natural and artificial recharge. Soil sampling is done to obtain parameters important for assessing the soil's conductivity to water, including permeability and infiltration rates. We also conduct groundwater quality analysis to locate potential recharge zones, as the differences in chemical composition can indicate productive recharge zones. These field observations provide a solid understanding of subsurface conditions and the recharge dynamics. By providing the field data in concert with more advanced map techniques, geospatial tools help optimize recharge potential analytics. Thematic maps produced through GIS (e.g. land use, soil type, and slope maps) are then integrated with field measurements to produce spatially explicit recharge potential maps. These maps are validated and improved using remote sensing data, including satellite imagery, for greater accuracy. Using machine learning techniques such as Random Forest and Support vector machine to detect patterns in the recharge area. This creates a bridge from ground-level observations to spatial analysis and maps providing holistic and digestible insights.

3.7 *Validation*

Validating recharge potential assessments adds to their reliability and accuracy. Field observations, including measured water levels and infiltration rates, are compared with the GIS-derived recharge zones to verify their consistency. Statistical models (only) assess the goodness of fit of predictions and numerically measure the degree of concordance of modeled data with the data observed. In addition to ensuring the credibility of analyses, this step also highlights potential improvements needed to ensure that final outputs are robust and usable in decision-making contexts in groundwater management. Integrating the ground data with the geospatial analysis can provide a solid and reliable assessment of the recharged potential zones, leading towards an efficient management of water resources.

4 Data Sources and Methodology

4.1 *Data Acquisition: Remote Sensing, GIS, and Ground-Based Observations*

Remote sensing, GIS tools, and ground-based observations each contribute diverse datasets that assist in identifying aquifer recharge potential zones (Santarosa et al. 2024). These datasets are combined to help fully understand the things influencing recharge potential (Fig. 1).

4.2 *Criteria for Identifying Recharge Zones*

Unfortunately, most popular recharge potential zones are established by considering several factors affecting the infiltration, storage, or flow of water (Yeh et al. 2009) (Fig. 2).

Each factor receives a weight to reflect its importance to recharge potential. The criteria are thereafter fused using analytical methods, to delineate recharge zone. Weighted Overlay Analysis is a commonly used technique in Geographic Information System (GIS)-based decision-making that integrates multiple raster layers, each representing a different criterion, to produce a final suitability map. In this method, each input layer is assigned a weight based on its relative importance to the overall decision objective. The cell values of the raster layers are reclassified to a common scale, typically ranging from least suitable to most suitable, and then combined using a weighted sum. This method is particularly useful in environmental management, land use planning, and site suitability analysis. While Multi-Criteria Decision Analysis (MCDA) refers to a set of techniques or approaches used to evaluate and prioritize multiple conflicting criteria in decision-making processes. MCDA provides

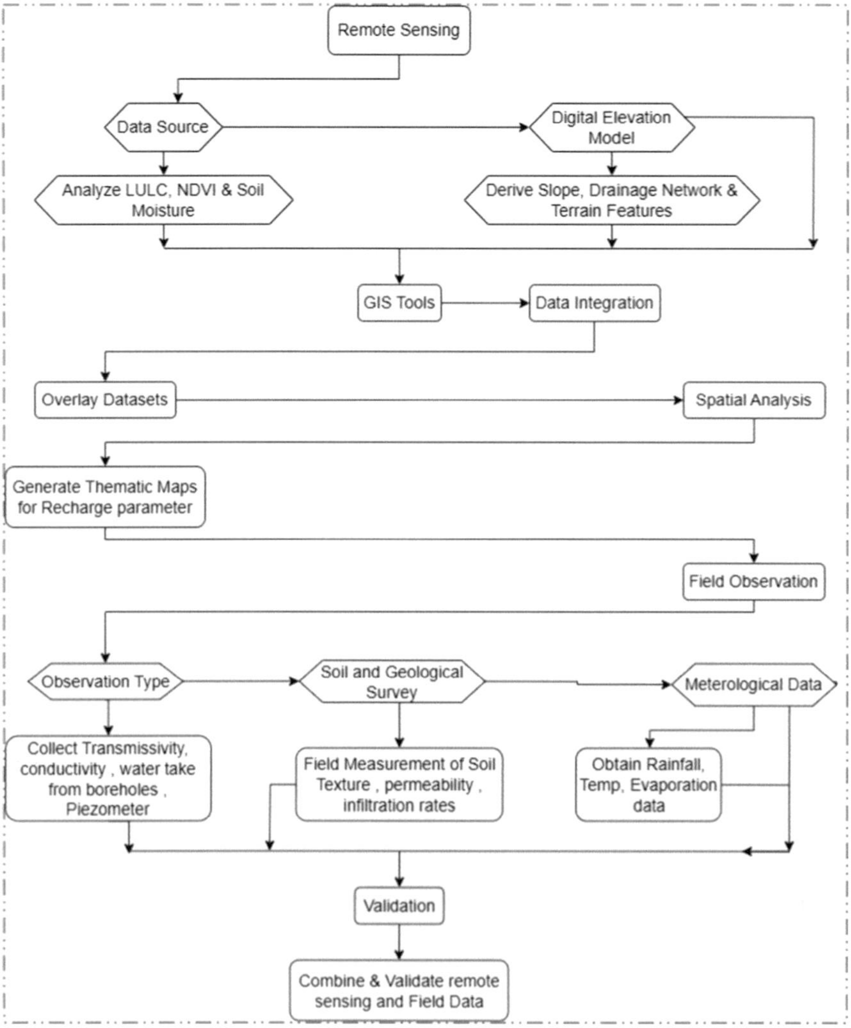

Fig. 1 Combined methodology of data acquisition: remote sensing, gis, and ground-based observations

a structured framework that allows decision-makers to incorporate both qualitative and quantitative data, along with stakeholder preferences, to identify the most suitable option among alternatives. In spatial analysis, MCDA is often integrated with GIS to support spatial decision-making, allowing for the inclusion of geographic and socio-environmental factors in complex decision scenarios To identify recharge zones, the following analytical techniques are employed (Drobne et al. 2009; Zabihi et al. 2019; Romano et al. 2015; Hill et al. 2005 and Huang et al. 2011) (Fig. 3).

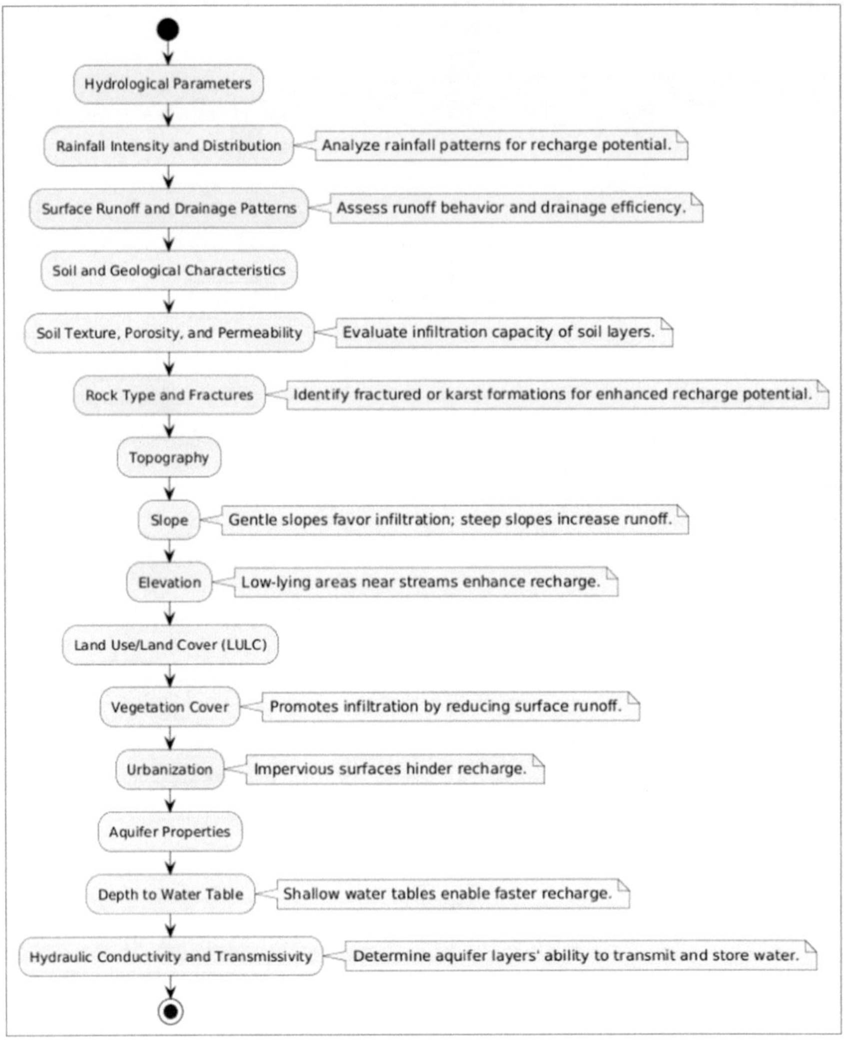

Fig. 2 Flow work of identifying Recharge Zones

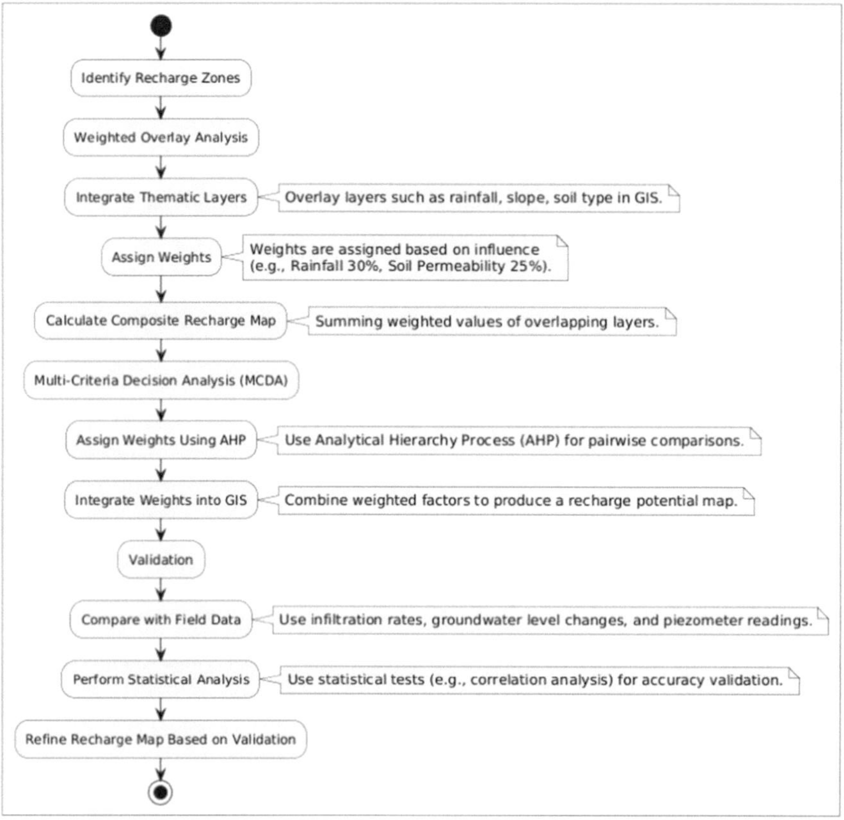

Fig. 3 Combined methodology for identifying recharge zones

5 Case Studies and Their Implications

5.1 Examples of Recharge Potential Mapping in Various Regions

Recharge potential mapping has been successfully applied in diverse geographical and climatic regions, demonstrating the effectiveness of integrated approaches combining remote sensing, GIS, and field data. Below are notable examples: Dar et al. (2021), Souissi et al. (2018), Acharya et al. (2019), Arshad et al. (2020) (Fig. 4 and Table 7).

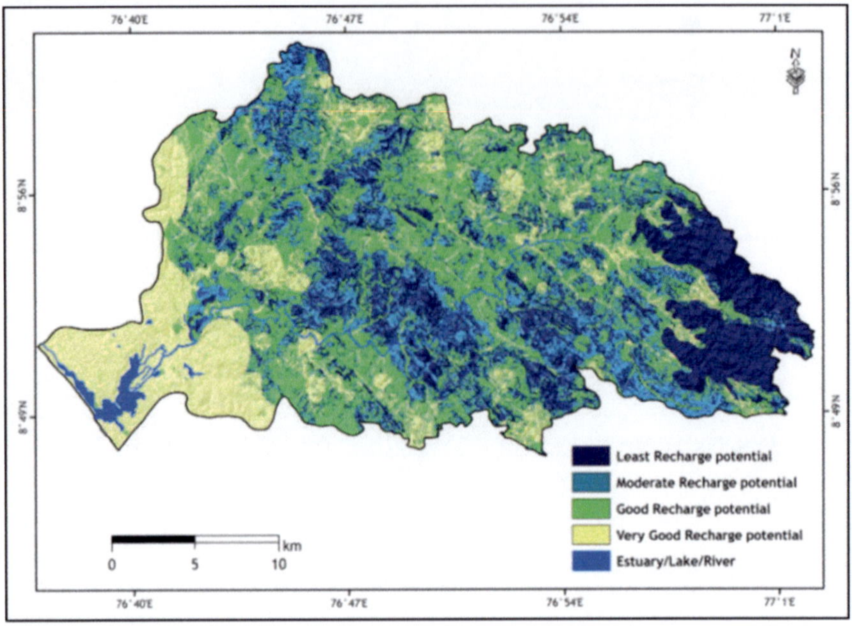

Fig. 4 Groundwater recharge potential zones of the IRB (*Source* Achu et al. 2020)

Table 7 Different case studies from different regions

Region	Objective	Methodology	Findings
Himalayan region, India	Identify groundwater recharge zone in mountainous terrain	GIS-based multi-criteria decision analysis (MCDA) using factors like rainfall, slope, soil type, and vegetation cover	High recharge potential in areas with moderate slope, forest cover, and fractured rock formation. Urbanized and steeply sloped areas had low capacity
Semi-arid zone, Africa	Enhanced water security by identifying high-potential recharge zones in drought prone areas	Remote sensing (Landsat Imagery) combined with GIS and field data on soil permeability and groundwater levels	Sandy soil regions near ephemeral rivers were high-potential zones, but agricultural activities reduced recharge due to soil compaction
Coastal region, Brazil	Prevent saltwater intrusion by enhancing recharge in coastal aquifers	Geophysical survey (ERT), GIS-based mapping, and hydrological modeling	High recharge potential in forested sandy soil regions; limited recharge in urbanized areas with impervious surfaces
Urban Area, Australia	Mitigate the impact of urbanization on groundwater recharge	Remote sensing (DEM and land use map) and in-situ measurement to map recharge zones	High recharge zones in urban parks and green spaces; impermeable urban centers had low recharge potential

5.2 Lessons Learned and Best Practices

Several valuable lessons and best practices have emerged from these case studies for assessing and managing recharge potential of groundwater:

i. **Integrated Approaches are a must: Integration of remote sensing, GIS**, and field-based methods provides more comprehensive and accurate recharge potential mapping.
ii. **Adapting Regional Setting**: These local factors include climate, topography, and land use, all needing to be considered in recharge assessment methods. The techniques that work for the arid regions may not be effective in humid or mountainous regions.
iii. **Community Engagement Matters**: Involving local communities in the planning and implementation of recharge interventions enhances sustainability and helps ensure long-term success.
iv. **Role of Policy and Governance**: Supportive policies and governance frameworks are required for effective management. As for most industry studies, the aim is to move scientific knowledge into action with policies that touch land use, water resources, and other decision networks.
v. **Addressing Data Gaps**: One of the pronounced challenges is limited data available where IoT is applied, particularly in remote or underdeveloped locations. This challenge can be solved using publicly available satellite data and enhanced field observations.
vi. **Validation and Monitoring**: Based on the interventions made, periodic assessment of various recharge zones using either piezometers, tracer studies, or geophysical surveys enables adaptive management to confirm their continued effectiveness
vii. **Embracing cutting-edge strategies**: New technologies such as machine learning and artificial intelligence can improve recharge zone mapping accuracy and efficiency.
viii. **Sustainability as a Priority**: Recharge projects must manage the ecological balance and need for water in a way that preserves the local ecosystem and biodiversity to the lowest degree possible.

6 Challenges in Identifying Recharge Zones

6.1 Limitations and Uncertainties

Groundwater recharge studies face the critical challenge of data availability and access, particularly in developing countries where detailed geological maps, soil characteristics, and long-term hydrological records remain elusive (Moeck et al. 2020). Ground-based data collection methods, such as piezometer measurements and borehole surveys, are typically expensive and time-consuming, leading to sparse data coverage (Uhlemann et al. 2016). Another major problem is the resolution of remote sensing data. Although the most common components of recharge potential assessments are extracted from satellite imagery and Digital Elevation Models (DEMs), their spatial and temporal resolutions might not be suitable for fine-scale assessments (Jarihani et al. 2015). Coarse-resolution DEMs might miss micro-topographical features critical to understanding infiltration patterns. Moreover, errors in input data silos, such as amounts of rainfall, soil permeability values, and land use classifications, can transmit errors through models, resulting in inaccurate predictions. Seasonal and interannual variations of important parameters, such as precipitation and evapotranspiration reinforce this uncertainty. The inherent heterogeneity of aquifers the variability of permeability and presence of fractures or karst features generate additional obstacles; there remains an inherent imprecision when mapping subsurface conditions. Finally, lack of strong field data- e.g., infiltration tests or tracer studies, etc., for testing predictive model efficacy determine recharge potential map verification (Neukum et al. 2008). These issues highlight the need for more data collection, remote sensing tools with better spatial resolutions, and field validation methods to improve recharge studies.

6.2 Environmental and Anthropogenic Constraints

However, the significant reduction in infiltration capacity is due to urbanization and land use (Patra et al. 2018). Increasingly urbanized sectors characterized by impervious areas such as roads and buildings hinder natural infiltration processes, making recharge zone identification difficult. Land use in rural areas, such as agriculture, especially soil compaction from heavy machinery—is an additional factor that reduces recharge. Climate change further complicates this issue by changing precipitation patterns, increasing temperatures, and exacerbating droughts, impacting recharge availability (Cook et al. 2018). Moreover, the potential for recharge is also compromised by higher evapotranspiration rates in arid and semi-arid areas, which makes the historical data less certain to predict the future. Environmental degradation also plays an important role, as do deforestation, overgrazing, and mining activities, lowering the quality of soil and vegetation cover and reducing infiltration rates. Industrializations, such as factories and agricultural sources like fertilizers, poisons, etc., can contaminate potential recharge areas, making it tougher to use these spaces to help fill that underground chamber. Competing land uses put extra pressure on high-potential recharge zones. Rapid urban development and intensive farming encroach on these vital areas, impairing their functionality. Regulatory and institutional challenges make recharge zone issues even more difficult. Weak enforcement of land-use policies leads to unrestrained encroachment and fragmented governance, with little coordination among agencies responsible for land, water, forests, and wildlife, which continues to pose serious challenges to effectively protecting and managing these ecosystems. Geological and topographical factors can complicate these projections:

areas of high slope or impervious bedrock naturally limit on-site recharge opportunities. In such regions, identifying suitable zones for artificial recharge is often carried out using advanced techniques, which tend to be resource-intensive. Furthermore, intricate subsurface environments, including multi-layered systems or deep aquifers, require advanced investigations, highlighting the necessity for consolidated strategies to tackle these diverse challenges.

7 Sustainable Water Supply Management

7.1 Strategies for Protecting Recharge Areas

Groundwater recharge zones can be protected and enhanced through land-use planning, solutions based on nature, engineered interventions, and pollution management (Tripp et al. 1979). The first is to set aside recharge areas as protected areas through sound land-use planning and zoning. High-potential recharge areas must not be used for activities such as mining, deforestation, and similar processes of intensive urbanization (DeMello et al. 2020). Nature-based solutions for enhancing recharge potential are more sustainable. Vertical trees and green infrastructure like rain gardens, bioswales, and wetlands encourage a natural recharge process of the hydrological cycle by enhancing infiltration and reducing surface runoff, feeding avenues for water back into the groundwater table. Artificial recharge techniques further enhance natural infiltration (Gale et al. 2002). Scientific analysis of the area identifies high-potential zones for structures such as check dams, percolation tanks, and recharge wells, which can raise groundwater levels when placed in high-potential zones. Also, pollution risk mitigation is important for rechargeable water quality. Creating buffer zones around these areas prevents contamination from industrial and agricultural activities, and the implementation of wastewater treatment and reduced waste disposal ensures that groundwater is safe and sustainable for future use. These approaches come together as a palette for protecting and improving recharge zones.

7.2 Community Participation and Policy Formulation

Strategies for Protecting and Improving Recharge Zones. Awareness is a key element to building a sense of responsibility in the communities. Awareness means educating people about recharge zones and sustainable water practices through educational campaigns and promoting engagement through workshops and training in local water conservation initiatives. Organizations with local stakeholders, including farmers, industries, and the government body, will assist in adhering to water application in the water recharge areas. Community-based water management groups supplement groundwater extraction/local recharge balancing and allow regular back-end monitoring infrastructure jobs to be easily approached, creating a bottom-up approach to sustainable management. Further, policy recommendations are key to institutionalizing recharge zone conservation. It is also important to formulate effective and comprehensive water management policies focusing on the recharge areas and offer incentives like subsidies or tax benefits to encourage adopting sustainable practices like rainwater harvesting and reducing groundwater extraction. Enhancing enforcement mechanisms can regulate land use and prevent encroachment upon these vital zones. Integrated governance is also a key aspect, as coordinated action is needed across agencies responsible for the management of water, land, and the environment.

Comprehensive planning and implementation is ensured by setting up regional and national systems to monitor and protect recharge zones. Integrating modern technology improves the effectiveness of recharge zone management (Regnery et al. 2013). Remote sensing and GIS tools allow for real-time mapping and monitoring of areas where recharge is possible, while spatial decision-support systems help policymakers make informed decisions. Sensors in piezometers and drones that map high-resolution geographical structures are some of these advanced monitoring techniques, providing time-series analytical data on groundwater levels and their interaction with current surface water systems. Hydrological models such as SWAT and MODFLOW provide insights into recharge scenarios of the future alongside climate projections with the incremental effects of land-use changes and climate variability. Utilizing advanced technologies such as machine learning and AI further increases the potential for identifying and managing recharge zones (Pourghasemi et al. 2020). Historically, machine learning algorithms can predict the genic potential zones, whereas recharging interventions can also be allotted through AI-driven decision-making frameworks. Local populations can be engaged through mobile applications and community tools to report groundwater levels and recharge structure functionality. This approach integrates crowdsourced data with official monitoring efforts to conserve recharge zones. These solutions make a holistic approach, empowering groundwater recharge and sewage management by linking education, stakeholders, policies, and technologies.

Hiware Bazar, a village located in the Ahmednagar district of Maharashtra, is a notable example of how community participation, combined with proactive policy formulation, can lead to sustainable water supply management. The region, once facing acute water scarcity and drought-like conditions, transformed itself through a community-led water conservation model. The process began in the early 1990s when the village adopted watershed management programs with full community involvement. Under the leadership of the village council and guided by local leader Popatrao Pawar, the community implemented water harvesting structures like check dams, contour trenches, and percolation tanks, all designed through participatory rural appraisal techniques (Agarwal and Narain 1999). Every household was involved in planning, execution, and monitoring of the water resource development activities, promoting a sense of ownership and accountability. To institutionalize water governance, the village established rules for water use, including the restriction of water-intensive crops and a ban on borewells. This was made possible through local-level policy formulation that aligned with the community's long-term vision of sustainability (Kerr 2002). Decisions were taken collectively in gram sabhas (village assemblies), where men and women from all social sections had equal say. Importantly, the village's efforts were supported by the state government's Jal Swarajya Yojana, a policy that aimed to decentralize water management by empowering communities. Hiware Bazar utilized these policy provisions to access funds and technical expertise, showing the critical linkage between top-down policy and bottom-up execution (Reddy and Reddy 2009). The result of this participatory model has been extraordinary. Within a decade, water availability increased, agricultural productivity improved, and the village was declared drought-free. Moreover, the model inspired other villages and became a blueprint for community-managed water systems across the country. This case illustrates how community participation, when supported by locally formulated policies and aligned with government frameworks, can ensure sustainable and equitable water supply management (Fig. 5).

Fig. 5 Community participation for the sustainable water supply management

8 Future Perspectives and Innovations

8.1 Role of Artificial Intelligence and Machine Learning in Hydrogeology

Groundwater recharge management is undergoing a paradigm shift with the emergence of Artificial Intelligence (AI) and Machine Learning (ML) techniques. Data-driven modeling represents one of the main applications, which consists of an analysis of historical and real-time data by artificial intelligence algorithms such as neural networks and random forests, allowing the identification of recharge potential zones (Elshorbagy et al. 2010). These models can simulate complex interactions between individual factors over time, such as soil permeability, rainfall patterns, and land use. They can lend explanatory support that traditional models cannot yet provide. Moreover, ML systems can identify trends in groundwater quality and changes in a water table by applying pattern recognition and anomaly detection (El-Shafeiy et al. 2023). By identifying anomalies, these systems can signal potential issues like contamination or over-extraction, enabling timely interventions. AI aids in optimization through simulation to find the effective location and design of artificial recharge structures such as check-dams and infiltration ponds (Singh et al. 2017). This maximizes aquifer recharge and conserves resources. Finally, machine learning (ML) models enable predictive analytics to assess the influence of future scenarios, including urbanization and climate change, on both recharge potential and groundwater sustainability (Banerjee et al. 2024). Such forecasts help stakeholders design nimble approaches to address changing environmental threats. In general, AI and ML are important tools for managing recharge in groundwater. These innovations empower policymakers and water resource managers to maintain sustainable and resilient groundwater systems by enhancing the accuracy of assessments, improving infiltration strategies, and predicting long-term consequences (Fig. 6).

8.2 Real-Time Monitoring with IoT Sensors

This is a new technological area in which IoT (Internet of Things) applications have been used for real-time monitoring and better resource utilization (Aderemi et al. 2021). Real-time data collection of groundwater levels, recharge rates, and water quality parameters is facilitated by IoT-enabled sensors that are placed in wells, recharge structures, and aquifers. This stream of data ensures that important changes are detected quickly. Sending big data instantly through wireless networks will allow remote monitoring of a remote or hard-to-reach location, eliminating the need for regular field visits and providing continuous monitoring. At the same time, IoT enables the creation of smart recharge management systems that automatically alert the device when there is a need to divert water at peak flow times or when the recharge efficiency declines. They improve operational efficiency and minimize

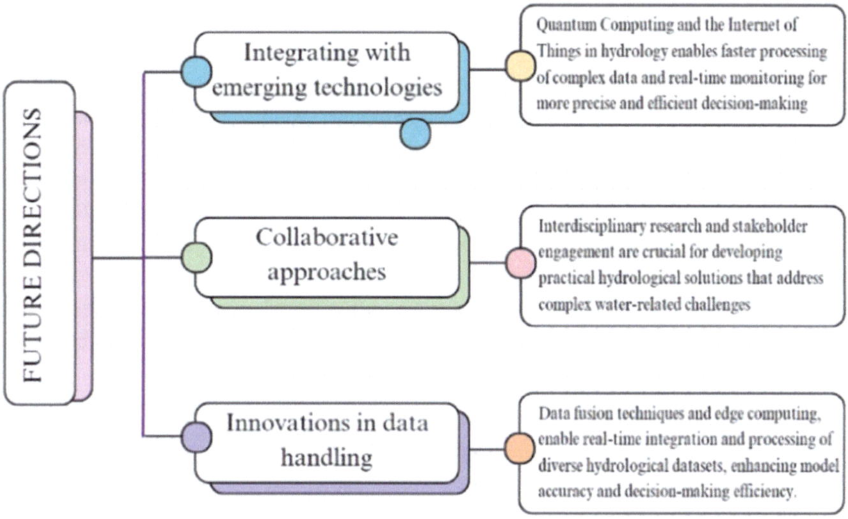

Fig. 6 Flow diagram related to future perspectives and innovations (*Source* Maity et al. 2024)

human intervention. After the IoT gadgets collect this data, it shows up in the form of centralized dashboards that let users visualize trends, how systems have performed, and what influencing decision support is found. Moreover, the availability of low-cost IoT devices enabling such advanced monitoring solutions to local communities and small organizations democratizes the management of water resources. Collectively, they represent IoT-driven innovations that lead to smarter, more sustainable strategies for sustaining groundwater recharge. Now, we have real-time tracking to make sure we get our processes done in a timely fashion while also being able to identify early risks & issues to resolve them before they snowball, affecting efficiency or effectiveness.

8.3 Climate Resilience and Adaptive Strategies

Building climate resilience in groundwater recharge involves integrating innovative models, sustainable infrastructure, and community-driven approaches. Dynamic recharge models are essential because they simulate future climate scenarios (e.g., diminished rainfall, amplified storms) to assess alterations in recharge potential and shape adaptive pathways (Davamani et al. 2024). Adjustable flow systems and modular recharge structures are examples of infrastructure designs that consider variable water availability and the need for responsiveness to different extreme weather conditions (Makropoulos et al. 2010). Solutions based on nature, from modifying land use to restoring wetlands and forests, increase groundwater recharge while offering benefits like flooding control and biodiversity protection. Adaptive management frameworks are, therefore, an essential means by which the sustainability of policy and practice can be considered over the long term through dynamically responsive adjustment to changing environmental conditions (Pahl-Wostl 2007). Community-based climate action is also a pillar, engaging communities on issues like water conservation, harvesting rain, and monitoring recharge areas to build resilience (Mukarram et al. 2023). Freshwater storage systems, like Managed Aquifer Recharge (MAR), can capture abundant water that comes with flooding conditions and store it for use during subsequent dry seasons, thereby addressing seasonal water scarcity. Finally, the integration of collaboration and policy within scientists, policymakers, and stakeholders is crucial. Finally, the integration of collaboration and policy within scientists, policymakers, and stakeholders is crucial.

9 Conclusion

This chapter investigates the evaluation of potential zones of aquifer recharge zones through a holistic, integrated framework, incorporating macro (remote sensing), meso (GIS), and micro (field observations) tools. An important conclusion from this research is that integrated methods can often more accurately identify recharge zones, as the combination of remote sensing and GIS with hydrogeological data creates an accurate framework for recharge point identification. Analytical techniques, such as weighted overlay analysis and multi-criteria decision analysis (MCDA), were emphasized for their capacity to convert varied datasets into viable insights. The chapter also addresses the influence of environmental and anthropogenic factors on recharge potential. Soil permeability, rainfall patterns, land use, and slope all affect infiltration and aquifer replenishment are key determinants. However, urbanization and climate change disrupt natural recharge processes, creating a need for focused mitigation strategies. Identify and manage our recharge zones—it will be said that technology has come to the rescue. Emerging technologies like AI, machine learning, and IoT-based systems improve the accuracy of mapping and allow real-time monitoring of recharge dynamics. Big Data Solution Harnessing ETL Tools With the help of hive, the big data solution can be handled very well to convert the traditional water source management- not perfect process- into a predictive one. It also highlights some of the challenges in implementing recharge zone mapping and preservation, including limited data availability, environmental degradation, and competing land-use demands that make it difficult to identify and preserve recharge zones. Addressing these challenges requires interdisciplinary partnerships and establishing effective governance systems focused on safeguarding critical replenishment zones. This holistic perspective helps pave the way for the sustainable management of water resources and protecting aquifer recharge areas. High-resolution recharge potential maps and modeling tools that can provide better tools for informed decision-making will aid effective groundwater management in a changing climate. This allows policymakers to make data-backed decisions in identifying the localized regions for artificial recharge projects and ensure judicious use of resources. Similarly, safeguarding high-potential recharge zones is crucial for sustaining aquifer health and ensuring long-term water supplies. The establishment of protected areas of groundwater recharge zones with the adoption of nature-based solutions (e.g., reforestation and wetland restoration) provides sustainable approaches to traditional engineering. Climate adaptation is also key; climate change makes groundwater resources less reliable. Integrating climate resilience into the recharge assessment ensures that strategies are effective amid changing environmental conditions. Manage risks through strategies with dynamic recharge modeling and flexible policy frameworks to ensure sustainable use of strategic capacity. This is further supported by active community and stakeholder involvement in groundwater management activities. With the involvement of local communities in monitoring and protecting recharge zones comes collective responsibility for the water in the region; conversely, governments, researchers, and practitioners should work together and across borders to achieve better scalability and longevity of successful strategies. Sustainable groundwater management requires these co-existing approaches to work together for the long-term viability of groundwater resources.

References

Acharya T, Kumbhakar S, Prasad R, Mondal S, Biswas A (2019) Delineation of potential groundwater recharge zones in the coastal area of north-eastern India using geoinformatics. Sustain Water Resour Manag 5:533–540

Achu AL, Ragunath R, Thomas J (2020) Mapping of groundwater recharge potential zones and identification of suitable site-specific recharge mechanisms in a tropical river basin. Earth Syst Environ 4(1):131–145

Aderemi BA, Olwal TO, Ndambuki JM, Rwanga SS (2021) A review of groundwater management models with a focus on IoT-based systems. Sustainability 14(1):148

Agarwal A, Narain S (1999) Community and household water management: the key to environmental regeneration and poverty alleviation. Centre for Science and Environment

Agarwal R, Garg PK (2016) Remote sensing and GIS based groundwater potential & recharge zones mapping using multi-criteria decision making technique. Water Resour Manag 30:243–260

Ambast SK, Tyagi NK, Raul SK (2006) Management of declining groundwater in the Trans Indo-Gangetic Plain (India): some options. Agric Water Manag 82(3):279–296

Arshad A, Zhang Z, Zhang W, Dilawar A (2020) Mapping favorable groundwater potential recharge zones using a GIS-based analytical hierarchical process and probability frequency ratio model: a case study from an agro-urban region of Pakistan. Geosci Front 11(5):1805–1819

Banerjee D, Ganguly S, Kushwaha S (2024) Forecasting future groundwater recharge from rainfall under different climate change scenarios using comparative analysis of deep learning and ensemble learning techniques. Water Resour Manag 1–19

Cook BI, Mankin JS, Anchukaitis KJ (2018) Climate change and drought: from past to future. Curr Clim Chang Rep 4:164–179

Dar T, Rai N, Bhat A (2021) Delineation of potential groundwater recharge zones using analytical hierarchy process (AHP). Geol Ecol Lands 5(4):292–307

Davamani V, John JE, Poornachandhra C, Gopalakrishnan B, Arulmani S, Parameswari E, Naidu R (2024) A critical review of climate change impacts on groundwater resources: a focus on the current status, future possibilities, and role of simulation models. Atmosphere 15(1):122

deMello K, Taniwaki RH, de Paula FR, Valente RA, Randhir TO, Macedo DR, Hughes RM (2020) Multiscale land use impacts on water quality: assessment, planning, and future perspectives in Brazil. J Environ Manage 270:110879

Döll P, Siebert S (2002) Global modeling of irrigation water requirements. Water Resour Res 38(4):8–1

Drobne S, Lisec A (2009) Multi-attribute decision analysis in GIS: weighted linear combination and ordered weighted averaging. Informatica 33(4)

El-Shafeiy E, Alsabaan M, Ibrahem MI, Elwahsh H (2023) Real-time anomaly detection for water quality sensor monitoring based on multivariate deep learning technique. Sensors 23(20):8613

Elshorbagy A, Corzo G, Srinivasulu S, Solomatine DP (2010) Experimental investigation of the predictive capabilities of data driven modeling techniques in hydrology-Part 1: concepts and methodology. Hydrol Earth Syst Sci 14(10):1931–1941

Feldman AD (2000) Hydrologic modeling system HEC-HMS: technical reference manual. US Army Corps of Engineers, Hydrologic Engineering Center

Gale I, Neumann I, Calow R, Moench M (2002) The effectiveness of artificial recharge of groundwater: a review

Graham DN, Butts MB (2005) Flexible, integrated watershed modelling with MIKE SHE. Watershed Models 849336090:245–272

Harbaugh AW (2005) MODFLOW-2005, the US geological survey modular ground-water model: the ground-water flow process, vol 6. US Department of the Interior, US Geological Survey, Reston, VA, USA

Hill MJ, Braaten R, Veitch SM, Lees BG, Sharma S (2005) Multi-criteria decision analysis in spatial decision support: the ASSESS analytic hierarchy process and the role of quantitative methods and spatially explicit analysis. Environ Model Softw 20(7):955–976

Huang IB, Keisler J, Linkov I (2011) Multi-criteria decision analysis in environmental sciences: ten years of applications and trends. Sci Total Environ 409(19):3578–3594

Idris MA, Garba ML, Kasim SA, Madabo IM, Dandago KA (2018) The role of geological structures on groundwater occurrence and flow in crystalline basement aquifers: a status review. Bayero J Pure Appl Sci 11(1):155–164

Jarihani AA, Callow JN, McVicar TR, Van Niel TG, Larsen JR (2015) Satellite-derived Digital Elevation Model (DEM) selection, preparation and correction for hydrodynamic modelling in large, low-gradient and data-sparse catchments. J Hydrol 524:489–506

Kerr J (2002) Watershed development, environmental services, and poverty alleviation in India. World Dev 30(8):1387–1400

Lesk C, Anderson W, Rigden A, Coast O, Jägermeyr J, McDermid S, Konar M (2022) Compound heat and moisture extreme impacts on global crop yields under climate change. Nat Rev Earth Environ 3(12):872–889

Maity R, Srivastava A, Sarkar S, Khan MI (2024) Revolutionizing the future of hydrological science: impact of machine learning and deep learning amidst emerging explainable AI and transfer learning. Appl Comput Geosci 24:1–16

Makropoulos CK, Butler D (2010) Distributed water infrastructure for sustainable communities. Water Resour Manag 24:2795–2816

McDonald RI, Weber K, Padowski J, Florke M, Schneider C, Green PA, Montgomery M (2014) Water on an urban planet: urbanization and the reach of urban water infrastructure. Glob Environ Chang 27:96–105

Mensah JK, Ofosu EA, Yidana SM, Akpoti K, Kabo-bah AT (2022) Integrated modeling of hydrological processes and groundwater recharge based on land use land cover, and climate changes: a systematic review. Environ Adv 8:100224

Moeck C, Grech-Cumbo N, Podgorski J, Bretzler A, Gurdak JJ, Berg M, Schirmer M (2020) A global-scale dataset of direct natural groundwater recharge rates: a review of variables, processes and relationships. Sci Total Environ 717:137042

Mukarram MMT, Kafy AA, Mukarram MMT, Rukiya QU, Almulhim AI, Das A, Chowdhury MA (2023) Perception of coastal citizens on the prospect of community-based rainwater harvesting system for sustainable water resource management. Resour Conserv Recycl 198:107196
Nandi S, Swain S (2024) Role of groundwater systems in fulfilling sustainable development goals: a focus on SDG6 and SDG13. Curr Opin Environ Sci Health 100576
Neukum C, Hotzl H, Himmelsbach T (2008) Validation of vulnerability mapping methods by field investigations and numerical modelling. Hydrogeol J 16:641–658
Nolan BT, Healy RW, Taber PE, Perkins K, Hitt KJ, Wolock DM (2007) Factors influencing groundwater recharge in the eastern United States. J Hydrol 332(1–2):187–205
Pahl-Wostl C (2007) Transitions towards adaptive management of water facing climate and global change. Water Resour Manag 21:49–62
Patra S, Sahoo S, Mishra P, Mahapatra SC (2018) Impacts of urbanization on land use/cover changes and its probable implications on local climate and groundwater level. J Urb Manag 7(2):70–84
Pourghasemi HR, Sadhasivam N, Yousefi S, Tavangar S, Nazarlou HG, Santosh M (2020) Using machine learning algorithms to map the groundwater recharge potential zones. J Environ Manage 265:110525
Reddy VR, Reddy MG (2009) How participatory is participatory irrigation management? Econ Pol Wkly 44(26–27):100–107
Regnery J, Lee J, Kitanidis P, Illangasekare T, Sharp JO, Drewes JE (2013) Integration of artificial recharge and recovery systems for impaired water sources in urban settings: overcoming current limitations and engineering challenges. Environ Eng Sci 30(8):409–420
Romano G, Dal Sasso P, Liuzzi GT, Gentile F (2015) Multi-criteria decision analysis for land suitability mapping in a rural area of Southern Italy. Land Use Policy 48:131–143
Santarosa LV, Pinto GVF, Blandon Luengas JS, Gastmans D (2024) Remote sensing to quantify potential aquifer recharge as a complementary tool for groundwater monitoring. Hydrol Sci J 1–11
Singh LK, Jha MK, Chowdary VM (2017) Multi-criteria analysis and GIS modeling for identifying prospective water harvesting and artificial recharge sites for sustainable water supply. J Clean Prod 142:1436–1456
Souissi D, Msaddek MH, Zouhri L, Chenini I, El May M, Dlala M (2018) Mapping groundwater recharge potential zones in arid region using GIS and Landsat approaches, southeast Tunisia. Hydrol Sci J 63(2):251–268
Tripp JT, Jaffe AB (1979) Preventing groundwater pollution: towards a coordinated strategy to protect critical recharge zones. Harv Envtl l Rev 3:1
Tyagi DS (1987) Domestic terms of trade and their effect on supply and demand of agricultural sector. Econ Polit Wkly A30–A36
Uhlemann S, Smith A, Chambers J, Dixon N, Dijkstra T, Haslam E, Mackay J (2016) Assessment of ground-based monitoring techniques applied to landslide investigations. Geomorphology 253:438–451
Yeh HF, Lee CH, Hsu KC, Chang PH (2009) GIS for the assessment of the groundwater recharge potential zone. Environ Geol 58:185–195
Zabihi H, Alizadeh M, Kibet LP, Karami M, Shahabi H, Ahmad A, Lee S (2019) GIS multi-criteria analysis by ordered weighted averaging (OWA): toward an integrated citrus management strategy. Sustainability 11(4):1009
Zhao J, Zhang N, Liu Z, Zhang Q, Shang C (2024) SWAT model applications: from hydrological processes to ecosystem services. Sci Total Environ 172605

Comprehensive Groundwater Recharge Assessment Using Integrated Approaches

Petros Chavula, Yusuf Umer, Elias Abdi, Agnes Uwimbabazi, Chebelo Habowa, George Bennah Mensah, Gilbert Lungu, and Fredrick Kayusi

Abstract An essential component of sustainable water resource management is groundwater recharge, a process that replenishes aquifers and ensures a reliable supply of water for various uses. This study presents a comprehensive approach that integrates hydrological, geological, and remote sensing techniques to evaluate groundwater recharge more effectively. By combining these methods, the study aims to overcome the limitations of using individual techniques in isolation and to provide a more accurate assessment of recharge rates across different landscapes. The integrated method begins with remote sensing, which is utilized to analyze land cover and soil moisture levels. This technology allows for large-scale monitoring of surface conditions that influence groundwater recharge, offering a detailed view of how different types of vegetation, land use, and soil characteristics affect water infiltration. Geological characterization follows, where the study identifies specific areas

P. Chavula (✉) · Y. Umer · E. Abdi
Africa Centre of Excellence for Climate-Smart Agriculture and Biodiversity Conservation, Haramaya University, Dire Dawa, Ethiopia
e-mail: chavulapetros@outlook.com

P. Chavula
World Agroforestry Centre, Kabulonga, Lusaka, Zambia

A. Uwimbabazi
Chair of Environment and Development, School of Natural Resources, Oliver R. Tambo Africa Research Chair Initiative (ORTARChl), The Copperbel University, Kitwe, Zambia

Department of Natural Conservation, Rwanda Polytechnic, Kitabi College, Huye, Rwanda

G. B. Mensah
Department of Legal Research, Regulatory Advisory and Management, Africa Institute for Regulatory Affairs LBG, Accra, Ghana

C. Habowa · G. Lungu
School of Natural Resources, Department of Agroforestry, The Copperbelt University, Kitwe, Zambia

F. Kayusi
Department of Environmental Sciences, Pwani University, Kilifi, Kenya

Department of Environmental Studies, Geography and Planning, Maasai Mara University, Narok, Kenya

A. Nath et al. (eds.), *Groundwater Depletion and Sustainability*, Advances in Geographic Information Science, https://doi.org/10.1007/978-3-032-09921-1_12

with favourable conditions for recharge, such as porous rock formations and fault zones that facilitate water percolation into underground aquifers. Finally, hydrological modelling is employed to estimate the actual recharge rates, taking into account factors such as precipitation, evapotranspiration, and surface runoff. This integrated approach was applied in a semi-arid region where groundwater serves as the primary water source for both human consumption and agricultural activities. Notably, the approach also uncovered previously unknown recharge zones, highlighting the potential of this method to improve groundwater management practices. The findings demonstrate that by using an integrated approach, researchers can gain a more comprehensive understanding of groundwater recharge dynamics. This approach addresses the limitations of relying on individual methods and provides more robust data for decision-making. The results of this study can serve as a foundation for developing sustainable groundwater management strategies, ensuring long-term water security for communities and ecosystems in the region.

Keywords Groundwater recharge · Integrated approaches · Hydrological modelling · Geological characterization · Remote sensing · Sustainable water resource management

1 Introduction

Groundwater recharge is a fundamental process within the Earth's hydrologic cycle, playing a crucial role in maintaining the health and sustainability of subsurface water resources (Chenini and Ben Mammou 2010). This natural phenomenon refers to the movement of water from the Earth's surface downward through various geological layers until it reaches the water table, effectively replenishing aquifers and sustaining groundwater systems (Hess and Ebi 2016; Zhang 2020). The recharge process begins when water from various sources, such as precipitation, surface water bodies, or irrigation, infiltrates the soil surface. As this water percolates through the unsaturated zone, it navigates through pores, fractures, and other openings in the soil and rock formations. The rate and volume of recharge are influenced by numerous factors, including soil type, vegetation cover, land use practices, topography, and climatic conditions. Once the infiltrating water reaches the saturated zone below the water table, it becomes part of the groundwater system, contributing to the overall volume of water stored in aquifers. This replenishment is essential for maintaining groundwater levels and ensuring the long-term availability of this vital resource for human use and ecosystem support (Fig. 1) (Fan et al. 2019; Sheehan and Fox 2020).

The significance of groundwater recharge extends beyond mere water quantity. It also plays a crucial role in maintaining water quality by diluting contaminants and supporting natural filtration processes (Roy and Sahu 2015; Yeh et al. 2016). Additionally, recharge helps to prevent land subsidence, maintain base flow in rivers and streams during dry periods, and preserve the overall balance of the hydrologic cycle. Understanding and quantifying groundwater recharge is critical for effective

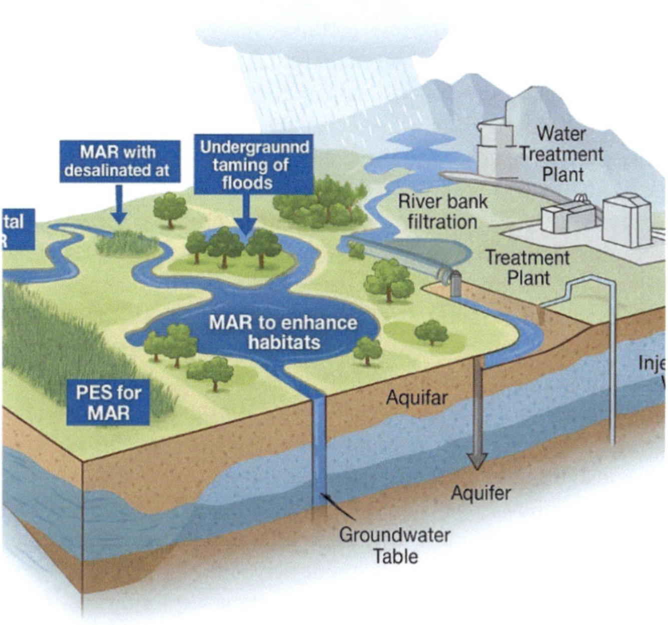

Fig. 1 Water recharge, adapted from Maven's book

water resource management, sustainable development planning, and the protection of groundwater-dependent ecosystems (Freitas et al. 2019). As such, it remains a key focus of hydrological research and water resource policy worldwide.

Traditional methods for assessing groundwater recharge often rely on simplistic approaches that fail to capture the full complexity of hydrological systems. These conventional techniques frequently include rainfall-runoff models that oversimplify intricate hydrological processes, potentially overlooking crucial factors influencing recharge rates. Point measurements of recharge rates, while providing localized data, may not accurately represent larger-scale recharge patterns across diverse landscapes (Hashemi et al. 2015). Furthermore, these methods often have limited consideration of spatial and temporal variability in recharge, failing to account for seasonal fluctuations, land use changes, and geological heterogeneity. Additional factors such as the impact of climate change on recharge patterns, interactions between surface water and groundwater systems, and the influence of vegetation and land cover on recharge processes are often inadequately addressed (Patil and Lokesh 2022). These limitations can result in inaccurate or incomplete assessments of groundwater recharge, significantly hindering the development of effective management strategies and protection measures for this vital water resource. Consequently, decision-makers may lack the comprehensive information necessary to implement sustainable groundwater policies and ensure long-term water security.

A comprehensive approach to groundwater recharge assessment is crucial for developing a thorough understanding of this vital hydrological process. Such an approach enables the accurate estimation of recharge rates and their spatial distribution across diverse landscapes, providing a more nuanced view of how water replenishes aquifers in different geological and climatic settings. It also facilitates the identification of key recharge areas and mechanisms, allowing for targeted conservation and management efforts (Girmay and Singh 2012).

Moreover, a holistic assessment methodology is essential for understanding the complex interplay between groundwater recharge and various external factors. This includes evaluating the impacts of climate change, which can alter precipitation patterns and evapotranspiration rates, significantly affecting recharge dynamics (Freitas et al. 2019; Hashemi et al. 2015). Land use changes, such as urbanization or deforestation, can also profoundly influence recharge processes, necessitating a comprehensive approach to capture these effects accurately. By adopting integrated assessment techniques, water resource managers and policymakers can gain valuable insights to inform sustainable groundwater management and protection strategies. This knowledge is critical for ensuring the long-term availability and quality of groundwater resources, which are vital for human consumption, agriculture, and ecosystem health. This chapter will delve into the cutting-edge approaches for comprehensive groundwater recharge assessment, exploring how the integration of various methodologies can overcome the limitations of traditional techniques. It will examine the combination of hydrological modelling, geological investigations, and advanced remote sensing technologies to provide a complete and more accurate picture of groundwater recharge processes (Patil and Lokesh 2022; Hosseini et al. 2019).The discussion will cover innovative methods such as isotope analysis, geophysical surveys, and satellite-based observations, which, when used in conjunction with traditional field measurements and modeling techniques, can offer unprecedented insights into recharge dynamics. By exploring these integrated approaches, the chapter aims to equip readers with a robust understanding of how to conduct thorough and reliable groundwater recharge assessments, ultimately contributing to more effective and sustainable management of this critical water resource.

2 Integrated Approaches for Groundwater Recharge Assessment

2.1 Overview of Integrated Approaches

Integrated approaches to groundwater recharge assessment represent a significant advancement in hydrological science, offering a more holistic and nuanced understanding of this complex process. These methodologies combine multiple disciplines and techniques, recognizing that recharge is not an isolated phenomenon but rather a result of intricate interactions between various environmental factors (EL-Bana

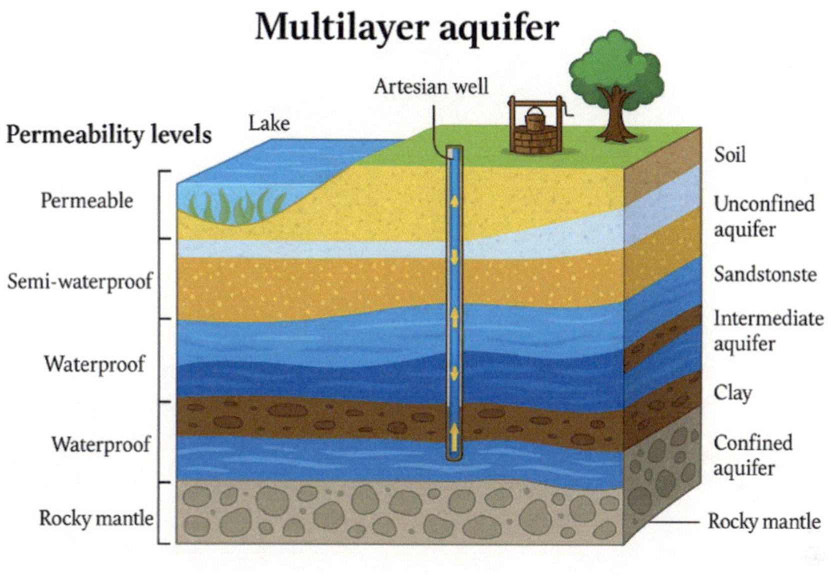

Fig. 2 Water processes, multilayer aquifer, adapted from UN water book

et al. 2024; Chung et al. 2010) (Fig. 2). At the core of these integrated approaches is the acknowledgement that groundwater recharge is influenced by a wide array of elements, each playing a crucial role in determining the quantity and quality of water that replenishes aquifers. These influencing factors can be broadly categorized into several key areas:

Hydrological Processes

The hydrological cycle forms the foundation of groundwater recharge. Precipitation, as the primary source of water input, initiates the recharge process. However, the path from rainfall to aquifer replenishment is far from straightforward (Woldearegay et al. 2024). Surface runoff, which is affected by topography, soil saturation, and rainfall intensity, can significantly impact the amount of water available for infiltration. Infiltration rates, in turn, are governed by soil properties, vegetation cover, and antecedent moisture conditions (Fig. 3) (Pande et al. 2020). Evapotranspiration further complicates this picture, as it can return a substantial portion of water to the atmosphere before it reaches the water table.

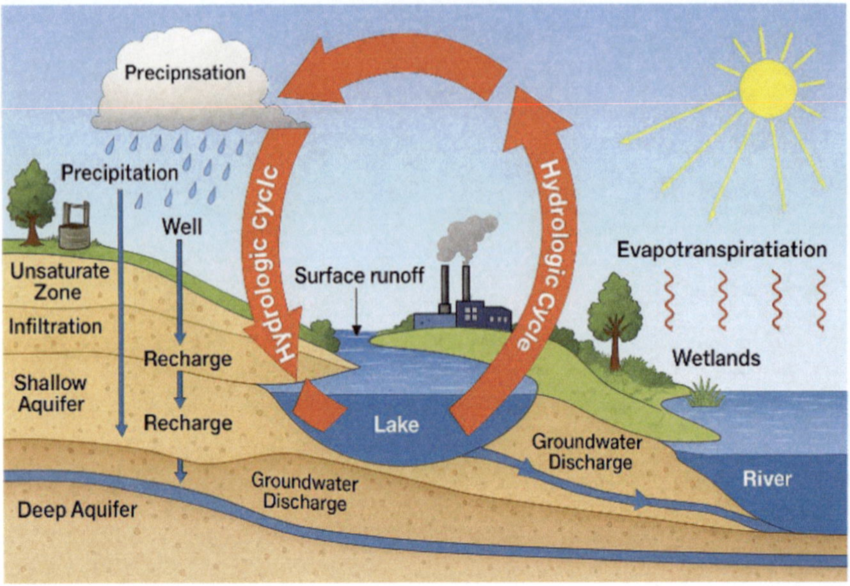

Fig. 3 Hydrological processes

Geological Characteristics

The geological framework within which recharge occurs plays a critical role in determining its rate and spatial distribution. Aquifer properties, such as porosity, permeability, and storage capacity, directly influence how much water can be absorbed and retained. The presence of confining layers or preferential flow paths can create complex recharge patterns that are not immediately apparent from surface observations (Al Atawneh et al. 2021). Soil type is another crucial factor, with variations in texture and structure affecting water retention and movement through the unsaturated zone.

Land Use and Land Cover Changes

Human activities have profound impacts on groundwater recharge processes. Urbanization, for instance, increases impervious surface area, potentially reducing infiltration and altering local hydrological cycles (Tesfaye et al. 2019; Mulat et al. 2018). Agricultural practices, including irrigation methods and crop selection, can significantly modify recharge patterns. Deforestation or afforestation can alter evapotranspiration rates and soil properties, thereby affecting the water balance and recharge dynamics (Fig. 4) (Seyum et al. 2019). Even seemingly minor changes in land management practices can have cumulative effects on groundwater systems over time.

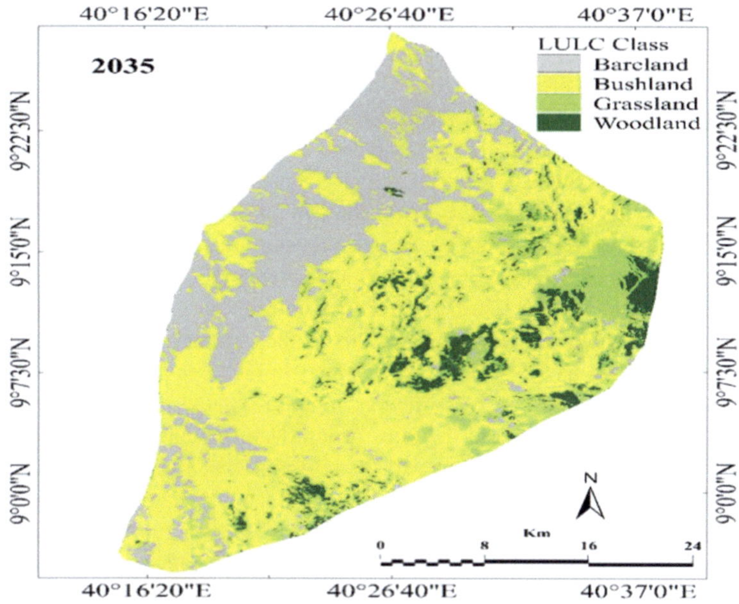

Fig. 4 Predicted land use and land cover

Climate Variability and Change

Climate exerts a profound influence on groundwater recharge, affecting both the timing and magnitude of recharge events. Short-term climatic fluctuations, such as seasonal changes in precipitation and temperature, naturally cause variability in recharge rates across regions (Holman 2006). However, long-term climate change presents more enduring and complex challenges. Shifts in precipitation regimes—including variations in the intensity, timing, and frequency of rainfall—can significantly alter groundwater recharge dynamics. Rising global temperatures further complicate this picture by increasing evapotranspiration, which can reduce the amount of water that infiltrates into the soil and ultimately reaches aquifers. In drought-prone areas, such as parts of sub-Saharan Africa or western India, increasingly frequent and intense droughts linked to climate change have already led to diminished recharge and mounting stress on groundwater systems (Nune et al. 2021).

To address the multifaceted nature of these influences, integrated approaches to groundwater recharge assessment are essential. These methods combine various tools and datasets to create a more accurate and holistic picture of recharge processes. For example, hydrological modeling, when combined with geological data, can simulate water movement through complex subsurface environments, providing insights into how water percolates through different soil and rock layers. Remote sensing technologies offer valuable data on land cover and vegetation, allowing for large-scale estimation of evapotranspiration and monitoring of changes in land use that influence recharge potential. In regions undergoing rapid land transformation, such as parts of Southeast Asia, remote sensing has been instrumental in identifying declining recharge zones due to urban sprawl.

Additionally, isotope analysis offers a means to determine the origin and age of groundwater, helping to validate recharge estimates and distinguish between recent and older recharge sources. For instance, in the Sahel, isotope techniques have been used to differentiate modern recharge from ancient fossil water. Geophysical surveys, such as electrical resistivity imaging, can also reveal underground features like fractures and porous zones that significantly influence recharge pathways—especially in geologically complex areas like fractured rock aquifers in Australia. Incorporating long-term climate data and future projections further strengthens these assessments by enabling scenario-based planning for changes in recharge patterns.

Through the synthesis of such diverse methodologies, integrated approaches improve both the reliability and depth of groundwater recharge assessments. This comprehensive understanding is critical for informed water resource management, guiding decisions on groundwater extraction, land use planning, and climate adaptation strategies (Wola 2023). Moreover, these approaches support the identification of key recharge areas, allowing for targeted conservation efforts. They also enable predictions of how changes in land use or climate may affect recharge in the future, thereby supporting proactive and adaptive management strategies. In essence, the adoption of integrated assessment techniques marks a significant advancement in our ability to sustainably manage groundwater. By accounting for the interplay of climatic, hydrological, geological, and anthropogenic factors, these approaches offer a more detailed and nuanced understanding of recharge processes, ultimately contributing to more resilient and future-ready water management practices.

2.2 *Benefits of Integrated Approaches*

Employing integrated approaches in groundwater recharge assessment yields substantial benefits by deepening our understanding of recharge mechanisms and supporting more effective resource management. These multidisciplinary methods combine tools and knowledge from hydrology, geology, climatology, remote sensing, and geographic information systems (GIS) to form a holistic perspective on groundwater recharge processes (Bae and Ryu 2015). This integration offers several practical and strategic advantages.

One of the primary benefits is the ability to capture spatial and temporal variability in recharge. For instance, in a case study conducted in the semi-arid region of Maharashtra, India, researchers utilized a combination of remote sensing, GIS, and hydrological modeling to delineate potential recharge zones. The study identified areas with high permeability and rainfall concentration as key recharge hotspots, which were not evident through traditional methods alone. This allowed local authorities to implement targeted recharge interventions, such as check dams and percolation tanks, in areas with the greatest potential for impact (Wola 2023; Bae and Ryu 2015).

Integrated approaches also enhance the accuracy of recharge estimates. In the United States, the High Plains Aquifer—one of the largest aquifer systems in the world—was assessed using a mix of hydrological models, geological characterization, and satellite data. This combination revealed significant variations in recharge rates across different soil types, with higher recharge observed in sandy regions and lower rates in clay-rich areas. By accounting for geological heterogeneity, this approach provided more precise estimations than those derived from uniform recharge assumptions (Bae and Ryu 2015; Jyrkama 2003; Pazhuparambil Jayarajan et al. 2022).

Another notable advantage is the ability to assess the effects of land use change and climate variability on groundwater systems. A study in South Australia employed remote sensing, field measurements, and climate data to evaluate recharge in a fractured rock aquifer. Researchers found that land clearing for agriculture had altered surface runoff patterns and reduced recharge potential in some areas. The integration of long-term climate data also enabled projections of how future temperature increases and changes in rainfall might influence recharge rates. These insights guided the development of adaptive water management policies for local communities (Jyrkama 2003).

Moreover, integrated assessments support the identification and prioritization of critical recharge areas for conservation. In Kenya's Central Highlands, researchers combined GIS-based mapping with community-led field surveys and hydrological data to locate vulnerable recharge zones. The resulting maps informed watershed management strategies and helped secure funding for reforestation and soil conservation initiatives that improved local groundwater levels (Pazhuparambil Jayarajan et al. 2022; Vishwakarma et al. 2021).

Ultimately, the comprehensive perspective offered by integrated approaches supports better-informed decision-making for groundwater extraction, infrastructure development, and ecosystem protection. By accounting for the interactions between climate, land use, geology, and hydrology, these methods reduce uncertainty and build a more resilient foundation for sustainable water management. Integrated approaches not only improve scientific understanding but also bridge the gap between research and policy, ensuring that groundwater resources are managed effectively for present and future generations.

Improved Accuracy and Reliability of Recharge Estimates

One of the primary benefits of integrated approaches is the substantial improvement in the accuracy and reliability of recharge estimates. By incorporating multiple data sources and methodologies, these approaches can cross-validate results and reduce the uncertainties associated with individual techniques (Holman 2006). For instance, combining physical measurements from field studies with satellite-based remote sensing data and hydrological modelling can provide a more robust estimate of recharge rates. This multi-pronged approach helps to overcome the limitations of single-method assessments, which may be subject to biases or errors inherent to specific techniques. Furthermore, integrated approaches allow for the consideration of a wider range of factors influencing recharge, including subtle interactions that might be overlooked in simpler assessments (Mulat et al. 2018; Nune et al. 2021). This comprehensive consideration of variables leads to more realistic and reliable estimates, crucial for effective water resource management and policy-making.

Enhanced Understanding of Recharge Mechanisms and Spatial Distribution

Integrated approaches facilitate a deeper understanding of the complex mechanisms driving groundwater recharge and their spatial distribution. By combining geological, hydrological, and climatological data, these methods can reveal intricate patterns of recharge that may not be apparent when using single-discipline approaches (Wola 2023; Bae and Ryu 2015). For example, the integration of high-resolution topographic data with soil moisture measurements and vegetation cover analysis can highlight preferential recharge pathways and identify critical recharge zones. This enhanced spatial understanding is particularly valuable for identifying areas of high recharge potential, which can inform conservation efforts and land use planning. It also allows for the recognition of areas where recharge may be impaired due to natural or anthropogenic factors, enabling targeted interventions to maintain or improve recharge rates.

Ability to Account for Uncertainty and Variability in Recharge Processes

Groundwater recharge is inherently variable across both time and space, influenced by a myriad of factors that can change rapidly or gradually. Integrated approaches excel in accounting for this uncertainty and variability by incorporating probabilistic methods and scenario analyses. By using ensemble modelling techniques or Bayesian approaches (Jyrkama 2003), these methods can provide a range of possible recharge estimates rather than a single value, offering a more realistic representation of the variability inherent in natural systems. This ability to quantify and communicate uncertainty is crucial for risk assessment and decision-making in water resource management. It allows stakeholders to make more informed choices, considering best-case, worst-case, and most likely scenarios in their planning processes.

Integration of Multiple Data Sources and Scales

One of the most powerful aspects of integrated approaches is their ability to seamlessly combine data from various sources and across different scales. This integration can span from highly localized field measurements to regional or even global satellite observations. For instance, point measurements of soil moisture or groundwater levels can be combined with large-scale climate models and land use maps to create a comprehensive picture of recharge dynamics. The multi-scale nature of these approaches allows for a more holistic understanding of recharge processes. Local-scale data can provide detailed insights into specific recharge mechanisms, while regional-scale analyses can reveal broader patterns and trends (Pazhuparambil Jayarajan et al. 2022). This scalability makes integrated approaches particularly valuable for addressing recharge questions at various levels of governance, from local water management districts to national or international water resource planning. Moreover, the integration of diverse data sources often leads to synergistic insights that might not be apparent when analyzing each dataset in isolation. For example, combining high-resolution satellite imagery with historical climate data and geological maps can reveal long-term changes in recharge patterns that may be linked to climate change or land use alterations (Vishwakarma et al. 2021).

Additional Benefits

Beyond the primary benefits, integrated approaches to groundwater recharge assessment offer several additional advantages that further enhance their value in hydrological research and water resource management. One significant benefit is their improved predictive capability. By accounting for the complex interactions between climate, land use, soil characteristics, and hydrogeology, these methods allow for more accurate forecasting of future recharge under varying scenarios, including those related to climate change and anthropogenic land modifications. This foresight is essential for long-term planning and adaptation strategies. Another key advantage is cost-effectiveness. Although integrated assessments may demand higher initial investments in terms of data collection and model development, they often prove more economical over time. The depth and reliability of the insights they provide can reduce the need for repeated or redundant studies, leading to better-informed and more efficient management decisions (Vishwakarma et al. 2021; Sajil Kumar et al. 2021).

Additionally, integrated approaches promote interdisciplinary collaboration by bringing together experts from fields such as hydrology, geology, remote sensing, climatology, and environmental science. This convergence fosters innovation and leads to a more holistic understanding of water systems, which is crucial for addressing the multifaceted challenges associated with groundwater management (Pazhuparambil Jayarajan et al. 2022). These approaches are also inherently adaptable. As new data become available or as analytical techniques evolve, integrated methods can be updated and refined, ensuring their continued relevance and accuracy in a changing environmental and technological landscape (Mensah et al. 2022).

In conclusion, the adoption of integrated approaches marks a significant step forward in both hydrological science and water governance (Sajil Kumar et al. 2021; Malmir et al. 2022). By offering improved precision, greater spatial and temporal insight, better uncertainty management, and the flexibility to evolve with emerging knowledge, these methodologies provide a solid foundation for sustainable groundwater management. In the context of increasing demand, climate variability, and resource degradation, integrated assessment techniques stand out as essential tools for informed and resilient water resource planning (Malmir et al. 2022).

2.3 Components of Integrated Approaches

Integrated approaches to groundwater recharge assessment typically involve a sophisticated combination of various methodologies, each contributing unique insights to create a comprehensive understanding of recharge processes (Sajil Kumar et al. 2021). These approaches leverage the strengths of different disciplines and technologies to overcome the limitations of individual methods (Fig. 5). The primary components of these integrated approaches include.

Hydrological Modelling

Hydrological modelling forms a cornerstone of integrated recharge assessment, providing a framework to simulate and analyze complex water movement processes.

Fig. 5 Components of integrated approaches

These models range from simple conceptual representations to complex, physically-based systems that account for numerous variables and their interactions. Rainfall-runoff models are commonly employed to estimate the portion of precipitation that becomes surface runoff versus the amount available for infiltration and potential recharge (Mensah et al. 2022). These models incorporate factors such as topography, soil properties, and land cover to simulate how water moves across and into the landscape. Soil moisture models are another crucial component, simulating water movement through the unsaturated zone. These models account for processes like infiltration, evapotranspiration, and percolation, providing insights into how water moves from the surface towards the water table (Malmir et al. 2022). More advanced hydrological models may integrate surface water and groundwater systems, allowing for a more comprehensive representation of the entire hydrological cycle. These coupled models can capture the dynamic interactions between rivers, lakes, and aquifers, providing a more realistic simulation of recharge processes.

Geological Geophysical Investigations and Geospatial Analysis

Understanding the geological context is essential for accurate recharge assessment, as subsurface characteristics significantly influence water movement and storage. Geological investigations involve a range of techniques to characterize aquifer properties and subsurface structures (Pazhuparambil Jayarajan et al. 2022). Aquifer characterization is a key aspect, involving the determination of properties such as porosity, permeability, and storage capacity. This may involve analyzing core samples, conducting pump tests, or using geophysical methods to estimate aquifer parameters over larger areas. Geophysical surveys, such as electrical resistivity tomography (ERT) or ground-penetrating radar (GPR), provide a non-invasive means of imaging subsurface structures. These techniques can reveal preferential flow paths, confining layers, or fracture networks that influence recharge patterns (Panda et al. 2019). Borehole logging and well data analysis offer direct insights into subsurface conditions, providing information on lithology, water levels, and aquifer properties. When combined with surface geophysical data, these methods can create detailed 3D models of subsurface hydrogeological conditions.

Remote sensing and Geographic Information Systems (GIS) have revolutionized recharge assessment by providing spatially continuous data over large areas. These technologies offer valuable information on various factors influencing recharge: Land use and land cover mapping using satellite imagery allows for the assessment of how different surface conditions affect infiltration and runoff. Changes in vegetation cover or urbanization can be tracked over time, providing insights into long-term recharge trends (Vishwakarma et al. 2021). Soil moisture estimation from satellite data, such as microwave remote sensing, offers large-scale information on near-surface water content (Fig. 6). This data can be used to validate hydrological models and identify areas of potential recharge. Digital elevation models (DEMs) derived from remote sensing provide crucial topographic information for hydrological modelling and the identification of potential recharge zones. Thermal infrared imagery can be used

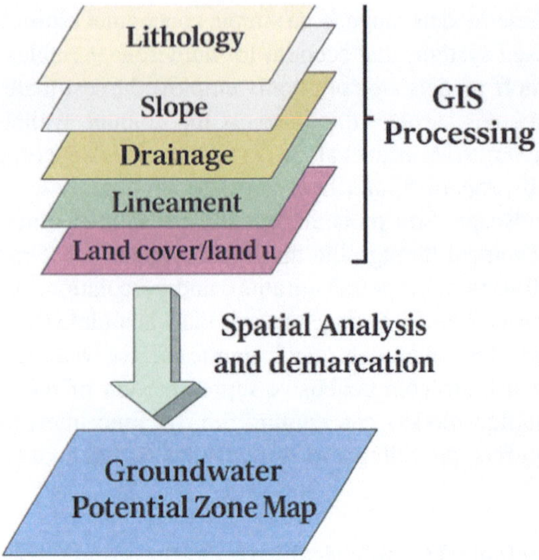

Fig. 6 Spatial analysis and demarcation

to estimate evapotranspiration rates, an important component of the water balance affecting recharge. GIS analysis integrates these various spatial datasets, allowing for complex spatial analyses and the creation of recharge potential maps. It also facilitates the integration of remote sensing data with other data sources, such as geological maps or climate data.

Field Measurements and Monitoring

While technological advancements have greatly enhanced recharge assessment capabilities, field measurements and monitoring remain crucial for ground-truthing and providing direct observations of recharge processes: Recharge rate measurements using techniques such as lysimeters or infiltration tests provide point-scale data on water movement through the unsaturated zone (Serur 2020). Groundwater level monitoring in observation wells offers direct evidence of aquifer responses to recharge events and long-term trends in groundwater storage. Soil moisture sensors deployed at various depths can track the movement of wetting fronts and provide data for calibrating soil moisture models. Stream gauging and spring flow measurements can provide information on groundwater-surface water interactions and baseflow contributions, which are often linked to recharge processes (Simiyu et al. 2022; TerAvest et al. 2015).Tracer studies, using environmental or artificial tracers, can provide insights into groundwater flow paths and residence times, helping to validate recharge estimates.

Integration of These Components

The true power of integrated approaches lies in the synergistic combination of these various components. For example, remote sensing data on land use and soil moisture can be used to parameterize and validate hydrological models. Geological information from field investigations can inform the structure of these models and help interpret geophysical survey results. Field measurements provide crucial calibration and validation data for both hydrological models and remote sensing-based estimates (Piemontese et al. 2021; Assegide et al. 2022). This integration often involves the use of advanced data assimilation techniques and machine learning algorithms to combine diverse datasets effectively. Geographic Information Systems (GIS) play a crucial role in managing and analyzing the spatial aspects of these integrated datasets. By leveraging the strengths of each component and addressing the limitations of individual methods, integrated approaches provide a more comprehensive, accurate, and spatially explicit understanding of groundwater recharge processes (Rath and Hinge 2024). This holistic view is essential for effective water resource management, especially in the face of changing climate and land use conditions.

2.4 Examples of Integrated Approaches

Several integrated approaches have been successfully applied to assess groundwater recharge, demonstrating the power and versatility of combining multiple methodologies and data sources. These approaches have significantly advanced our understanding of recharge processes and have proven invaluable in supporting the sustainable management of groundwater resources. Here are some prominent examples of integrated approaches.

Hydrological-Geological Frameworks and Integrated Hydrological Models

These frameworks combine hydrological processes with geological characteristics to provide a comprehensive view of groundwater recharge. They typically integrate surface water hydrology, unsaturated zone processes, and aquifer dynamics within a geological context. One example is the development of hydrogeological landscape units (HLUs), which divide a region into areas with similar recharge characteristics based on factors such as geology, topography, and climate (Yadav and Telila 2020; Alfieri et al. 2012). These units serve as a basis for more targeted recharge assessments and management strategies. Another application involves coupling geological models with distributed hydrological models. This approach allows for a more realistic representation of subsurface heterogeneity and its impact on recharge processes. For instance, 3D geological models derived from borehole data and geophysical surveys can be integrated with hydrological models to simulate water movement through complex aquifer systems (Kbrom et al. 2017). Hydrological-geological frameworks often incorporate GIS technologies to manage and analyze spatial data, allowing for the identification of critical recharge zones and the assessment of spatial variability in recharge rates.

These models simulate various components of the hydrological cycle simultaneously, providing a holistic view of water movement within a catchment or region. Two widely used examples are SWAT (Soil and Water Assessment Tool) and MODFLOW: SWAT is a semi-distributed watershed model that simulates surface runoff, soil water dynamics, and groundwater recharge. It incorporates land use, soil properties, and management practices, making it particularly useful for assessing the impacts of land use changes on recharge.

MODFLOW is a modular groundwater flow model that can be coupled with surface water models to simulate the entire hydrological system. When integrated with packages like UZF (Unsaturated Zone Flow), it can provide detailed simulations of recharge processes from the surface through the unsaturated zone to the water table. These integrated models often incorporate climate data, allowing for the assessment of climate change impacts on recharge. They can also be coupled with water quality modules to evaluate the transport of contaminants through recharge processes.

Remote Sensing-Based Recharge Estimation

Remote sensing technologies have revolutionized recharge assessment by providing spatially continuous data over large areas. Integrated approaches leveraging remote sensing typically combine satellite data with ground-based measurements and modelling techniques (Fig. 7).

One common application is the use of GRACE (Gravity Recovery and Climate Experiment) satellite data to estimate changes in terrestrial water storage, which can be used to infer recharge rates at regional scales (Feyisa et al. 2014; Yadav and Telila 2020). This approach often integrates GRACE data with land surface models and ground-based observations for improved accuracy. Another example is the combination of optical and radar satellite imagery to map soil moisture, vegetation cover, and land use changes. These data can be integrated into water balance models or used to parameterize hydrological models for recharge estimation. Thermal infrared imagery is also used to estimate evapotranspiration rates, a crucial component in recharge calculations. When combined with precipitation data and soil moisture estimates, it allows for the closure of the water balance and improved recharge estimates.

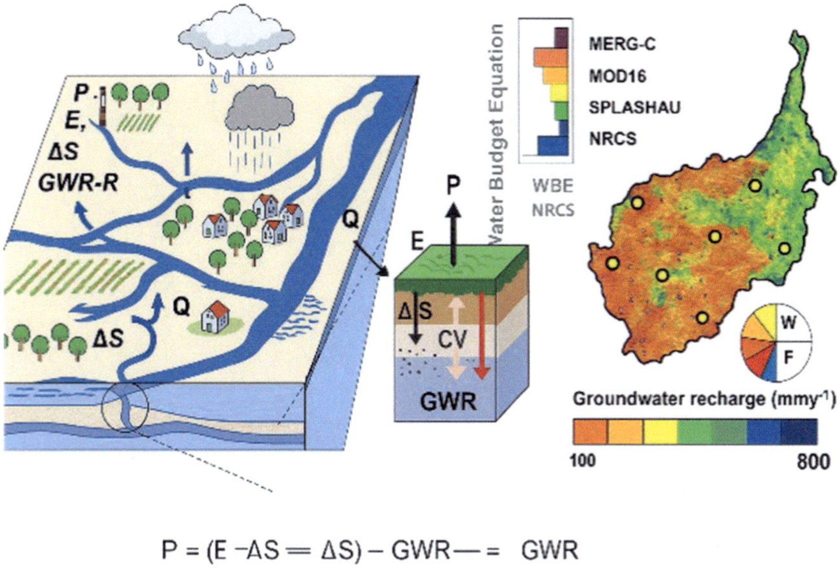

Fig. 7 Application of remote sensing

Machine Learning and Data-Driven Approaches

The increasing availability of large datasets and computational power has led to the development of machine learning and data-driven approaches for recharge assessment. These methods can identify complex patterns and relationships that might be overlooked by traditional methods.

One application is the use of artificial neural networks (ANNs) to predict recharge rates based on various input parameters such as rainfall, temperature, soil properties, and land use. These models can be trained on existing data and then used to estimate recharge in areas with limited measurements (Alfieri et al. 2012). Random forest algorithms have been employed to identify the most important factors influencing recharge in different environments, helping to prioritize data collection efforts and inform management strategies. Data-driven approaches often integrate multiple data sources (Fig. 8), including remote sensing data, field measurements, and model outputs, to create more robust and accurate recharge estimates.

Integration and Synergies

The true strength of integrated approaches to groundwater recharge assessment lies in their synergistic combination of diverse methodologies. When hydrological and geological frameworks are combined, they provide a robust structural basis for integrated models, ensuring that key parameters and boundary conditions align

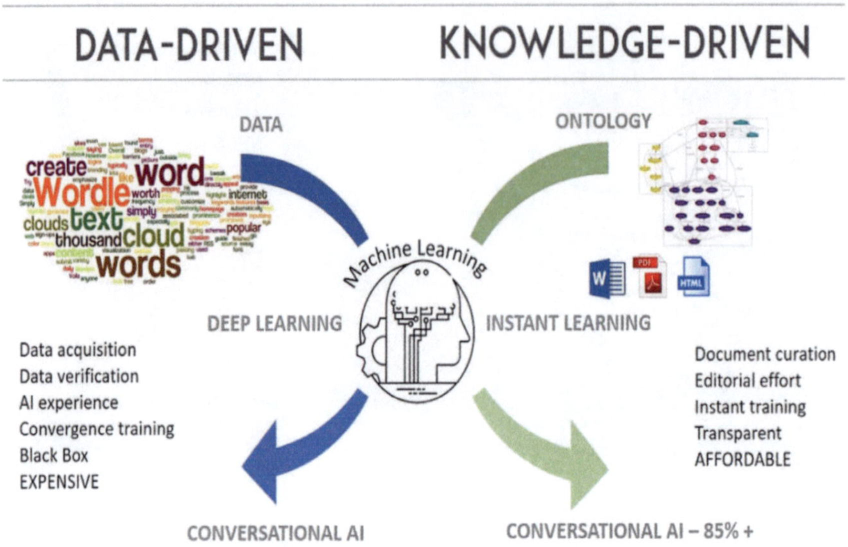

Fig. 8 Machine learning techniques

with the physical characteristics of the subsurface. Remote sensing data enhances these models by offering accurate spatial and temporal parameters, which improve the calibration and validation of simulations. In regions where data are limited, machine learning techniques can analyze outputs from hydrological models and remote sensing imagery to detect patterns and make informed predictions (Turyasingura et al. 2023). Geographic Information Systems (GIS) often serve as the unifying platform for managing, analyzing, and visualizing the various datasets and outputs from these integrated tools.

The effectiveness of these integrated approaches is evident in their ability to significantly enhance the accuracy and reliability of recharge estimates across different environmental and geographical contexts. They allow for a more refined understanding of recharge dynamics, such as the delineation of critical recharge areas, assessment of temporal variations, and evaluation of the impacts of climate change and land use alterations (Turyasingura et al. 2023; Surendra Singh Choudhara and Garg 2012). This comprehensive perspective supports more informed and evidence-based decisions in groundwater management. With better data and understanding, stakeholders can implement targeted strategies for groundwater conservation, protect vital recharge zones, and plan adaptive responses to environmental shifts (Chavula et al. 2023; Lubembe et al. 2022). As technological capabilities expand and our comprehension of hydrological processes continues to grow, integrated approaches will remain indispensable tools for addressing the increasing challenges of groundwater resource management (Chavula et al. 2023).

3 Hydrological, Geological and Geospatial Applications in Recharge Assessment

Hydrological approaches are essential components of integrated groundwater recharge assessments. These approaches focus on understanding the movement of water through the hydrologic cycle, from precipitation to infiltration and recharge. One key aspect of hydrological approaches is precipitation-runoff modelling (Sivakumar et al. 2022), which simulates the partitioning of precipitation into runoff and infiltration. This helps estimate the amount of water available for recharge. Soil moisture monitoring is another crucial aspect, as it provides insights into the amount of water stored in the soil profile and available for recharge. Evapotranspiration estimation is also important, as it accounts for water losses from the soil and vegetation, allowing for more accurate recharge estimates.

Hydrological models, such as SWAT and MODFLOW, are widely used to simulate recharge processes. These models consider factors like precipitation, soil type, land use, and topography to estimate recharge rates and spatial distribution. Hydrological approaches also involve field measurements, such as monitoring water levels and recharge rates in wells and piezometers. These measurements help calibrate and validate hydrological models, ensuring that they accurately represent real-world recharge processes (Wang and Zhang 2011). By combining these approaches, hydrologists can gain a comprehensive understanding of groundwater recharge dynamics and identify areas and mechanisms of recharge. The benefits of hydrological approaches include improved accuracy and reliability of recharge estimates, enhanced understanding of recharge mechanisms, and identification of recharge areas. However, these approaches also have limitations, such as reliance on high-quality input data and simplification of complex hydrological processes. Nevertheless, hydrological approaches remain a crucial component of integrated groundwater recharge assessments, providing valuable insights into this critical process.

Geological approaches play a vital role in understanding groundwater recharge by providing insights into the subsurface environment. Aquifer characterization is a key aspect, involving the identification of aquifer properties, such as permeability, porosity, and thickness. This information helps determine the potential for recharge and the movement of water through the aquifer (Mengistu et al. 2022; Turyasingura et al. 2023). Geological mapping and structural analysis are also essential, as they reveal the spatial distribution of recharge areas and the relationships between geological features and recharge processes. Geophysical methods, like electrical resistivity tomography (ERT) and ground-penetrating radar (GPR), offer non-invasive ways to investigate subsurface structures and identify recharge zones. Hydrogeological modelling, such as MODFLOW, integrates geological and hydrological data to simulate groundwater flow and recharge. Field investigations, including drilling and coring, provide direct observations of subsurface conditions and help calibrate geological models.

Geological approaches help identify recharge areas, such as fractures, faults, and lithological contacts and understand the role of geological structures in controlling recharge. By combining geological and hydrological data, researchers can develop a comprehensive understanding of recharge processes and improve the accuracy of recharge estimates. However, geological approaches also have limitations, such as the need for high-quality data and the complexity of interpreting geological information. Nevertheless, geological approaches are essential for understanding groundwater recharge and managing this vital resource sustainably.

Remote sensing and GIS technologies have revolutionized groundwater recharge assessment by providing spatially explicit data on land surface processes. Satellite-based sensors can monitor precipitation, soil moisture, and evapotranspiration, while GIS analysis integrates these data with geological and hydrological information. Land use/land cover analysis, using datasets like Landsat or Sentinel-2, helps identify areas with high recharge potential. Using sensors like SMOS or SMAP, soil moisture monitoring provides insights into infiltration and recharge processes.

Remote sensing-based recharge estimation models, such as the Soil Water Balance (SWB) model, use satellite data to simulate recharge rates. GIS-based mapping and analysis enable the visualization of recharge patterns and the identification of recharge hotspots. Integration of remote sensing and GIS with hydrological models, like SWAT or MODFLOW, enhances the accuracy of recharge estimates. Field measurements, such as soil moisture monitoring, validate remote sensing-based recharge estimates.

Spatially explicit data on land surface processes are essential for accurate recharge estimation, providing critical insights into the variability of infiltration and evapotranspiration across landscapes. Developing cost-effective and efficient data collection methods is paramount, enabling comprehensive monitoring without prohibitive expenses. These approaches should leverage advanced technologies such as remote sensing and automated sensor networks. The ability to monitor recharge processes at various scales, from local to regional, is crucial for understanding the complex interactions between climate, geology, and land use. Seamless integration of this multiscale data with existing hydrological and geological information is vital for a comprehensive recharge assessment (Surendra Singh Choudhara and Garg 2012). This integration allows for a holistic understanding of groundwater systems, incorporating surface water interactions, subsurface flow patterns, and aquifer characteristics (Fig. 9). By synthesizing diverse datasets, researchers can develop more accurate and nuanced models of recharge dynamics, ultimately supporting more effective and sustainable groundwater management strategies.

4 Limitations

Remote sensing techniques for groundwater recharge estimation heavily rely on the availability of high-quality satellite data, necessitating continuous improvements in sensor technology and data processing methods. The development of robust algorithms and models is crucial to accurately interpret complex spectral signatures and derive meaningful hydrological information from satellite imagery (Vishwakarma

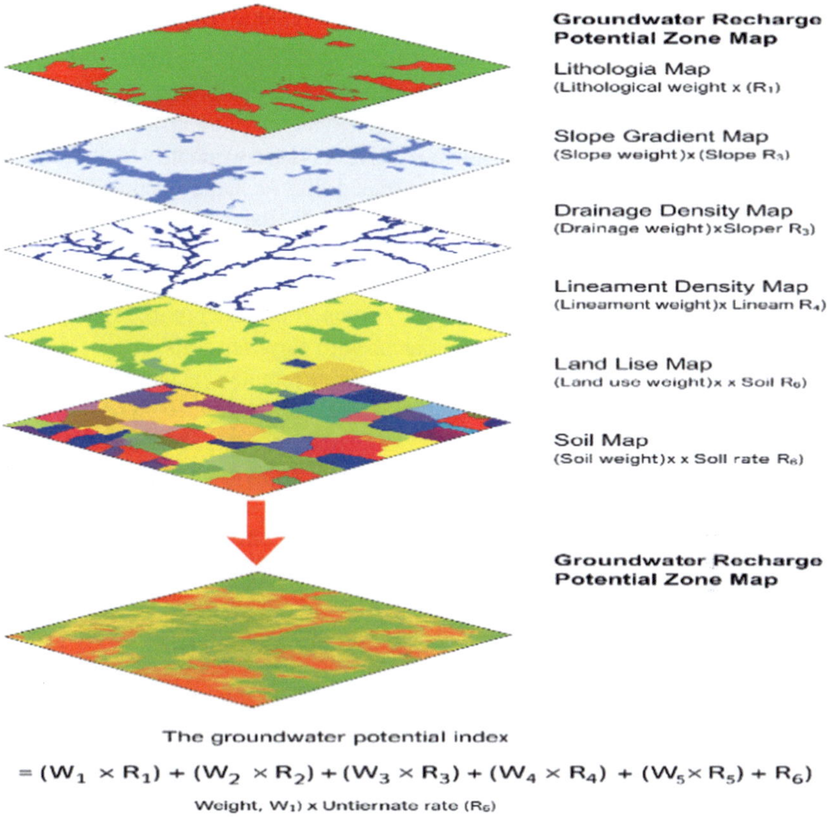

Fig. 9 GIS and Remote sensing in water resource mapping

et al. 2021). These models must account for diverse landscapes and varying atmospheric conditions to ensure reliable results. A significant challenge lies in the integration of remote sensing data with field measurements and other data sources, requiring sophisticated data fusion techniques to reconcile differences in spatial and temporal resolutions. Overcoming these integration challenges is essential for validating remote sensing-based recharge estimates and enhancing their accuracy. Despite these hurdles, the potential of remote sensing to provide cost-effective, large-scale recharge assessments makes it an invaluable tool in groundwater management, particularly in data-scarce regions where traditional monitoring methods are limited or impractical.

Despite these limitations, remote sensing and GIS technologies have significantly improved groundwater recharge assessment, enabling researchers and managers to identify recharge areas, estimate recharge rates, and develop effective management strategies.

5 Case Studies of Integrated Recharge Assessment

Several case studies demonstrate the effectiveness of integrated approaches for groundwater recharge assessment. In the semi-arid region of India, a combination of remote sensing, GIS, and hydrological modelling was used to identify recharge zones and estimate recharge rates. The study revealed that recharge occurred primarily in areas with high rainfall and permeable soils. In the United States, an integrated approach combining hydrological modelling, geological characterization, and remote sensing was used to assess recharge in the High Plains Aquifer. The study found that recharge rates varied significantly across the aquifer, with higher rates in areas with sandy soils and lower rates in areas with clay-rich soils.

In Australia, a case study used a combination of remote sensing, GIS, and field measurements to assess recharge in a fractured rock aquifer (Fig. 10). The study identified areas with high recharge potential and estimated recharge rates using a soil water balance model. These case studies demonstrate the benefits of integrated approaches, including improved accuracy and reliability of recharge estimates, identification of recharge areas, and enhanced understanding of recharge mechanisms.

These studies also highlight the importance of considering local hydrogeological conditions, land use, and climate variability when assessing groundwater recharge. By adopting an integrated approach, researchers and managers can develop effective strategies for managing groundwater resources, ensuring their sustainability for future generations. The case studies provide valuable insights into the application of integrated approaches for recharge assessment, offering lessons for similar studies in other regions.

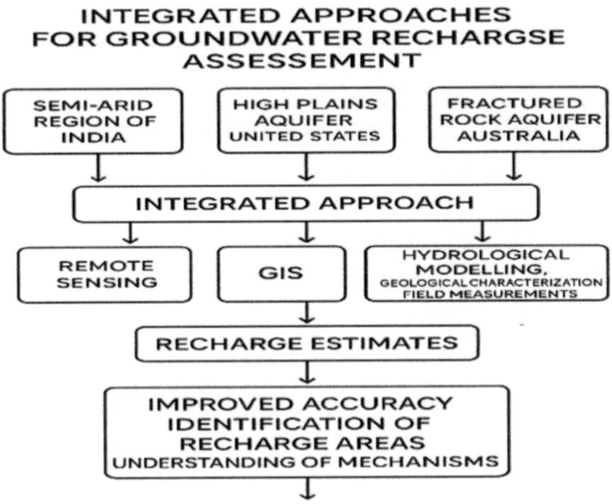

Fig. 10 Integrated recharge assessment

6 Data Integration and Analysis for Recharge Assessment

Effective data integration and analysis are crucial for comprehensive groundwater recharge assessment. This involves combining diverse data sources, such as hydrological, geological, and remote sensing data, to provide a complete understanding of recharge processes. Data integration techniques, like data fusion and machine learning algorithms, help merge data from different sources, scales, and formats. This enables the identification of patterns and relationships between variables, leading to improved recharge estimates.

Analysis of integrated data involves applying statistical and numerical models to simulate recharge processes. Techniques like regression, time series, and spatial analysis help identify key factors controlling recharge and estimate recharge rates. Data visualization tools, such as maps and graphs, facilitate the communication of complex data insights to stakeholders.

Data integration and analysis offer several benefits, including:

The study aims to significantly improve the accuracy and reliability of recharge estimates, providing water resource managers with more precise and dependable data for long-term planning and reducing uncertainty in hydrological models. This enhanced understanding of recharge mechanisms and controlling factors will offer deeper insights into the processes that facilitate groundwater replenishment, including a clearer comprehension of environmental and geological influences, as well as seasonal variations impacting recharge patterns (Chavula et al. 2023; Chavula 2022). By identifying specific recharge areas and hotspots, the research will enable precise mapping of locations where significant groundwater replenishment occurs, recognizing critical zones for prioritized protection and management, and improving spatial understanding of recharge distribution across the region. Ultimately, these findings will contribute to the development of effective groundwater management strategies, allowing for the formulation of data-driven policies for sustainable water resource utilization, the creation of targeted conservation measures for high-recharge areas, the implementation of balanced extraction practices based on recharge capacity, and design of adaptive management approaches to address changing climate and land use patterns (Freitas et al. 2019). This comprehensive approach will ensure more sustainable and efficient use of groundwater resources in the long term.

However, data integration and analysis also pose challenges, such as:

Handling large and diverse datasets presents a significant challenge in groundwater recharge analysis, requiring robust data management systems and advanced computational tools. Researchers must address data quality and uncertainty issues through rigorous validation processes and error assessment techniques to ensure reliable results. Selecting appropriate analysis techniques is crucial, as different methods may be suitable for various geological settings and data types (EL-Bana et al. 2024). Ensuring data consistency and compatibility across different sources and formats is essential for meaningful integration and interpretation. Temporal and spatial variability in recharge processes necessitates careful consideration when aggregating or interpolating data. Researchers must also account for potential biases in historical records and reconcile discrepancies between various data collection methods. Effective visualization techniques are vital for communicating complex spatial and temporal patterns in recharge data. Collaboration between hydrogeologists, statisticians, and data scientists can enhance the robustness of analysis approaches. Regular updates to methodologies are necessary to incorporate new scientific understanding and technological advancements (Shekar and Mathew 2023). Finally, transparent documentation of data processing steps and assumptions is critical for reproducibility and peer review of research findings (Umer et al. 2024; Ali et al. 2024).

By overcoming these challenges, researchers and managers can harness the power of integrated data analysis to advance groundwater recharge assessment and ensure sustainable management of this vital resource.

7 Uncertainty Analysis and Validation of Recharge Assessment

Uncertainty analysis and validation are essential components of groundwater recharge assessment, ensuring the reliability and accuracy of recharge estimates. Uncertainty arises from various sources, including data quality, model assumptions, and parameter uncertainty. Techniques like sensitivity analysis, Monte Carlo simulations, and Bayesian inference help quantify uncertainty in recharge estimates.

Validation involves comparing recharge estimates with independent field measurements, such as groundwater levels, soil moisture, and streamflow (Yeh et al. 2016). This ensures that recharge assessments accurately represent real-world processes. Validation data can also be used to calibrate recharge models, improving their performance and reducing uncertainty.

7.1 Uncertainty Analysis and Validation Offer Several Benefits

The quantification of uncertainty in recharge estimates is a critical aspect of groundwater research, providing essential context for the interpretation and application of results. By systematically identifying sources of uncertainty, such as measurement errors, model assumptions, and natural variability, researchers can better understand the limitations of their estimates (Chenini and Ben Mammou 2010). This process leads to improved accuracy and reliability of recharge assessments, as potential biases and errors are explicitly accounted for and minimized where possible. Advanced statistical techniques and sensitivity analyses play a crucial role in quantifying the range of plausible recharge values, offering decision-makers a more nuanced understanding of the data. Enhanced confidence in recharge estimates results from this rigorous approach, enabling water resource managers to make more informed decisions based on a clear understanding of the associated uncertainties (Yusuf et al. 2025; Mume et al. 2024). Furthermore, communicating these uncertainties effectively to stakeholders fosters transparency and realism in water management strategies. As methodologies for uncertainty quantification continue to evolve, incorporating new data sources and analytical tools, the reliability of recharge estimates is expected to improve further, contributing to more sustainable and resilient groundwater management practices (Fig. 11).

7.2 Uncertainty Analysis and Validation also Pose Challenges

The limited availability of validation data poses a significant challenge in groundwater recharge studies, often hindering the thorough assessment of model accuracy. This scarcity of benchmark data necessitates innovative approaches to model validation and calibration. The complexity of uncertainty analysis techniques further compounds these challenges (Alam et al. 2022), requiring advanced statistical skills and computational resources to implement effectively. Researchers must navigate sophisticated methodologies to quantify and communicate uncertainties in recharge estimates reliably. There is a pressing need for robust and reliable recharge models that can account for diverse geological settings, climate variability, and anthropogenic influences (Chavula and Turyasingura 2021). These models must balance complexity with practicality, providing accurate results while remaining accessible to water resource managers. Developing such models demands interdisciplinary collaboration and continuous refinement based on emerging data and an improved understanding of hydrological processes. Ultimately, addressing these challenges is crucial for enhancing the credibility and applicability of recharge estimates in sustainable groundwater management. By addressing these challenges, researchers and managers can develop robust and reliable recharge assessments, supporting informed decision-making and sustainable groundwater management (Vishwakarma et al. 2021). Uncertainty analysis and validation are critical components of comprehensive recharge assessments, ensuring that recharge estimates are accurate, reliable, and useful for managing this vital resource.

8 Conclusion

This study has demonstrated the importance of using integrated approaches for a comprehensive assessment of groundwater recharge. By combining hydrological modelling, remote sensing, and field-based measurements, we have provided a more accurate and holistic understanding of groundwater recharge dynamics. Our results highlight the significant spatial and temporal variability in recharge rates, driven by both natural factors such as precipitation patterns, soil characteristics, and land

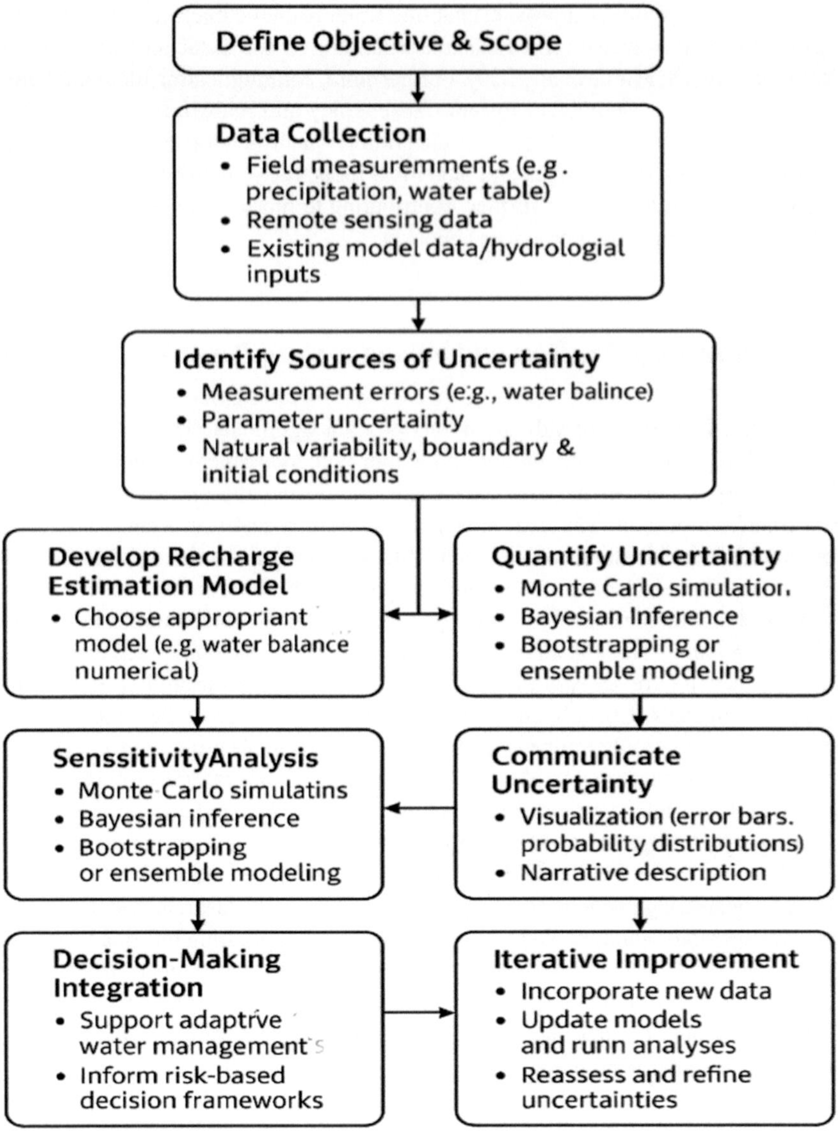

Fig. 11 Flow diagram of assessing uncertainty with validation

cover, as well as anthropogenic influences like land-use changes and water extraction practices. The integrated methodology employed in this study offers a robust framework for addressing the complexities of groundwater recharge estimation. It not only improves the precision of recharge estimates but also enhances the predictive capabilities for managing groundwater resources sustainably. This approach can be particularly useful for regions experiencing water scarcity or high variability in recharge rates due to climate change and human activities. Furthermore, the insights gained from this assessment underscore the need for adopting multidisciplinary approaches in groundwater management. Future research should focus on refining these methods to address specific local and regional challenges, including the impacts of climate change, increased urbanization, and agricultural intensification on groundwater recharge. The combination of these strategies will provide policymakers and stakeholders with critical information for developing effective groundwater management policies that ensure long-term water security. In conclusion, integrated approaches to groundwater recharge assessment are essential for capturing the complexity of the process and ensuring sustainable water management practices in an era of growing environmental pressures.

References

Abdi E, Mwewa T, Adame AA, Umer Y, Kayusi F (2025) Promoting a human-rights-centered just energy transition for communities in Zambia's mineral resource areas: a critical review of policies, challenges, and pathways

Al Atawneh D, Cartwright N, Bertone E (2021) Climate change and its impact on the projected values of groundwater recharge: a review. J Hydrol 601:126602

Alam F et al (2022) Identification of potential natural aquifer recharge sites in Islamabad, Pakistan, by integrating GIS and RS techniques. Remote Sens 14(23):6051

Alfieri L, Salamon P, Pappenberger F, Wetterhall F, Thielen J (2012) Operational early warning systems for water-related hazards in Europe. Environ Sci Policy 21:35–49. https://doi.org/10.1016/j.envsci.2012.01.008

Ali EA et al (2024) Monitoring the evolution of a wetland pattern near Addis Ababa with the use of Landsat data (1986–2019). Nov Geod 4(1):166

Assegide E, Alamirew T, Bayabil H, Dile YT, Tessema B, Zeleke G (2022) Impacts of surface water quality in the Awash River Basin, Ethiopia: a systematic review. Front Water 3:790900

Bae J, Ryu Y (2015) Land use and land cover changes explain spatial and temporal variations of the soil organic carbon stocks in a constructed urban park. Landsc Urban Plan 136:57–67

Chavula P (2022) Effects of climate change on water paucity. Int J Ground Sediment Water 15:959–971. https://doi.org/10.5281/zenodo.5835190

Chavula P, Turyasingura B (2021) Review on climatic change upshot on livestock husbandry in Ethiopia. Reports 1(4):47–53. https://doi.org/10.11648/j.reports.20210104.14

Chavula P, Shepande C, Feyissa S, Sileshi M (2023) Factors influencing climate-smart agriculture practices adoption and crop productivity among smallholder farmers in Nyimba District, Zambia

Chenini I, Ben Mammou A (2010) Groundwater recharge study in arid region: an approach using GIS techniques and numerical modelling. Comput Geosci 36(6):801–817

Chung I-M, Kim N-W, Lee J, Sophocleous M (2010) Assessing distributed groundwater recharge rate using integrated surface water-groundwater modelling: application to Mihocheon watershed, South Korea. Hydrogeol J 18(5):1253

1Asst. Surendra Singh Choudhara DSKG, Dr. Garg PK (2012) Mapping of agriculture drought using remote sensing and GIS. Int J Sci Eng Technol 1(4):149–157

Dersseh MG et al (2019) Potential of water hyacinth infestation on lake Tana, Ethiopia: a prediction using a GIS-based multi-criteria technique. Water 11(9):1921

EL-Bana EMM, Alogayell HM, Sheta MH, Abdelfattah M (2024) An integrated remote sensing and GIS-based technique for mapping groundwater recharge zones: a case study of SW Riyadh, Central Saudi Arabia. Hydrology 11(3):38

Edmondson JL, Davies ZG, McCormack SA, Gaston KJ, Leake JR (2014) Land-cover effects on soil organic carbon stocks in a European city. Sci Total Environ 472:444–453

Fan L, Li J, Pan Y, Wang S, Yan C, Yao D (2019) Research and application of smart grid early warning decision platform based on big data analysis. In: 2019 4th International conference on intelligent green building and smart grid (IGBSG), pp 645–648

Feyisa GL, Meilby H, Fensholt R, Proud SR (2014) Automated water extraction index: a new technique for surface water mapping using Landsat imagery. Remote Sens Environ 140:23–35

Freitas L, Afonso MJ, Pereira AJSC, Delerue-Matos C, Chaminé HI (2019) Assessment of sustainability of groundwater in urban areas (Porto, NW Portugal): a GIS mapping approach to evaluate vulnerability, infiltration and recharge. Environ Earth Sci 78(5):140

Girmay G, Singh BR (2012) "Changes in soil organic carbon stocks and soil quality: land-use system effects in northern Ethiopia. Acta Agric Scand Sect B-Soil Plant Sci 62(6):519–530

Hashemi H, Uvo CB, Berndtsson R (2015) Coupled modeling approach to assess climate change impacts on groundwater recharge and adaptation in arid areas. Hydrol Earth Syst Sci 19(10):4165–4181

Hess JJ, Ebi KL (2016) Iterative management of heat early warning systems in a changing climate. Ann N Y Acad Sci 1382(1):21–30

Holman IP (2006) Climate change impacts on groundwater recharge-uncertainty, shortcomings, and the way forward? Hydrogeol J 14:637–647

Hosseini SM, Parizi E, Ataie-Ashtiani B, Simmons CT (2019) Assessment of sustainable groundwater resources management using integrated environmental index: case studies across Iran. Sci Total Environ 676:792–810

Jyrkama MI (2003) A methodology for estimating groundwater recharge

Kbrom AG, Abebe FB, Mehari G (2017) Towards optimal irrigation water abstraction from Haramaya dry lake basin. Ethiop J Environ Stud Manag 10(9):1221–1233

Lubembe SI, Turyasingura B, Chavula P (2022) Reflection on impacts of climate change on fisheries and aquaculture: sub-Sahara Africa. Weather 6(10):62–72

Malmir M, Javadi S, Moridi A, Randhir T, Saatsaz M (2022) Integrated groundwater management using a comprehensive conceptual framework. J Hydrol 605:127363

Mengistu TD, Chung IM, Kim MG, Chang SW, Lee JE (2022) Impacts and implications of land use land cover dynamics on groundwater recharge and surface runoff in East African Watershed. Water (Switzerland) 14(13). https://doi.org/10.3390/w14132068

Mensah JK, Ofosu EA, Yidana SM, Akpoti K, Kabo-bah AT (2022) Integrated modeling of hydrological processes and groundwater recharge based on land use land cover, and climate changes: a systematic review. Environ Adv 8:100224

Mulat Y, Kibret K, Bedadi B, Mohammed M (2018) Soil organic carbon stock under different land use types in Kersa Sub Watershed, Eastern Ethiopia. African J Agric Res 13(24):1248–1256

Mume AA et al (2024) Impact of climate change on the environment: a synthesis study. Asian J Res Agric For 10:86–96

Nune R, George BA, Western AW, Garg KK, Dixit S, Ragab R (2021) A comprehensive assessment framework for attributing trends in streamflow and groundwater storage to climatic and anthropogenic changes: a case study in the typical semi-arid catchments of southern India. Hydrol Process 35(8):e14305

Panda B et al (2019) An integrated novel approach to understand the process of groundwater recharge in mountain and riparian zone aquifer system of Tamil Nadu, India. Aquat Geochem 25:137–159

Pande CB, Moharir KN, Singh SK, Varade AM (2020) An integrated approach to delineate the groundwater potential zones in Devdari watershed area of Akola district, Maharashtra, Central India. Environ Dev Sustain 22:4867–4887

Patil VBB, Lokesh KN (2022) Integrated approach for groundwater recharge assessment—a review. Groundw Water Qual Hydraul Water Resour Coast Eng 93–103

Pazhuparambil Jayarajan SK, Schneider M, Lakshmanan E (2022) Estimation of natural groundwater recharge in Chennai River basin using multiple approaches. Hydrol Sci J 67(8):1165–1184

Piemontese L, Kamugisha RN, Tukahirwa JMB, Tengberg A, Pedde S, Jaramillo F (2021) Barriers to scaling sustainable land and water management in Uganda: a cross-scale archetype approach. Ecol Soc 26(3)

Rath S, Hinge G (2024) Groundwater sustainability mapping for managed aquifer recharge in Dwarkeswar River basin: integration of watershed modeling, multi-criteria decision analysis, and constraint mapping. Groundw Sustain Dev 26:101279

Roy S, Sahu AS (2015) Investigation for potential groundwater recharge area over the Kunur river basin, Eastern India: an integrated approach with geosciences. J Geomatics 9(2):165–177

Sajil Kumar PJ, Schneider M, Elango L (2021) The state-of-the-art estimation of groundwater recharge and water balance with a special emphasis on India: a critical review. Sustainability 14(1):340

Serur AB (2020) Modeling blue and green water resources availability at the basin and sub-basin level under changing climate in the Weyb River basin in Ethiopia. Sci African 7:e00299. https://doi.org/10.1016/j.sciaf.2020.e00299

Seyum S, Taddese G, Mebrate T (2019) Land use land cover changes on soil carbon stock in the Weshem Watershed, Ethiopia. For Res Eng Int J 3(1):24–30

Sheehan MC, Fox MA (2020) Early warnings: the lessons of COVID-19 for public health climate preparedness. Int J Heal Serv 50(3):264–270

Shekar PR, Mathew A (2023) Assessing groundwater potential zones and artificial recharge sites in the monsoon-fed Murredu river basin, India: an integrated approach using GIS, AHP, and Fuzzy-AHP. Groundw Sustain Dev 23:100994

Simiyu BM, Amukhuma HS, Sitoki L, Okello W, Kurmayer R (2022) Interannual variability of water quality conditions in the Nyanza Gulf of Lake Victoria, Kenya. J Great Lakes Res 48(1):97–109

Sivakumar R, Prasanth BRSV, Ramaraj M (2022) An empirical approach for deriving specific inland water quality parameters from high spatio-spectral resolution image. Wetl Ecol Manag 30(2):405–422

TerAvest D, Carpenter-Boggs L, Thierfelder C, Reganold JP (2015) Crop production and soil water management in conservation agriculture, no-till, and conventional tillage systems in Malawi. Agric Ecosyst Environ 212:285–296. https://doi.org/10.1016/j.agee.2015.07.011

Tesfaye MA, Gardi O, Bekele T, Blaser J (2019) Temporal variation of ecosystem carbon pools along altitudinal gradient and slope: the case of Chilimo dry afromontane natural forest, Central Highlands of Ethiopia. J Ecol Environ 43(1):17

Tsehaye MT, Assayie AAY, Besha ATY, Tufa RAY, Gebreyohannes AYY (2022) Membrane technology for water and wastewater treatment in Ethiopia: current status and future prospects. J Membr Sci Res 8(1)

Turyasingura B et al (2023) A review of the effects of climate change on water resources in Sub-Saharan Africa. African J Clim Chang Resour Sustain 2(1):84–101

Turyasingura B et al (2023) Progressive efforts in the implementation of integrated water resources management (IWRM) in Uganda. In: Disaster risk reduction in agriculture. Springer, pp 543–558

Umer Y et al (2024) Small-scale irrigation farming as a climate-smart agriculture practice; its adoption and impact on food security for Ethiopian smallholder farmers: a review. Asian Res J Curr Sci 6(1):163–180

Vishwakarma A, Goswami A, Pradhan B (2021) Prioritization of sites for Managed Aquifer Recharge in a semi-arid environment in western India using GIS-Based multicriteria evaluation strategy. Groundw Sustain Dev 12:100501

Wang S, Zhang Z (2011) Effects of climate change on water resources in China. Clim Res 47(1–2):77–82. https://doi.org/10.3354/cr00965

Wola AW (2023) Review on the above and below ground biomass and carbon stock estimation options for trees and forest resources: within and around Mago National Park, Southern Region of Ethiopia. South Reg Ethiop 24

Woldearegay K et al (2024) Watershed management, groundwater recharge and drought resilience: an integrated approach to adapt to rainfall variability in northern Ethiopia. Int Soil Water Conserv Res 12(3):663–683

Yadav BK, Telila FK (2020) Delineating potential groundwater recharge zones in the Haramaya Watershed of Ethiopia using GIS based Multi-criteria Assessment. Main Spons. This issue NABARD

Yeh H-F, Cheng Y-S, Lin H-I, Lee C-H (2016) Mapping groundwater recharge potential zone using a GIS approach in Hualian River, Taiwan. Sustain Environ Res 26(1):33–43

Yusuf U, Chavula P, Abdumalik M, Adame AA, Ahmed S, Eliyas A (2025) Determinants of household-level food and nutritional security in urban vegetable producers: a case study. Nov Geod 5(1):285

Zhang W (2020) Geological disaster monitoring and early warning system based on big data analysis. Arab J Geosci 13(18):946

Modelling of Groundwater Condition for Developing the Aquifer System

Priya, Namratha Bhandari, and Santosh L. Gaonkar

Abstract The variables encompassing climate change, demographic expansion, and unsustainable practices of water extraction render the management and preservation of freshwater resources an imperative global concern. This chapter investigates the advancements and requirements in the development of analytical tools for groundwater management, offering a thorough examination of the methodologies for identifying strategies for aquifer exploitation and conservation. This chapter explains the key unanswered scientific questions that necessitate groundwater modelling, summarizes the current state of modelling efforts, and provides a thorough framework for comprehending the crucial role of groundwater within the Earth's system by combining hydrological, climatic, and socioeconomic models. This chapter accentuates the importance of policy evaluation tools that incorporate contemporary simulation methodologies and underscores the development of hydrologic economic models that effectively represent water systems in conjunction with human behaviours. There exists an integration of sophisticated modelling techniques alongside pertinent background information essential for global modelling concerning groundwater processes and hydrogeology. This chapter sets the standards for the crucial next steps that the community has to take in order to address and lessen the long-term, systemic effects of groundwater resource degradation and depletion.

Keywords Climate change · Groundwater management · Groundwater modelling · Subsurface water · Aquifer replenishment

Priya · B. Namratha
Department of Chemistry, The Yenepoya Institute of Arts, Science, Commerce and Management, Yenepoya (Deemed to be University), Mangaluru, Karnataka, India

S. L. Gaonkar (✉)
Department of Chemistry, Manipal Institute of Technolgy, Manipal Academy of Higher Education, Manipal, Karnataka, India
e-mail: sl.gaonkar@manipal.edu

A. Nath et al. (eds.), *Groundwater Depletion and Sustainability*, Advances in Geographic Information Science, https://doi.org/10.1007/978-3-032-09921-1_13

1 Introduction

A vital component of the hydrological cycle, groundwater is essential to maintaining the balance of ecological changes. Water, as a finite natural resource, is indispensable for sustaining life and facilitating the vast majority of economic endeavours (Hoff 2009). Groundwater serves as a key factor in the adaptability and resilience of humans to extreme and unforeseen global environmental shifts, particularly as surface water systems increasingly exhibit unsustainability in response to population growth and climatic changes. It is irreplaceable and non-expandable through mere human intent, rendering it particularly vulnerable and susceptible to successive utilizations (Scanlon et al. 2023a; Loucks et al. 2017). On a global scale, groundwater-dependent ecosystems are confronted by challenges such as coastal aquifer salinization, industrial and urban contamination and mineral-origin pollution contamination, resource fragilities, and considerations regarding long-term sustainability (Gorelick and Zheng 2015a). Over one-third of the water utilized globally is sourced from underground reserves. Groundwater is crucial for preserving the natural balance of the planet and meeting societal demands for food production and drinkable water in dry and semiarid mid-latitude areas that do not receive enough surface water from rivers and reservoirs. Groundwater demand is rapidly increasing in line with population expansion, and climate change increases water resource stress and the probability of extreme drought events (Durán-Sánchez et al. 2020).

Groundwater represents a crucial freshwater resource that satisfies the needs of a significant proportion of the Earth's populace, various industries, and ecosystems. India, being an agriculturally dependent nation, heavily relies on groundwater for purposes such as drinking, agriculture, and irrigation, among others (Suhag 2016) This nation accounts for over one-fourth of the total global groundwater consumption, approximately 230 km^3 per year, thereby supporting roughly 85%, 66%, and 50% of rural drinking, irrigation, urban drinking, and industrial requirements, respectively (Amanambu et al. 2020).

Water security is severely hampered by the demands that climate change, population increase, and human activities are putting on the world's water resources. Groundwater resources face threats from over extraction and contamination, particularly in non-renewable aquifers with limited recharge capacities, which consequently leads to issues such as land subsidence, saltwater intrusion, and ecological degradation (Scanlon et al. 2023b). Changes in water storage are caused by a variety of factors, but human activities—particularly irrigation—are the main cause of these changes. With rivers in most of the regions already been exploited worldwide and instances of groundwater over usage and extensive contamination of ground water is prevalent in numerous areas, human demands and climatic extremes have already placed significant strain on many water sources, with readily accessible supplies being extensively exploited (Scanlon et al. 2023c) Integrated Water Resources Management (IWRM), which promotes a coordinated approach across many sectors to boost social and economic wellbeing without compromising ecological integrity, is one strategy that advocates for sustainable water resource management in response to the growing global water requirements (Al Atawneh et al. 2021).

Increasingly, groundwater resources play more significant role in supplying freshwater for the all the activities (Fig. 1). To effectively address these challenges, the application of contemporary modelling tools in groundwater management is deemed essential (Gorelick and Zheng 2015b) Over the past 20 years, the total water storage system has shown declining, steady, and increasing patterns in all geographical regions, according to Gravity Recovery and Climate Experiment (GRACE) satellites (Wu et al. 2020). In order to investigate important topics ranging from pollution cleanup to agricultural irrigation management, groundwater management modelling has focused on combining simulation and optimization methodologies throughout the past 50 years. Due to pressures from around the world, these models must be continuously improved for groundwater management to be sustainable. This phenomenon has precipitated a global challenge regarding the sustainability of groundwater resources, wherein trans-boundary disputes and environmental hazards emerge from the utilization and shared nature of aquifers (Wu et al. 2020).

A new generation of water table protection plans that concentrate on worldwide consequences of aquifer storage dynamics and land subsidence has been made possible by the complex issues currently faced by water resource planners. The resilience of water resources can be enhanced through the diversification of management strategies (Scanlon et al. 2023b). These tactics include grey solutions, which include enhancing storage in surface reservoirs and depleted aquifers, facilitating water transportation, and augmenting supply through techniques like desalination and wastewater reuse, as well as green solutions, which include the preservation of forests and wetlands (Mendieta-Mendoza et al. 2021).

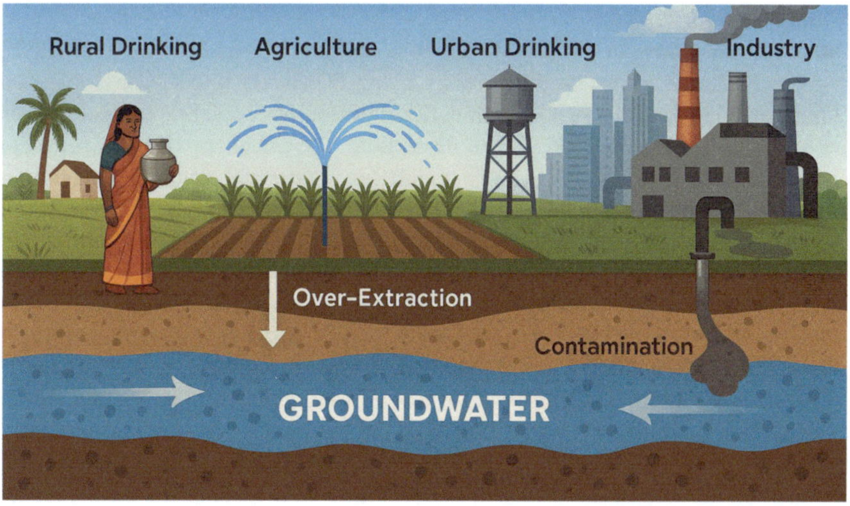

Fig. 1 Ground water-its uses

A thorough understanding of the implications of climatic shift on groundwater recharge (GWR) is imperative for the effective strategic planning and administration of groundwater supplies. The advancement of analytical tools for groundwater management is directed towards the identification of aquifer exploitation and preservation strategies that adhere to organizational, fiscal, ecological, and legal parameters (Gorelick and Zheng 2015b). These models make it easier to quantitatively evaluate the effects of suggested policy tools, such as management mechañisms, quotas, taxation, and hydrological rights policies (Hora 2022).

The scope of research has increased to provide governance and policy evaluation methods that integrate multi-agent system simulation methods with contemporary simulation techniques and a variety of optimization methodologies. Hydrologic economic models have depicted human behavior patterns and water systems. For groundwater resources to be produced sustainably or managed to reduce short-term risks, a thorough grasp of the complex relationships between hydrologic responses and human activity is essential (Srinivasan et al. 2017) Aquatic science and water resource management are falling behind in tackling the urgent need to mitigate and resolve systemic long-term consequences associated with the degradation of subsurface water reserves, despite the fact that aquifer simulation and management tools are available and are gradually being improved to accommodate more complex dynamics (Salvo 2023).

2 Effects of Climate Variability on Groundwater

2.1 Climatic Variables Affecting Groundwater

One important source of water for both public consumption and commercial activity is groundwater. This reserve is currently being negatively impacted by pollution, climate change, and intense abstraction. Societal structures and biophysical systems are about to be impacted by climate disruption. The effects of climate disruption on groundwater systems, however, have only lately attracted scholarly attention, in contrast to other aspects of the climate system (Taylor et al. 2013a). The recognition of groundwater as a vital freshwater resource that is essential to the world's food and water sustainability, ecosystem sustainability, and human adaptation to climatic instability and change is the source of this new attention. Determining the immediate impacts of climate disruption on the groundwater system as a whole is crucial (Lewandowski et al. 2020).

As societies adjust to changes in freshwater availability, climate change has an indirect impact on subsurface water networks through changes in groundwater extraction and direct effects through changes in the hydrological balance. Human-driven actions, such as modifications to land use and land cover (LULC), can further impact climate change and terrestrial water balances. Sea level rise is largely caused by the sea-level rise due to thermal expansion and the release of freshwater from long-term storage in glacial ice sheets, both of which are triggered by global temperature rise (Gu et al. 2020).

2.2 *Direct Consequences of Climate Disruption on Groundwater*

Aquifer replenishment constitutes a principal determinant of the hydrological equilibrium and sustainability of hydrological systems. The process of water table restoration is influenced, in part, by means of atmospheric processes such as precipitation and evaporation, and thus, any alterations in the planetary climate may exert considerable effects on groundwater, both qualitatively and quantitatively (Alley et al. 2002) Aquifer replenishment is the volume of water that infiltrates a specified region into water tables, as well as the slow escape occurring beneath adjacent water bodies, including lakes and ponds (Gebrechorkos et al. 2020).

Due to its significant challenges, indirect strategies, such as operational and experimental modelling approaches, may be used for aquifer replenishment measurement in situ. However, when evaluating the effects of climate disruption on aquifer replenishment, the use of these approaches adds more uncertainty. Climate change may alter the amounts of ground absorption, subsurface flow, and aquifer recharge within the hydrological cycle. Additionally, a rise in temperature increases the evaporative demand across land surfaces, which limits the amount of water that can be used to replenish groundwater. Climate extremes like droughts and floods have a significant impact on aquifer replenishment over long periods of time. However, increased groundwater recharge is not always the outcome of wetter weather (Taylor et al. 2013b). The sustainability of freshwater supplies may be threatened by the effects of climate change on groundwater storage (GWS). The phenomena is typified by a decrease in meltwater and an increase in evapotranspiration, which together produce different reactions in GWS changes in different aquifer systems (Yuan et al. 2015).

Natural mechanisms that aid in groundwater replenishment are directly impacted by climate disruption. This replenishment can happen all over a landscape through surface water body leaks and direct precipitation. That being said, global projections of how climate alteration would affect aquifer replenishment are quite uncertain (Fig. 2). The main source of this uncertainty is the limitations in how groundwater dynamics and Global Circulation Models (GCMs) depict climate alteration, where the ability of excess water to be transferred to an underlying aquifer is determined by the underlying soil and geological features (Yang et al. 2021).

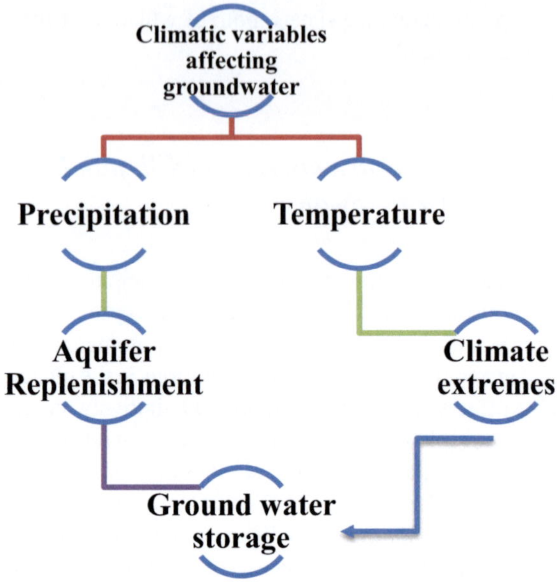

Fig. 2 Climatic variables affecting groundwater

2.3 *Long-Term Changes in Groundwater Recharge and Storage*

Global climate shift is generally defined as a shift in Earth's climate that lasts longer than a decade. It also refers to changes in human-induced and natural activities that affect the world's energy equilibrium. Examples of these changes include increasing sea levels, glacier melting, ozone layer depletion, and changes in rainfall and air wrmth. The pronounced climatic transformations witnessed in the twenty-first century have detrimentally impacted groundwater storage (GWS) and subsequently altered the hydrological cycle (Harik et al. 2023).

An understanding of the ramifications of climate alteration on aquifer replenishment is essential for the formulation of effective long-standing groundwater supply design and governance strategies. To date, no comprehensive study has encapsulated recent modelling endeavours to furnish a global perspective on anticipated long-term changes in groundwater recharge. Different projections are shown for Oceania, North America, Europe, and Asia depending on emission scenarios and seasonal factors (Ma et al. 2024).

2.4 *Vulnerability of Aquifers to Climate Change*

A semi-arid aquifer's vulnerability to climate change was assessed by comparing the effectiveness of the DRASTIC model and its modified versions. Climate change forecasts show that the highest recharge zones are decreasing at the same time that aquifer vulnerability is rising. As a result, the problems relating to replenishment deficiencies from aquifer replenishment continue to exist in the present and may worsen in the upcoming days. The economic resilience and social protection of local inhabitants that depend on this resource for agricultural irrigation, as well as the broader domains of water, land, and food security, will ultimately be threatened by these consequences, which will also make regional economic activities more vulnerable (Davamani et al. 2024; Condon et al. 2021).

The regional depletion of groundwater represents a pervasive global dilemma, wherein numerous groundwater systems are experiencing depletion rates that significantly exceed their natural replenishment, jeopardizing their continued utilization by both human societies and ecological systems. Since the mid-twentieth century, the global reliance on groundwater for irrigation has escalated tenfold. Groundwater volume and condition have declined as a result of increased use of this resource due to anthropogenic activities at the terrestrial surface. Groundwater vulnerability, as used in hydrogeology, is defined as an aquifer's susceptibility to contamination. Aquifer water quality degradation may also result from the mobilisation of naturally occurring contaminants like trace metals, the buildup of organic constituents like total dissolved organic carbon and nitrates from riparian environments, and the inter-aquifer movement of contaminants made possible by the short-circuiting of pathways to deep groundwater reservoirs. Additionally, the degradation may extend beyond the simple application of pollutants at the surface (Azar 2020; Sarkar et al. 2020).

3 Demographic Expansion and Unsustainable Water Extraction

3.1 Population Growth and Water Demand

India is a nation characterized by a burgeoning population, whereby the demographic count escalates every minute. To adequately meet the hydration needs of this substantial populace, the primary endeavour revolves around water storage, necessitating a rigorous scientific analysis of the quality and quantity of the groundwater sourced from subsurface reservoirs (Liu et al. 2021).

Groundwater is an invaluable resource integral to the sustenance of agriculture, industry, and domestic applications on a worldwide extent, even in dry or semi-arid regions. The Indian populace currently consumes approximately 230 km^3 of groundwater, thereby signifying a marked surplus exploitation of this resource. Groundwater serves as the principal source for irrigation, accounting for 80% of usage; between 1970 and 1999, the area designated for irrigation experienced a threefold increase to 33,100,000 hectares; it also constitutes 60% of drinking water sources. The most densely populated nations, including Middle East, China, North Africa and India, exhibit excessive groundwater utilization (Chen et al. 2014).

However, the excessive consumption of groundwater may precipitate drought conditions and significant resource scarcity, thereby threatening the integrity of ecosystems, social structures, and economic stability. Regions experiencing population surges will inevitably confront groundwater shortages if the current global depletion rates of groundwater and the escalating populace are not judiciously balanced. Agriculture has emerged as the primary consumer of groundwater, utilizing 90%, with some areas witnessing declines in groundwater levels of up to 100 m due to this overexploitation (Chaturvedi et al. 2020; Hoogesteger 2022). The recharging of groundwater is an exceedingly slow process, and the accelerating deterioration of this resource may pose a significant risk to affected regions, as restoring previous groundwater levels may prove challenging in the foreseeable future. Groundwater depletion represents a critical concern not only in the densely populated plains but also the mountainous regions, where snowmelt plays a pivotal role, contrasting with the plains (Cleland 2013).

3.2 Over-Extraction and Aquifer Degradation

Although the increased availability of subsurface water has significantly changed lives and economic structures, international attempts to manage this resource have largely failed. Numerous recent studies have questioned the enduring viability of current aquifer system use, since extraction rates much outpace natural groundwater recharge rates in many parts of the world, such as China, Mexico, India, and the United States. According to recent satellite measurements, unsustainable extraction methods are present in over one-third of the world's 37 major aquifers, many of which are located beneath important agricultural regions. Non-renewable groundwater abstraction rates have increased thrice in the last few decades, rising from 75 km^3 annually in 1960 to more than 234 km^3 annually in 2000 (Huggins et al. 2023).

It has been shown that unsustainable subsurface water extraction causes groundwater tables to permanently drop, which could eventually result in less subsurface flow into surface waters. Earth subsidence results from this, and saltwater intrusion exacerbates the salinisation of groundwater supplies. In India, where groundwater extraction rates have increased at an unparalleled rate over the past 50 years, the need to reform groundwater resource management is especially pressing. Groundwater extraction has increased dramatically since the 1960s, spurred by what has been called the "silent revolution," as millions of Indian farmers have been able to use mechanised pumps to collect groundwater resources. This unregulated extraction of groundwater has culminated in a complex series of feedback loops, yielding numerous adverse environmental and socio-economic repercussions.

A fundamental inquiry regarding groundwater and its interaction with society pertains to the extent of groundwater that can be extracted to satisfy water consumption needs. The goal of subsurface water management in relation to pumping rate

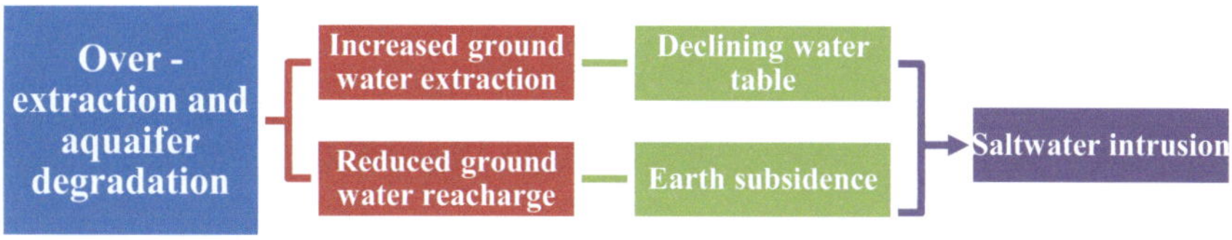

Fig. 3 Over extraction and aquifer degradation

establishment has historically been to limit regional pumping volumes to levels lower than the natural recharge rate. The renewable resource that a region can safely harvest (renewably) without causing a net depletion in long-term groundwater storage is known as the natural recharge rate. Most importantly, it has been discovered that groundwater stress assessments in India over-allocate the groundwater resources available for extraction while failing to adequately account for the environmental needs for water, even though they play a crucial role in directing support and interventions (Sadath et al. 2021) (Fig. 3).

Aquifer resource management has been complicated, with overutilization of groundwater reported in many parts of the world. Groundwater use patterns are having a negative impact on ecological systems and endangering future generations' ability to access this strategically important resource. India, which has the highest rates of groundwater withdrawal in the world, is currently at the centre of this problem, with millions of households' livelihoods and national food security depending more and more on the overutilization of aquifer resources.

3.3 *Urbanization and Groundwater Contamination Risks*

The demand for water in the households and public activities is never ending. However, the contemporary world has witnessed a steady rise in the necessity for water due to agricultural activities and industrial processes. Unfortunately, the available quantities of water continue to remain constrained, which undermines sustainable long-term economic development. Consequently, the quality of the existing water resources is increasingly compromised in various regions due to the augmented volume of industrial and other waste products resulting from human activities being released into the environment. The attributes of surface and groundwater play essential part in the overall advancement of environment and economy of a society. Population growth, urbanisation and agricultural expansion are the primary activities that are noted to adversely impact the water quality (Fig. 4). Anthropogenic activities such as urbanisation and industrialisation have considerably contributed to environmental pollution and deterioration, which negatively affects water resources including surface and subsurface that are necessary for life (Sarker et al. 2021).

The escalation of urbanization and population growth in numerous regions globally has intensified the contamination of groundwater, primarily through the release of

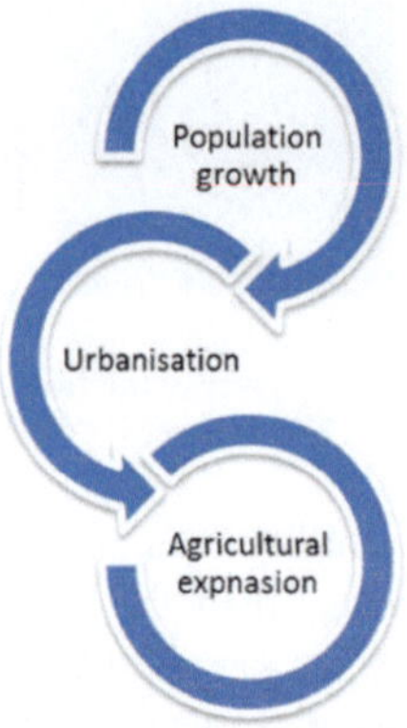

Fig. 4 Prominent major activities impacting subsurface water quality

domestic and industrial effluents into the groundwater system. Human health is highly vulnerable to exposure to metals and metalloids, even at minuscule concentrations, due to their persistence in environmental media and their acute toxicity.

4 Modelling Groundwater Systems

4.1 Hydrological Models for Groundwater Systems

A water resource model is characterised by the chemical and physical properties at the Earth's surface and subsurface that simulates the dynamics of water storage and water fluxes. It is mainly based on the water balance equation.

Understanding Aquifer Dynamics

The inherent ability of physics-based hydrological models to hold true across a wide range of sizes and environmental situations is well-known. These are the models that explain the intricate connections between the carbon flux, air circulation, and water cycle. They are able to assess hydroclimatic risks and forecast the amount and characteristics of water resources in the future in response to changes in the climate and terrestrial use (Beven 2012).

Managing groundwater resources is quite complex and challenging. Notably, groundwater numerical modelling is the most efficacious method that the researchers are following for simulating and evaluating subsurface water dynamics under various conditions. In order to evaluate conceptual ideas produced from field observations of the groundwater system, such models offer a deeper understanding of the system. As a result, these numerical models require the support of sufficient field data collection, accurate natural system conceptualisation, and careful computer code and solver selection. Each model has distinctive application and unique features. This is compelling modellers to identify the most suitable techniques to achieve specific objectives (Bizhanimanzar et al. 2020).

The integrated surface–subsurface hydrological models can resolve the subsurface Richardson-Richards equation in conjunction with a variation of the shallow water equation that represents surface flow. Integrated Surface–Subsurface Hydrological Models (ISSHMs) include a wide range of spatial resolutions, from a few meters to several hundred meters. ISSHMs are typically used in a small number of spatial domains (Maxwell and Miller 2005). The framework of earth system modelling that considers the energy and mass balance processes at the interface between the earthly surface and the atmosphere includes global hydrological and land surface models. But, such models often exhibit deficiencies in representing phenomena in highly heterogeneous regions that are extremely susceptible to hydro-climatic threats, such as mountainous and coastal areas (Camporese and Girotto 2022). Global hydrologic models along with integrated surface and subsurface hydrological models (ISSHMs) depend upon a precise representation of hydrological physical processes, meteorological forcing data and parameters like land cover, soil type, soil characteristics, etc. There is significant level of uncertainties and challenges pertaining to modelled hydrological states and fluxes. This is particularly considerable when extrapolating a model to encompass larger spatial domains (Clark et al. 2008).

The MODFLOW-2005 groundwater flow model has been integrated with the MODPATH advective particle tracking model, supplemented by geostatistical analyses derived from hydrochemical and hydrogeological assessments. This modelling approach can verify geogenic origin of NO^{3-} within the semi-confined aquifer within a multi-aquifer groundwater flow system. On the other hand, the uppermost unconfined aquifer is under threat of NO^{3-} contamination due to human activities (Letcher and Giupponi 2005).

Groundwater-Surface Water Interactions

There are various aspects to comprehending the interactions between surface water and subsurface water systems. One of the numerous factors contributing to this shift is that groundwater-surface water interactions encompass a diverse array of scientific disciplines. Scholars from hydrology, biogeochemistry, microbiology, biology, physics, and chemistry engage in the intricate interactions of processes, necessitating the consideration of pertinent elements from various scientific domains. Furthermore, interactions between surface and subsurface water occur across a multitude of marine and freshwater ecosystems; however, the potential for transferring technologies, methodologies, and resultant knowledge across fields remains underutilized (Yaseen 2021).

Groundwater-surface water interactions will only gain regular consideration in management and conservation practices when straightforward and robust methodologies and analytical procedures are readily accessible. Consequently, we urge all researchers operating within this domain to connect and actively disseminate their insights with practitioners and policymakers to cultivate the requisite focus on subsurface water-surface water relations in the framework of freshwater management and conservation. A physically based model (MIKE SHE) and an externally linked surface water-groundwater model (MOBIDIC-MODFLOW) can be used to capture the integrated hydrologic responses of the Thomas Brook catchment in Canada. In order to facilitate the sequential interchange of baseflow discharges and groundwater recharge at each computational time interval, MOBIDIC and MODFLOW were coupled. The hydrological processes in MOBIDIC-MODFLOW are simplified in the study based on its prediction accuracy (Al-Hashimi et al. 2021).

The coefficients for the gravitational and capillary reservoirs in MOBIDIC were systematically standardised using identical sets of hydrogeological features for both models in order to accurately predict the hydrological budget of the watershed represented by MIKE SHE. According to this simulation investigation, the two models accurately reproduce the water table oscillations that were observed at two specific monitoring wells. MOBIDIC-MODFLOW framework fails to comprehensively capture the immediate response of the water table when the case is locations that are exceptionally shallow water table. Additionally, the reductionist theory of the unsaturated flow in MOBIDIC-MODFLOW caused the groundwater recharge to be overestimated in the spring and underestimated in the summer. Furthermore, MOBIDIC-MODFLOW's computational efficiency in comparison to MIKE SHE and its lower input data requirements support its applicability for subsurface water-surface water interaction modelling at the regional scale (Bizhanimanzar et al. 2020).

4.2 Climate Models Integrated with Hydrological Models

A challenge that parallels short-term hydrological forecasting pertains to the transference of seasonal hydrological forecasts into actionable warnings and responses. Indeed, an impeccably accurate forecast or prompt early warning becomes redundant unless decision-makers possess a clear comprehension of the warning and subsequently engage in actions aligned with the forecasted predictions.

Understanding the mechanism of action linked to changes in the hydrological cycle within a changing environmental setting has been the main focus of hydrological research in recent decades. Significantly, the water flow cycle has changed due to climate alteration, as evidenced by an increase in extreme hydrological events. Besides, the socio-economic changes have substantially impacted human activities, resulting in a pronounced escalation in water demand across numerous global regions, thereby precipitating acute water scarcity. To promote the viable progress of human society, it is vital to enhance the comprehension of critical processes through an integrated examination of the water cycle (Singh 2014).

The technological progression has resulted in the rapid propagation of seasonal hydrologic prediction data for consumers; however, it is incumbent upon scientists to contemplate strategies for effectively communicating with authorities. This aids in better understanding of the forecast outcomes. Strategic decadal planning for infrastructure adaptation and capacity enhancement is vital for the management of water resources amid a varying climate, where seasonal hydrologic forecasting (SHF), are crucial for the mitigation of hydrological hazards by bolstering preparedness. This type of forecasting can differentiate between the uncontrollable land surface hydrological problems like drought and the short-term flood (Faisal et al. 1997). Traditional Supervised Hydrologic Forecasting (SHF) methodologies predominantly rely on temporal series modelling, establishing correlations between extensive remote climate indices. With the advent of conceptual and distributed hydrological models, the use of physical modelling for seasonal hydrology prediction has gained significant traction. These models effectively address both surface and subsurface water dynamics while ensuring the conservation of water balance in both vertical and horizontal dimensions (Zhou et al. 2023).

There are ample opportunities for research and development in the realm of Climate-Modelling Supervised Hydrologic Forecasting (CM-SHF) due to the recent scientific endeavours taking place in climate forecasting models that involve ocean–atmosphere teleconnections in association with the land surface hydrological modelling across different scales (Teshager et al. 2016). Current and potential stream flow changes in the Awash basin have been assessed using the Soil Water Assessment Tool (SWAT) model, more especially the ArcGIS version of SWAT (ArcSWAT). SWAT is a physically-grounded, spatially-distributed, process-based continuous-time model that simulates surface and subsurface water quantity on a daily temporal scale. Alongside, it gives an idea of environmental ramifications of climate alteration and land use planning. This model is adept at functioning efficiently in expansive basins with diverse management practices, optimizing the operational efficiency facilitating the assessment of long-term impacts without necessitating extensive temporal spans. Hydrological Response Units (HRUs), which are classified by certain soil types, terrain cover, and slope classes, are used by SWAT to represent the spatial variability that exists within a given watershed. The literature claims that this model is effective in predicting sediment output and hydrological variables for a variety of watersheds with different characteristics (Amin and Nuru 2020).

The evolution of global-scale glacier modelling has witnessed recent advancements; however, there remains a deficiency in physically-driven glacier behaviour and terminal ablation dynamics. Consequently, it is imperative to integrate the glacier mass balance model with hydrological models. To obtain more precise results, the remote sensing observations of snow cover must be combined with the understanding of inversion technology. Only these findings can be used to counteract the impact of snow cover on runoff and heat transport at the land surface in hydrological models. Furthermore, there is a necessity to bolster research efforts aimed at understanding snowmelt flood disasters. In the context of a number of simulations and prognostications concerning cryospheric hydrological regime shifts, there is an urgent demand for multi-dimensional coupled cryospheric water models.

4.3 Socio-Economic Models Are Intrinsically Linked with Groundwater Management Paradigms

Groundwater is interconnected with numerous socio-environmental dilemmas and constitutes a resource that is profoundly integrated within a global crisis. Nevertheless, it is frequently deprioritized or excluded from political and social priorities. Concurrently, there are appeals for innovative approaches and enhanced methodological experimentation in the domain of groundwater research. Examples of human-centric systems associated with groundwater encompass groundwater-irrigated agriculture, the water security of groundwater stakeholders, and the cultural relevance linked to surface manifestations of different sources of groundwater (Khan et al. 2024). A number of stresses, such as anthropogenic influences and climate alteration, affect these regions' environmental health. A significant threat to shoreline farming and crop productivity may result from anticipated climate variation, which is expected to increase the frequency and length of heat waves and cause spatiotemporal variations in precipitation patterns (Li et al. 2024).

5 Analytical Tools for Groundwater Management

Prominent analytical tools and technologies in groundwater management encompass the monitoring of water levels, pollution tracking, and the utilization of Geographic Information System (GIS) and aerial observation for resource assessment.

5.1 Current Analytical Tools and Technologies

Hydrogeophysical techniques such as electrical resistivity tomography, ground penetrating radar, and electromagnetic surveys are employed to delineate subsurface water and identify contaminants. Chemical and spectroscopic analyses evaluate contaminants and assess water quality. Stable isotope analysis facilitates the tracing of water sources and movement dynamics. Sensor networks and IoT technologies permit continuous monitoring of water quality and automated alerts for contamination (Alam et al. 2022).

The emergence of artificial intelligence (AI) has positioned it as an effective device for addressing practical difficulties, garnering substantial consideration due to its purposes across diverse arenas. Recently, AI methodologies have been applied in water purification and desalination processes to enhance efficiency and provide viable solutions for water pollution and scarcity. Data analytics and machine learning techniques forecast groundwater levels, contamination dispersal, and recharge patterns, while environmental DNA (eDNA) methodologies identify microbial signatures associated with pollutants. Groundwater flow modelling software such as MODFLOW simulates hydrological movement, aiding in the assessment of management strategies. Collectively, these tools enhance monitoring capabilities, quality evaluation, and sustainable groundwater management practices (Kumar and Singh 2023).

5.2 Emerging Trends in Groundwater Modelling

Studies reveal that subsurface water has been marginalized in global models. Lately, efforts have been put at the global level to nurture groundwater modelling. This advancement has yielded several crucial initial strides. However, further efforts are mandatory to begin rational global groundwater framework that intermingles with the global circulation models (Gaffoor et al. 2020).

Global land surface and Earth system models have been successfully assimilated with the groundwater representations. This is conjunction with the progress of groundwater modelling analyses and evaluation that is taking place at the regional to global scale. It should be noted that uncertainty quantification at spatial and temporal scales groundwater flow modelling incurs significant computational costs (Lykkegaard et al. 2020). Global high-resolution modelling endeavours necessitate the utilization of thousands of computational cores and the production of petabytes of data output. Global modelling is an initiative that depends on the successful tackling of computational and data-related challenges. The advancement of high performance computing will be requisite in realizing this objective, which will in turn demand significant software innovations within the hydrogeologic discipline (Engdahl 2023).

In contrast to the atmospheric and oceanographic scientific communities, which have been greatly aided by predictive analysis centres, it can be argued that the groundwater industry has been comparatively slow to adopt High Performance Computing (HPC) technologies (Asch et al. 2018). Although the deployment of such technology necessitates huge expenditure, scientific gains, social benefits and economic rewards of establishing an International Groundwater Network are undoubtedly greater.

5.3 Integration of Big Data and Artificial Intelligence in Groundwater Modelling

Artificial intelligence (AI) has developed as a formidable instrument for addressing practical challenges across diverse domains. AI methodologies are roped in to optimize processes and deliver pragmatic solutions to water scarcity concerns and contamination problems. The implementation of machine intelligence is anticipated to diminish the running costs associated with water treatment by lowering expenses and refining chemical usage (Duan et al. 2019).

Artificial intelligence (AI) represents a pivotal and well-established branch of computer science dedicated to the progress of intelligent systems that resolve issues in a way analogous to human cognitive processes. Enhancing computational capabilities relevant to human cognition is the main goal of incorporating AI applications into a system. AI is a rapidly evolving domain that manifests real-world applications across a multitude of sectors, including healthcare, intelligent urban design, education, travel, digital commerce and banking. Machine learning, deep learning, and data analytics are other subcategories of AI. These approaches are mostly used for cloud computing, blockchain applications, the Internet of Things, and intelligent decision-making. The proliferation of AI technologies can largely be attributed to their distinctive capabilities to learn from historical data and facilitate decision-making processes. The relevance of AI is continuously escalating, driven by the incorporation of AI-driven systems characterized by intelligence, adaptability, and intentionality within their algorithmic frameworks. AI has also stepped into water management to successfully overcome the shortcomings of traditional methodologies (Fig. 5). In contemporary times, water industry stakeholders are increasingly investing in artificial intelligence (Zhou et al. 2023).

Big data analytics (BDA) is a new paradigm that focuses on using large amounts of diverse data with sophisticated analytical methods to promote knowledge extraction. Facilitating sustainable subsurface water management is one important area where big data may be helpful. It is possible to fully digitise subsurface water well field data, gathering real-world data from sensor-equipped monitoring infrastructures and other relevant sources. This data can then be processed using sophisticated analytical algorithms to provide well field managers with insightful information that helps them make decisions. There are two ways to look at how big data analytics contributes to groundwater management. In the first place, big data analytics can help alleviate data scarcity by combining data from many sources, including conventional and non-traditional data sets. Second, especially at the local level, big data analytics can help turn unprocessed data into useful information that can improve groundwater management techniques (Kamyab et al. 2023).

Despite the many advantages that artificial intelligence offers, some restrictions prevent the widespread application of these techniques in real-world hydration treatment situations. Significant obstacles that require attention include the selection and availability of data, poor reproducibility, and a lack of empirical support for applications in real-world water refinement settings. There have been suggestions made to ensure the successful integration of AI in upcoming water-related technology. (Zhang et al. 2021).

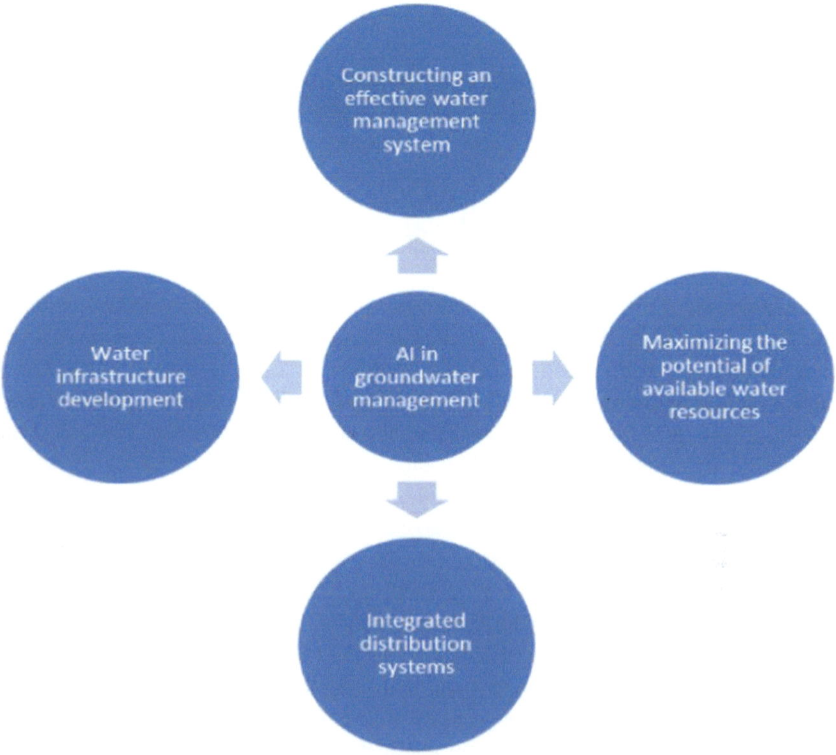

Fig. 5 Applications of artificial intelligence in groundwater management

6 Policy Evaluations and Decision-Making

6.1 Simulation-Based Policy Models

In general, quantitative simulation models of groundwater articulate the dynamics of groundwater resources and their management. Contemporary simulation models typically exhibit high computational demands due to their capacity to accurately represent the complexities of real-world systems. Recently, numerous studies have utilized both quantitative and qualitative simulations to ascertain hydraulic parameters, hydro-chemical characteristics, and aquifer management strategies. To this end, the majority of studies documented in the literature have employed MODFLOW and MT3D (MT3DMS) for quantitative and qualitative simulations, respectively. Numerical models, particularly MODFLOW models, are significant within the corpus of quantitative groundwater simulation models due to their user-friendliness, operational ease, and versatility (Theesfeld 2010).

A holistic simulation of hydrological processes enhances the precision of estimating evapotranspiration and recharge values in relation to one another. Consequently, quantitative estimations concerning recharge and evapotranspiration values constitute fundamental inputs in the modelling of groundwater flow.

6.2 Hydrologic Economic Models

The evaluation of groundwater resources with the objective of ensuring sustainable utilization presents a considerable challenge in applied hydrology, especially within arid regions where this resource is vital for socioeconomic development and, in certain instances, human survival. Comprehending the intricate hydrological and economic interactions occurring in these remote areas represents the initial step in enlightening stakeholders regarding the opportunities, risks, and significance of their natural resources. This understanding emerges from a conceptual model that encapsulates the essential characteristics of aquifer functionality in conjunction with its economic utilization (Mays 2013).

The integrated groundwater management model, encompassing both hydrologic and economic subsystems, effectively transforms into the best control model that intends to maximize cumulative discounted benefits throughout the planning horizon, while adhering to all pertinent physical and institutional constraints. Although the conceptual design of the optimal groundwater management model appears straightforward, its practical implementation encounters various theoretical and practical challenges (Knüppe 2011).

6.3 Risk Assessment Tools for Sustainable Water Use

The escalating concerns regarding freshwater scarcity and unsustainable water use manifest prominently in numerous regions globally. The rising demand for water, compounded by climatic fluctuations, has the potential to precipitate unsustainable freshwater utilization, culminating in water scarcity. To achieve a balance between the financial and social imperatives of water and to address competing water demands, it is imperative to sustain and effectively manage critical water systems (Abbass et al. 2022).

The complexity inherent in water systems, as well as the diversity of water users and sources, dictates that methodologies for assessing the efficacy of water management may range from simplistic approaches (e.g., indicator techniques) to more intricate frameworks (e.g., multiple models). It is posited that the condition of local infrastructure and the financial resources available are more pivotal to the sustainability of water management than the degree of regional development. (Juwana et al. 2012). The escalating issues of freshwater scarcity and the unsustainable utilization of water resources represent significant challenges in numerous regions globally.

The surging demand for water, compounded by climatic variability, can precipitate the unsustainable exploitation of freshwater, culminating in increased water scarcity (Baldassarre et al. 2019; Mseli et al. 2024).

One of the fundamental limitations of most large-scale studies is the omission of a groundwater component. Subsurface water use has increased significantly over the past 50 years in countries like China, India, and the United States. However, the lack of data prevents the subsurface water element from being sufficiently included in many large-scale assessments. One workable approach that might be incorporated into extensive evaluations is the use of GRACE data as an indirect indicator of groundwater (Famiglietti and Rodell 2013).

6.4 The Importance of Governance in Groundwater Conservation

For aquifer water resources to be managed effectively, legal frameworks are necessary. By outlining the rights and obligations of users and regulatory agencies in the management and preservation of these vital resources, they offer a framework that supports water policy and promotes sustainable and equitable water use. The persistent lack of focus on groundwater is evident in fragmented legislative measures, ineffective institutional frameworks, and inadequate enforcement of groundwater regulations in numerous regions worldwide. Increasingly, there is recognition of the imperative role that legal frameworks play in enhancing groundwater governance, necessitating their alignment with scientific and policy objectives (Kiparsky et al. 2017).

Modern water laws take a comprehensive approach to surface and groundwater resources, outlining clear rights and obligations, developing management tools, protecting the amount and characteristic features, involving investors, and guaranteeing alignment with strong institutions in charge of enforcing them. These regulations usually cover a wide range of topics, such as the requirements for groundwater access, protecting aquifers from contamination and depletion, using planning and monitoring techniques, striking a balance between public and private interests, and involving stakeholders in management and decision-making procedures. These important issues are typically covered by explicit legislation (Mechlem 2016). The complexity inherent in water systems, as well as the diversity of water users and sources, dictates that methodologies for assessing the efficacy of water management may range from simplistic approaches (e.g., indicator techniques) to more intricate frameworks (e.g., multiple models). It is posited that the condition of local infrastructure and the financial resources available are more pivotal to the sustainability of water management than the degree of regional development (Juwana et al. 2012).

The escalating issues of freshwater scarcity and the unsustainable utilization of water resources represent significant challenges in numerous regions globally. The surging demand for water, compounded by climatic variability, can precipitate the unsustainable exploitation of freshwater, culminating in increased water scarcity (Joseph et al. 2020).

The evaluation of water availability in minor water system assessments includes both aquifer replenishment and surface flow components. However, there are significant practical difficulties in incorporating groundwater data into global hydrologic models, especially when done on coarse grid resolutions ($0.5° \times 0.5°$). When the groundwater component is left out of large-scale studies, the total amount of water available is underestimated, leading to inaccurate conclusions about water shortage (Gleeson et al. 2021). It is imperative to identify, classify, and assess the characteristics and factors influencing groundwater. These recommendations were directed towards researchers, policymakers, and decision-makers, offering insights for further inquiry into sustainable groundwater management. The overarching objective is to safeguard and judiciously utilize groundwater resources (Taylor et al. 2013c).

7 Conclusion

In light of the world's changing climate, sustainable groundwater management is essential, and addressing this issue requires a dual focus on adjustment and reduction strategies. To increase groundwater's effectiveness in supporting climate adaption and sustainable development, more stringent regional and worldwide monitoring and supervision are needed. Furthermore, adding new variables to Representative Concentration Pathway scenarios—like shifting land use and aquifer properties—can improve prediction accuracy, leading to a more comprehensive understanding of groundwater dynamics and useful information for developing sustainable water resource organization plans.

In studies pertaining to the evaluation, preservation, and restoration of water resources, aquifer models have grown in importance. In order to assess and prepare for the effects of climate variation as they become more noticeable, hydrological models will be essential. More complex models that take into account the dynamic interactions between terrestrial use, climate, and water systems may be created in the upcoming days. Given that groundwater is essential to achieving sustainable development goals, it is necessary to create strategies that address both climate alteration and groundwater management, both of which must be implemented locally to support sustainable development. Planning and decision protocols for subsurface water management have been shown to benefit greatly from the use of groundwater modelling. Furthermore, to guarantee that hydrological models are successfully incorporated into real decision-making frameworks, cooperation between researchers, policy makers, and specialists will be crucial. These developments will help manage water resources sustainably in the face of changing environmental problems by improving our understanding of water systems.

References

Bizhanimanzar M, Leconte R, Nuth M (2020) Catchment-scale integrated surface water-groundwater hydrologic modelling using conceptual and physically based models: a model comparison study. Water 12(2):363. https://doi.org/10.3390/w12020363

Abbass K, Qasim MZ, Song H, Murshed M, Mahmood H, Younis I (2022) A review of the global climate change impacts, adaptation, and sustainable mitigation measures. Environ Sci Pollut Res 29(28):42539–42559. https://doi.org/10.1007/s11356-022-19718-6

Al Atawneh D, Cartwright N, Bertone E (2021) Climate change and its impact on the projected values of groundwater recharge: a review. J Hydrol 601:126602. https://doi.org/10.1016/j.jhydrol.2021.126602

Alam G, Ihsanullah I, Naushad M, Sillanpää M (2022) Applications of artificial intelligence in water treatment for optimization and automation of adsorption processes: recent advances and prospects. Chem Eng J 427:130011. https://doi.org/10.1016/j.cej.2021.130011

Al-Hashimi O, Hashim K, Loffill E, Marolt Čebašek T, Nakouti I, Faisal AA, Al-Ansari N (2021) A comprehensive review for groundwater contamination and remediation: occurrence, migration and adsorption modelling. Molecules 26(19):5913. https://doi.org/10.3390/molecules26195913

Alley WM, Healy RW, LaBaugh JW, Reilly TE (2002) Flow and storage in groundwater systems. Science 296(5575):1985–1990. https://doi.org/10.1126/science.1067123

Amanambu AC, Obarein OA, Mossa J, Li L, Ayeni SS, Balogun O, Oyebamiji A, Ochege FU (2020) Groundwater system and climate change: present status and future considerations. J Hydrol 589:125163. https://doi.org/10.1016/j.jhydrol.2020.125163

Amin A, Nuru N (2020) Evaluation of the performance of SWAT model to simulate stream flow of Mojo River watershed: in the upper Awash River Basin, in Ethiopia. Hydrology 8(1):7–18. https://doi.org/10.11648/j.hyd.20200801.12

Asch M, Moore T, Badia R, Beck M, Beckman P, Bidot T, Bodin F, Cappello F, Choudhary A, De Supinski B, Deelman E (2018) Big data and extreme-scale computing: pathways to convergence-toward a shaping strategy for a future software and data ecosystem for scientific inquiry. Int J High Perform Comput Appl 32(4):435–479. https://doi.org/10.1177/1094342018778123

Azar BB (2020) Energy-water-food security challenge for middle-east and North Africa (MENA). University of Technology Sydney (Australia)

Beven KJ (2012) Rainfall-runoff modelling: the primer, 2nd ed. Wiley-Blackwell. https://doi.org/10.1002/9781119951001

Camporese M, Girotto M (2022) Recent advances and opportunities in data assimilation for physics-based hydrological modeling. Front Water 4:948832. https://doi.org/10.3389/frwa.2022.948832

Chaturvedi RK, Jallu PR, Rajesh SBV, Zare N, Ashutosh S, Lakhchaura P, Ghosh S, Rao V (2020) Mapping climate change hotspots in Indian forests based on observed climate change and high resolution climate model projections. *A* Study done by BITS Pilani Goa campus and Forest Survey of India, Ministry of Environment, Forest and Climate Change, Government of India

Chen J, Li J, Zhang Z, Ni S (2014) Long-term groundwater variations in Northwest India from satellite gravity measurements. Global Planet Change 116:130–138. https://doi.org/10.1016/j.gloplacha.2014.02.007

Clark MP, Rupp DE, Woods RA, Zheng X, Ibbitt RP, Slater AG, Schmidt J, Uddstrom MJ (2008) Hydrological data assimilation with the ensemble Kalman filter: use of streamflow observations to update states in a distributed hydrological model. Adv Water Resour 31(10):1309–1324. https://doi.org/10.1016/j.advwatres.2008.06.005

Cleland J (2013) World population growth; past, present and future. Environ Resource Econ 55:543–554. https://doi.org/10.1007/s10640-013-9675-6

Condon LE, Kollet S, Bierkens MF, Fogg GE, Maxwell RM, Hill MC, Fransen HJH, Verhoef A, Van Loon AF, Sulis M, Abesser C (2021). Global groundwater modeling and monitoring: opportunities and challenges. Water Resour Res 57(12):e2020WR029500. https://doi.org/10.1029/2020WR029500

Davamani V, John JE, Poornachandhra C, Gopalakrishnan B, Arulmani S, Parameswari E, Naidu R (2024) A critical review of climate change impacts on groundwater resources: a focus on the current status, future possibilities, and role of simulation models. Atmosphere 15(1):122. https://doi.org/10.3390/atmos15010122

Di Salvo C (2023) Groundwater hydrological model simulation. Water 15(4):822. https://doi.org/10.3390/w15040822

Di Baldassarre G, Sivapalan M, Rusca M, Cudennec C, Garcia M, Kreibich H, Konar M, Mondino E, Mård J, Pande S, Sanderson MR (2019) Sociohydrology: scientific challenges in addressing the sustainable development goals. Water Resour Res 55(8):6327–6355. https://doi.org/10.1029/2018WR023901

Duan Y, Edwards JS, Dwivedi YK (2019) Artificial intelligence for decision making in the era of Big Data–evolution, challenges and research agenda. Int J Inf Manage 48:63–71. https://doi.org/10.1016/j.ijinfomgt.2019.01.021

Durán-Sánchez A, Álvarez-García J, González-Vázquez E, Del Río-Rama MDLC (2020) Wastewater management: bibliometric analysis of scientific literature. Water 12(11):2963.https://doi.org/10.3390/w12112963

Engdahl NB (2023) Next generation computers warrant next generation groundwater models. Groundwater 62(1):7–14. https://doi.org/10.1111/gwat.13325

Faisal IM, Young RA, Warner JW (1997) Integrated economic-hydrologic modelling for groundwater basin management. Int J Water Resour Dev 13(1):21–34. https://doi.org/10.1080/07900629749908

Famiglietti JS, Rodell M (2013) Water in the balance. Science 340(6138):1300–1301. https://doi.org/10.1126/science.1236460

Gaffoor Z, Pietersen K, Jovanovic N, Bagula A, Kanyerere T (2020) Big data analytics and its role to support groundwater management in the southern African development community. Water 12(10):2796. https://doi.org/10.3390/w12102796

Gebrechorkos SH, Bernhofer C, Hülsmann S (2020) Climate change impact assessment on the hydrology of a large river basin in Ethiopia using a local-scale climate modelling approach. Sci Total Environ 742:140504. https://doi.org/10.1016/j.scitotenv.2020.140504

Gleeson T, Wagener T, Döll P, Zipper SC, West C, Wada Y, Taylor R, Scanlon B, Rosolem R, Rahman S, Oshinlaja N, Maxwell R, Lo M-H, Kim H, Hill M, Hartmann A, Fogg G, Famiglietti JS, Ducharne A, de Graaf I, Cuthbert M, Condon L, Bresciani E, Bierkens MFP (2021) GMD perspective: the quest to improve the evaluation of groundwater representation in continental-to global-scale models. Geosci Model Dev 14(22):7545–7568. https://doi.org/10.5194/gmd-14-7545-2021

Gorelick SM, Zheng C (2015) Global change and the groundwater management challenge. Water Resour Res 51(5), 3031–3051.https://doi.org/10.1002/2014WR016825

Gorelick SM, Zheng C (2015b) Global change and the groundwater management challenge. Water Resour Res 51(5):3031–3051. https://doi.org/10.1002/2014WR016825

Gu L, Chen J, Yin J, Xu CY, Zhou J (2020) Responses of precipitation and runoff to climate warming and implications for future drought changes in China. Earth's Future 8(10):e2020EF001718. https://doi.org/10.1029/2020EF001718

Harik G, Alameddine I, Zurayk R, El-Fadel M (2023) An integrated socio-economic agent-based modeling framework towards assessing farmers' decision making under water scarcity and varying utility functions. J Environ Manage 329:117055. https://doi.org/10.1016/j.jenvman.2022.117055

Hoff H (2009) Global water resources and their management. Curr Opin Environ Sustain 1(2):141–147. https://doi.org/10.1016/j.cosust.2009.10.001

Hoogesteger J (2022) Regulating agricultural groundwater use in arid and semi-arid regions of the Global South: challenges and socio-environmental impacts. Curr Opin Environ Sci Health 27:100341. https://doi.org/10.1016/j.coesh.2022.100341

Hora T (2022) Addressing groundwater over-extraction in India: assessments, monitoring methods and interventions

Huggins X, Gleeson T, Castilla-Rho J, Holley C, Re V, Famiglietti JS (2023) Groundwater connections and sustainability in social-ecological systems. Groundwater 61(4):463–478. https://doi.org/10.1111/gwat.13305

Joseph N, Ryu D, Malano HM, George B, Sudheer KP (2020) A review of the assessment of sustainable water use at continental-to-global scale. Sustain Water Resour Manag 6:1–20. https://doi.org/10.1007/s40899-020-00379-7

Juwana I, Muttil N, Perera BJC (2012) Indicator-based water sustainability assessment—a review. Sci Total Environ 438:357–371. https://doi.org/10.1016/j.scitotenv.2012.08.093

Kamyab H, Khademi T, Chelliapan S, SaberiKamarposhti M, Rezania S, Yusuf M, Farajnezhad M, Abbas M, Jeon BH, Ahn Y (2023) The latest innovative avenues for the utilization of artificial Intelligence and big data analytics in water resource management. Results Eng 101566.https://doi.org/10.1016/j.rineng.2023.101566

Khan T, Samiullah M, Rouf II, Sultana S, Rahman S, Rahman MB, Khanum R (2024) The Nexus of water scarcity and climate change: understanding interconnected challenges and formulating resilient strategies. Int J Environ Sci 7(3):57–68. https://doi.org/10.54473/IJTRET.2024.9102

Kiparsky M, Milman A, Owen D, Fisher AT (2017) The importance of institutional design for distributed local-level governance of groundwater: the case of California's sustainable groundwater management act. Water 9(10):755. https://doi.org/10.3390/w9100755

Knüppe K (2011) The challenges facing sustainable and adaptive groundwater management in South Africa. Water Sa 37(1). https://doi.org/10.4314/wsa.v37i1.64110

Kumar A, Singh R (2023) Artificial intelligence applications in water treatment and desalination: a comprehensive review. Environ Technol Rev 12(1):45–67. https://doi.org/10.1080/21622515.2023.1234567

Letcher RA, Giupponi C (2005) Policies and tools for sustainable water management in the European Union. Environ Model Softw 20(2):93–98. https://doi.org/10.1016/j.envsoft.2004.01.003

Lewandowski HJ, Meinikmann K, Krause S (eds) (2020) Groundwater-surface water interactions. MDPI. https://doi.org/10.3390/w12010296

Li Y, Chen Y, Liu Y, Zhang H (2024) Effects of climate variability on agricultural productivity: a spatiotemporal assessment. Environ Res 245:117007. https://doi.org/10.1016/j.envres.2024.117007

Liu B, Zou X, Yi S, Sneeuw N, Cai J, Li J (2021) Identifying and separating climate-and human-driven water storage anomalies using GRACE satellite data. Remote Sens Environ 263:112559. https://doi.org/10.1016/j.rse.2021.112559

Loucks DP, van Beek E, Loucks DP, van Beek E (2017) Water resources planning and management: an overview. Water resource systems planning and management: an introduction to methods, models, and applications, pp 1–49

Lykkegaard MB, Dodwell TJ, Moxey D (2020) Accelerating uncertainty quantification of groundwater flow modelling using a deep neural network proxy. arXiv:2007.00400. https://doi.org/10.48550/arXiv.2007.00400

Ma Z, Wang W, Chen X, Gehman K, Yang H, Yang Y (2024) Prediction of the global occurrence of maize diseases and estimation of yield loss under climate change. Pest Manag Sci. https://doi.org/10.1002/ps.8309

Maxwell RM, Miller NL (2005) Development of a coupled land surface and groundwater model. J Hydrometeorol 6: 233–247. https://doi.org/10.1175/JHM422.1

Mays LW (2013) Groundwater resources sustainability: past, present, and future. Water Resour Manage 27:4409–4424. https://doi.org/10.1007/s11269-013-0436-7

Mechlem K (2016) Groundwater governance: the role of legal frameworks at the local and national level—established practice and emerging trends. Water 8(8):347. https://doi.org/10.3390/w8080347

Mendieta-Mendoza A, Hanson RT, Renteria-Villalobos M (2021) Potential adverse impacts on vulnerability and availability of groundwater from climate-change and land use. J Hydrol 594:125978. https://doi.org/10.1016/j.jhydrol.2021.125978

Mseli ZH, Sankaranna G, Mwegoha WJ (2024) Toward sustainable groundwater management: a comprehensive framework for resource protection and utilization. Environ Qual Manage 34(1):e22308. https://doi.org/10.1002/tqem.22308

Sadath PVR, Kartheeshwari MR, Elango L (2021) Sustainable groundwater management under global climate change: mitigation and adaptation measures. In: Proceedings of the earth and environmental sciences international webinar conference. Springer International Publishing, Cham 187–206. https://doi.org/10.1007/978-3-031-34783-2_10

Sarkar T, Kannaujiya S, Taloor AK, Ray PKC, Chauhan P (2020) Integrated study of GRACE data derived interannual groundwater storage variability over water stressed Indian regions. Groundw Sustain Dev 10:100376. https://doi.org/10.1016/j.gsd.2020.100376

Sarker B, Keya KN, Mahir FI, Nahiun KM, Shahida S, Khan RA (2021) Surface and ground water pollution: causes and effects of urbanization and industrialization in South Asia. Sci Rev 7(3):32–41. https://doi.org/10.32861/sr.73.32.41

Scanlon BR, Fakhreddine S, Rateb A, de Graaf I, Famiglietti J, Gleeson T, Zheng C (2023a) Global water resources and the role of groundwater in a resilient water future. Nat Rev Earth Environ 4(2):87–101

Scanlon BR, Fakhreddine S, Rateb A, de Graaf I, Famiglietti J, Gleeson T, Zheng C (2023b) Global water resources and the role of groundwater in a resilient water future. Nat Rev Earth Environ 4(2):87–101. https://doi.org/10.1038/s43017-022-00378-6

Scanlon BR, Fakhreddine S, Rateb A, de Graaf I, Famiglietti J, Gleeson T, Zheng C (2023) Global water resources and the role of groundwater in a resilient water future. Nat Rev Earth Environ 4(2):87–101. https://doi.org/10.1038/s43017-022-00378-6

Singh A (2014) Simulation and optimization modeling for the management of groundwater resources. II: Combined applications. J Irrig Drain Eng 140(4):04014002. https://doi.org/10.1061/(ASCE)IR.1943-4774.0000689

Srinivasan V, Konar M, Blöschl G, Sivapalan M, Thompson SE (2017) Predictions in a socio-hydrological world. Hydrol Sci J 62(3):394–406. https://doi.org/10.1080/02626667.2016.1253844

Suhag R (2016) Overview of ground water in India. PRS on standing committee on water resources, legislative research, p 12

Taylor RG, Scanlon BR, Döll P, Rodell M, van Beek R, Wada Y, Edmunds M (2013b) Ground water and climate change. Nat Clim Chang 3(4):322–329. https://doi.org/10.1038/nclimate1744

Taylor RG, Scanlon B, Döll P, Rodell M, Van Beek R, Wada Y, Longuevergne L, Leblanc M, Famiglietti JS, Edmunds M, Konikow L (2013c) Ground water and climate change. Nat Clim Chang 3(4):322–329. https://doi.org/10.1038/nclimate1744

Taylor RG, Scanlon B, Döll P, Rodell M, van Beek R, Wada Y, Edmunds M (2013) Ground water and climate change. Nat Clim Change 3(4):322–329. https://doi.org/10.1038/nclimate1744

Teshager AD, Gassman PW, Secchi S, Schoof JT, Misgna G (2016) Modeling agricultural watersheds with the soil and water assessment tool (swat): calibration and validation with a novel procedure for spatially explicit hrus. Environ Manage 57:894–911. https://doi.org/10.1007/s00267-015-0636-4

Theesfeld I (2010) Institutional challenges for national groundwater governance: policies and issues. Groundwater 48(1):131–142. https://doi.org/10.1111/j.1745-6584.2009.00624.x

Wu W-Y, Lo M-H, Wada Y, Famiglietti JS, Reager JT, Yeh P-F, Ducharne A, Yang Z-L (2020) Divergent effects of climate change on future groundwater availability in key mid-latitude aquifers. Nat Commun 11(1):3710. https://doi.org/10.1038/s41467-020-17581-y

Yang D, Yang Y, Xia J (2021) Hydrological cycle and water resources in a changing world: a review. Geogr Sustain 2(2):115–122. https://doi.org/10.1016/j.geosus.2021.05.003

Yaseen ZM (2021) An insight into machine learning models era in simulating soil, water bodies and adsorption heavy metals: review, challenges and solutions. Chemosphere 277:130126. https://doi.org/10.1016/j.chemosphere.2021.130126

Yuan X, Wood EF, Ma Z (2015) A review on climate-model-based seasonal hydrologic forecasting: physical understanding and system development. Wiley Interdiscip Rev Water 2(5):523–536. https://doi.org/10.1002/wat2.1088

Zhang X, Lin G, Sun S, Li G (2021) Deep neural network surrogate-assisted likelihood-free Bayesian uncertainty quantification of groundwater flow models. Comput Methods Appl Mech Eng 381:113848. https://doi.org/10.1016/j.cma.2021.113848

Zhou Y, Liu Y, Wang H, Zhang X (2023) Coupling deep learning and physically based hydrological models for monthly streamflow prediction. Water Resour Res 59(5):e2023WR035618. https://doi.org/10.1029/2023WR035618

Groundwater Depletion and Recharging Strategies Augmented by Digital Technology

Channarayappa

Abstract Groundwater is an essential natural resource that provides drinking water worldwide. It occurs naturally in the Earth's subsurface, which is stored mainly in the aquifers. Due to overuse, groundwater is depleted, and water pollution threatens all life. Safeguarding groundwater is crucial to ensure its availability. Conservation efforts such as reducing water usage and recharging aquifers have led to sustainable water management. Recently, personal computers and space technology for land and water management have integrated contour mapping, watersheds, remote sensing (RS), geographic information systems (GIS), and Artificial intelligence (AI). These techniques have fundamentally changed our thoughts on managing natural resources and groundwater management. The spatial variations in the groundwater distribution in arid areas can be studied using topology, GIS, and AI techniques. Estimating water sources, spatial distribution, and mapping of groundwater hotspots by using the above technologies is promising. The info-maps for all the area specifications can be prepared using the Inverse Distance Weighted (IDW) interpolation technique of GIS. The established spatial database would help monitor and manage groundwater levels to apply efficient recharge techniques in the study area. Protecting groundwater is of global concern, and humans need to sustain it.

Keywords Groundwater · Depletion · Topography · Contours · Aquifers · Watersheds · Remote sensing · GIS · AI

1 Introduction

Life on Earth is water-based since the biological functions in cells work in an aqueous environment. Without water, no life can exist or be sustained. But not all types of water support life. Out of the total water resources available on the Earth, 97% is salt water, 3% is fresh water, and 0.6% is groundwater (TWB 2010). The clean water

Channarayappa (✉)
Research and Development, Department of Biotechnology and Genetics, MS Ramaiah College of Arts, Science, and Commerce, Bengaluru, India
e-mail: channarayappa@gmsil.com

A. Nath et al. (eds.), *Groundwater Depletion and Sustainability*, Advances in Geographic Information Science, https://doi.org/10.1007/978-3-032-09921-1_14

that supports living beings comes from the groundwater and is a natural resource, accounting for 30% of the world's freshwater reserves. Groundwater is replenished by rainfall, snowmelt, and surface water that percolate into the ground. Groundwater is stored in underground reservoirs called aquifers, typically composed of porous rocks and sediments. Understanding groundwater measures is essential for its conservation and sustainable use. The demand for groundwater is increasing due to the rapid increase in population, urbanization, industrialization, and agriculture. Hence, understanding the importance of groundwater for managing and protecting the groundwater resource is essential. Overuse of groundwater has led to depletion and pollution, making water undrinkable and affecting the environment. We are extracting more than is being recharged. Therefore, periodic observation and management of groundwater status are vital to ensure its sustainability for future generations.

The occurrence and distribution of groundwater are controlled by lithology, structure, geomorphology, and rainfall patterns. Most underground water is produced by rainfall, snowmelt, and leaks in lakes or rivers. The higher surface area of a saturated zone is called a water table. Underground water is normally stored and moves slowly through the cracks between rocks, spaces in soil, and sand, called aquifers. Aquifers are rocks and sediments that store underground water, filled after passing through the unsaturated soil regions (CGWB 2015). The recharging of groundwater reserves is accomplished by surface water and rainfall. Groundwater resources are badly threatened in most parts of the world by quantity and quality (Moeck et al. 2020). In an arid region, it is more important as it supplies domestic needs, agriculture, and industrial uses. Avoiding groundwater depletion, optimizing water use, and applying artificial recharging have been strongly recommended.

The satellite technology has been suggested for tracking groundwater resources. To begin these studies, the demarcation of suitable locations for the artificial recharging of aquifers is necessary. Rainwater harvesting, artificial recharge, and water-use efficiency are practiced by government and private departments. Several researchers (Samadder et al. 2011) have applied unified geographic information systems (GIS) and remote sensing techniques to identify potential groundwater zones, Inverse Distance Weighting (IDW) Techniques for Spatial Data Analysis, and topology maps, for the location of aquifers. Detailed investigations on controlling parameters are required for groundwater modelling and management. It requires reliable input data and the standard point sampling methods for input parameters. However, these are often biased because of heterogeneity in subsurface layers and structures, and restrictions on samples from harsh terrain. The remote sensing and GIS techniques provided ample scope to generate input data to study groundwater in an area more efficiently, at a lower cost, and in less time. Geospatial knowledge related to groundwater has been used to elucidate thematic layers such as geomorphology, drainage pattern, contour, rainfall pattern, and gradient (Brown and Molly 2015). Integration of these thematic tiers with post-monsoon water levels and depth of weathering was found to be important and suitable sites for artificial recharge within the aquifer system. With precise field operations, AI technology can produce higher productivity levels with minimal effects on the environment (Ibrahim et al. 2024).

Groundwater is of paramount importance for humans, and a major source of fresh water, and is used largely in agriculture and industries. The FAO describes groundwater as "the invisible ingredient in food". The US Geological Survey says groundwater is the world's most important natural resource. Around 70% of the world's groundwater is used for food, livestock, and industrial processes. Roughly a third of the water used in crop irrigation is groundwater. People in drier parts of the world depend almost exclusively on groundwater for their water supplies. India is the world's largest groundwater user, accounting for a quarter of a whole world, according to the World Bank. North America and South Asia are most reliant on them for irrigation. Our natural ecosystems also need groundwater, as it ensures water flows to wetlands and streams. Many industrial processes use groundwater, typically where water quality is important, but access to surface water is limited. These include manufacturing, mining, oil and gas, energy generation, engineering, and construction.

Many previous studies (Rodell et al. 2018) have highlighted groundwater depletion. Unfortunately, slowing or reversing groundwater losses has received less attention. To improve groundwater resources efficiently, applications of modern technology are essential. Transition of farming from traditional to industrial requires, and then to tech-based systems, for several reasons, such as increasing population, hunger, poverty, and climate change. A tech-based farming system, such as digital technologies, ICT, GIS, IDW, AI, etc., maintains farming operations at a precise scale and is sustainable.

2 Identification of Groundwater Sources

When people think of a water source, they immediately think of lakes, rivers, and streams, in other words, surface water. However, most freshwater (~97%) comes from groundwater. Groundwater, found beneath the earth's surface, is primarily sourced from rainfall that seeps into the ground. Other sources include rainfall, snowmelt, and water from rivers and lakes that leak into the soil. Groundwater is stored in underground layers of permeable rock and soil.

2.1 *Sources*

(a) **Aquifers**: An aquifer is a porous rock or sediment layer that can hold water. Aquifers are the most common source of groundwater and can be found worldwide. They are typically made up of porous rock or sediment, such as sandstone or gravel, and are often located beneath the Earth's surface. Aquifers can be divided into two categories: unconfined and confined. Unconfined aquifers are located near the Earth's surface and are not covered by impermeable rock or sediment. Confined aquifers are located deep beneath the Earth's surface and protected by impermeable rock or sediment. The depth of water aquifers indicates the level at which groundwater is available from the surface (Lorenzo-Lacruz et al. 2017).

(b) **Springs**: A spring is a natural source of groundwater that flows from the Earth's surface. Springs are typically found where the water table is close to the Earth's surface, such as in valleys or on hillsides. Springs can be classified into two categories: hot springs and Cold springs. Hot springs are formed when groundwater is heated by geothermal activity and rises to the surface. Cold springs are formed when groundwater flows to the surface without being heated.

(c) **Geysers**: Hot springs where heated groundwater erupts above the floor.

(d) **Wells**: Drilled tubes that tap into aquifers, bringing groundwater to the surface for consumption, irrigation, and business, and are artificial.

2.2 Identification

Identification of groundwater sources is important to find where a particular groundwater reserve is and the routes it has taken to reach a specific location. This is helpful for understanding water source, quality, water cycle dynamics, and for taking effective regulatory plans. The major techniques used include geophysical exploration, geochemical analysis, isotopic tracing, numerical modelling, pumping, aquifer tests, data-driven models, and ensemble learning etc.

3 Groundwater Extraction Methods

Water is taken out from the underground by various methods for different purposes, called as groundwater withdrawal. Judicious use of water resources is highly recommended. Over-extraction can lead to the depletion of aquifers and can cause water shortages. It is important to monitor water levels and use sustainable extraction practices to preserve groundwater resources, thus no land subsidence for future generations. A few main extraction methods are generally followed:

1. **Bore and Dug wells:** Groundwater is typically removed from aquifers through bore-wells. Bore wells are holes drilled into the ground to access the water table. The kind of bore well used will be determined by the depth of the water table, the kind of soil, and the amount of water needed. Whereas, dug wells are shallow and typically only reach the water table. Dug wells are often lined with bricks or stones to prevent collapse. Driven wells are usually used in sandy soils and can reach depths of up to 50 feet. They are created by driving a pipe into the ground and installing a screen to allow water to enter the pipe. Drilled wells reach deeper water tables and can be several hundred feet deep.

2. **Pumps**: Once a well has been drilled and water has been accessed, a pump extracts the water from the ground. Generally, two types of pumps are used: submersible pumps and jet pumps. Submersible pumps are placed directly into the well and are designed to push water to the surface. Jet pumps are used for shallow wells and are placed above ground. The jet pumps use suction to pull water up from the well.

4 Surface Water Versus Groundwater

The interrelation between surface water and groundwater constantly changes due to the different spatial and temporal distributions of the exchange between surface water and groundwater. The chemistry of groundwater is often more predictable than surface water composition, because groundwater doesn't change much over time or space as quickly as surface water does. However, surface water has fewer dissolved particles because it includes a major component of precipitation water that hasn't had a long time in contact with minerals from the soil. Groundwater is usually higher in pH (alkaline), and the hardness of surface water may vary depending on the source. In contrast, groundwater is soft or hard, depending on the stored water age. Surface water is rich in nutrients, dissolved oxygen, and the number of organisms because it is exposed to the sunlight, which activates Autotrophs to promote primary production, and is the base of the food web.

5 Influence of Climate on Groundwater

Groundwater is a vital resource that is severely affected by the climate. The factors that lead to groundwater depletion come mostly from human activities, and also from changes in our climate. Climate change has considerable impacts on the groundwater resources. Changes in precipitation patterns, temperature, and evapotranspiration rates all impact groundwater recharge and availability (Ebi et al. 2021). Here are a few ways climate change affects groundwater: As temperatures rise, evapotranspiration rates increase, so less water is available to recharge groundwater. Additionally, precipitation patterns can lead to less water reaching the ground and being available for recharge. This leads to decreased groundwater levels and reduced availability. Rising sea levels and vagaries in precipitation patterns can cause saltwater to intrude into coastal aquifers, making the water unusable for human consumption and agriculture. Raising air temperatures leads to higher ground temperatures, causing a rise in groundwater temperature. As surface water becomes scarcer, there may be increased demand for groundwater as an alternative source. This can lead to over-extraction and depletion of groundwater. In conclusion, changes in climatic conditions significantly threaten groundwater resources, impacting their quantity and quality (Fig. 1).

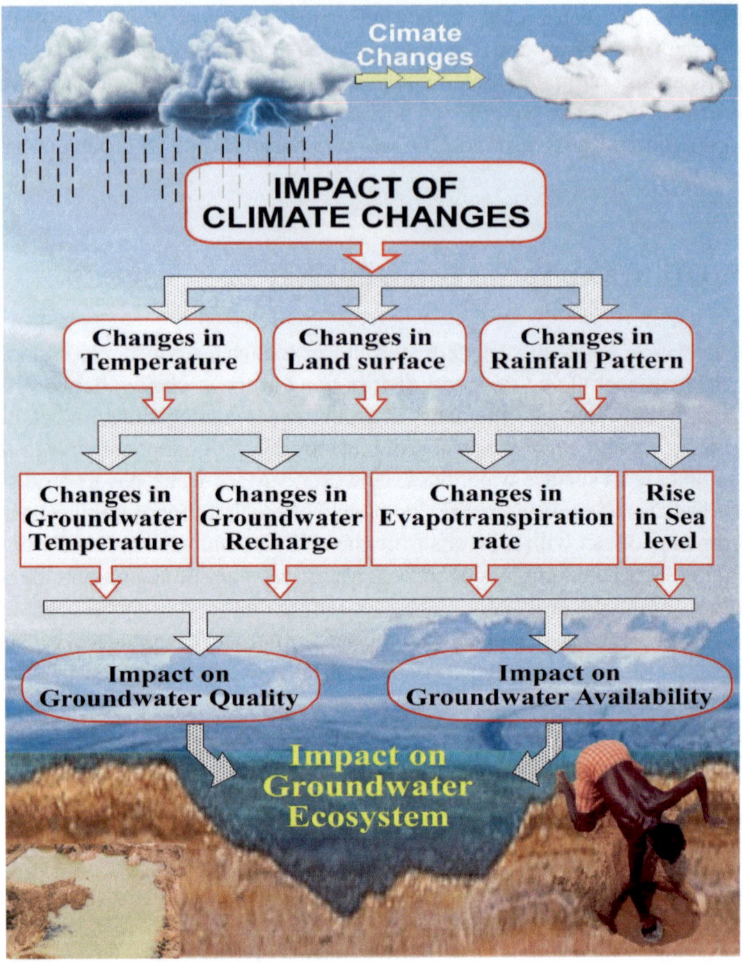

Fig. 1 Conceptual diagram illustrating natural effects on the groundwater ecosystem

6 Role of the Groundwater Cycle

Any dynamic system includes the inputs, outputs, and storage as prominent observable features. In watershed hydrology, various processes that influences water movement and storage are expressed as precipitation, interception, surface storage, evapotranspiration, infiltration, runoff, interflow, and groundwater flow (Fig. 2). The surface water, once it is heated by the sun, it vaporises and goes into the atmosphere as evaporation. The water vapour condenses and forms clouds, then creates precipitation, water drops fall as rain and snow. The rainwater is absorbed into the Earth and flows into the water bodies such as rivers, lakes, and oceans. Groundwater is also recharged by rainwater flowing from the building roof surfaces. Rainwater seeps into the ground, filling the spaces between soil particles and rocks, which will be stored in underground reservoirs called aquifers. The water cycle is a continuous process involving water movement between the Earth's surface and the atmosphere. The water cycle includes the processes of evaporation, transpiration, precipitation, and infiltration. Groundwater is an integral part of this cycle as it helps maintain the water balance on Earth (Grogan et al. 2022). Groundwater is replenished through infiltration. Rainwater recharges the aquifers. This process is essential for maintaining the water levels in the aquifers and providing base flow to rivers and streams. During dry periods, rivers and streams rely on groundwater to maintain flow. This is particularly important in arid regions where surface water is scarce. Overuse of groundwater can lead to depletion of the aquifers, which can have severe consequences for the environment and communities that rely on it. It is very critical to manage groundwater resources sustainably to ensure their long-term availability. Groundwater has multiple uses, including domestic, agricultural, industrial, and environmental.

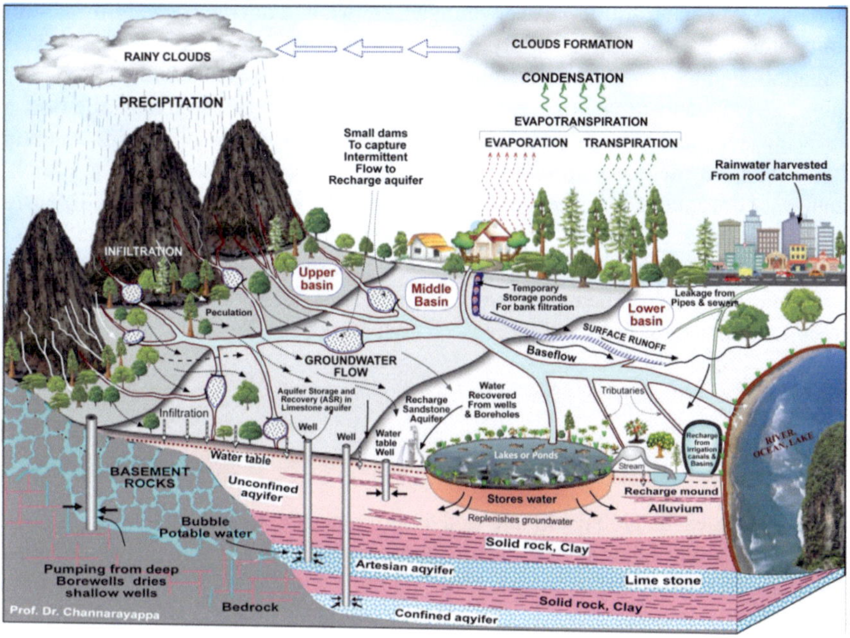

Fig. 2 The hydrologic cycle of groundwater and landscape: The diagram illustrates the watershed dynamics of the groundwater ecosystem

7 Groundwater Depletion and Its Impact on Society

Water is a highly exploited resource and, consequently, highly depleted. This is due to an increasing population. To realize the huge water requirements for domestic and agricultural uses, a system of groundwater pumping is used in vast areas, including in apartments, communities, and industries. Intensive industrialization and urbanization have a greater impact on the existing freshwater resources. Groundwater depletion can severely affect human activities that rely on it. Here are some of the effects of groundwater depletion:

Wells are essential for all societies, but due to groundwater depletion, wells are drying. Wells provide water for homes, irrigation, and industries. In deserts, wherever surface water is scarce, people use wells to get underground water. Groundwater is pumped out of the aquifer, and the soil above it can sink and settle, a process known as subsidence. This can cause damage to infrastructure such as buildings, roads, and bridges. In some cases, subsidence can also increase the risk of flooding and landslides. When an aquifer is over-pumped, saltwater from the ocean infiltrates and contaminates the freshwater supply. This can make the water unusable for drinking, irrigation, and all other purposes. Groundwater is often connected to surface water sources (rivers and streams). When groundwater depletes, it can reduce the flow of surface water sources, negatively impacting aquatic ecosystems and the communities that depend on them. As groundwater becomes scarcer, the usefulness of land use changes; consequently, farmers and other land users switch to less water-intensive crops. This can economically impact communities that rely on agriculture and water-dependent industries. Pumping groundwater requires more energy. This can have economic and environmental impacts, as increased energy use can lead to higher greenhouse gas emissions and increased consumer costs (Fig. 3).

8 Causes for Groundwater Depletion

Due to changes in global climate and interference of human activities, groundwater quality and quantity are drastically changing. The unpredictable precipitation and overuse of groundwater from aquifers without recharging cause water depletion. Continuous overuse can reduce long-term groundwater levels, threatening groundwater-dependent ecosystems such as springs and plant communities (Dehghani et al. 2022).

Climate change: The problems faced with freshwater shortage come from changes in the climate. Climate change escalates the global water cycle dynamics and subjects the world's masses to water-related crises. It is anticipated to increase severity in the coming years, due to the erratic rainfall patterns, frequent and severe floods and droughts, and rising sea levels. These changes threaten the ecosystem and can impede economic and social development. Earth's water resources are very

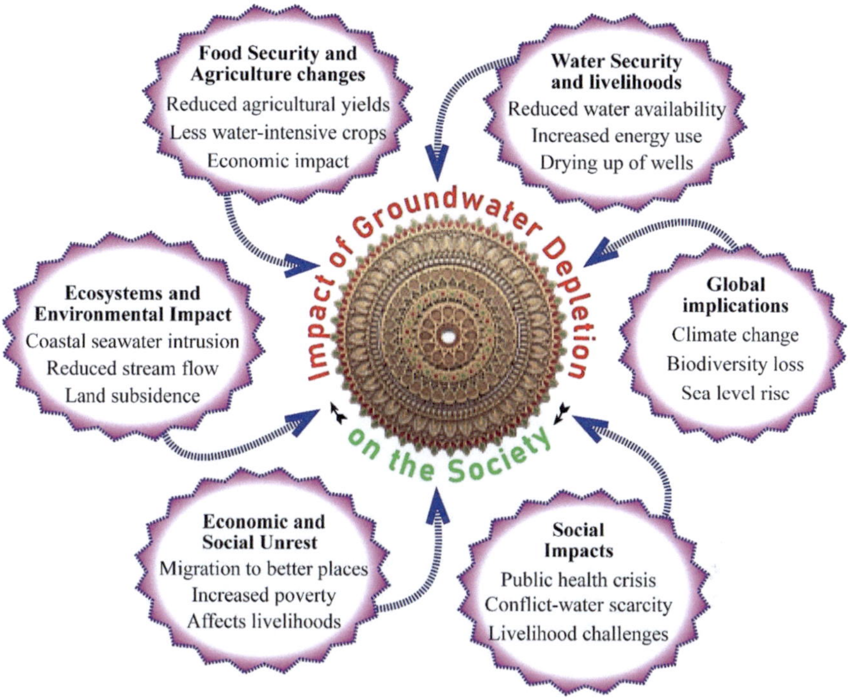

Fig. 3 The negative impacts of groundwater depletion on society, where water is used primarily for drinking, agriculture, and industrial purposes

much connected with climate dynamics. Therefore, it is crucial to have integrated approaches to have effective climate-resilient water management.

Excess withdrawal of groundwater: Overuse of groundwater increases rapidly in many areas, and is extracted faster than nature can replenish it. Frequent pumping out of excess groundwater from underground sources may deplete of groundwater levels. It adversely affects the availability for domestic and industrial uses, affecting agriculture and food security. The biological equilibrium and ecological cycle are also affected. This can counterbalance the water reserves generated by demand-side measures or the water increase produced by supply-side measures, which leads to further depletion.

Agricultural expansion and intensification (Food production): Agriculture needs a large volume of groundwater. Access to sufficient groundwater resources allows farmers to boost farming production significantly. Without the replenishment of aquifers, usage of groundwater for irrigation could lead to a decline in water tables, making agriculture less productive.

Groundwater contamination: Chemical and organic fertilizers used in agriculture can pollute groundwater and threaten human and ecosystem health. Contamination occurs when harmful substances are introduced into the groundwater (Hartmann

2022). Contamination can occur naturally or due to human activities such as industrial waste disposal, agricultural activities, and improper waste management. The World Bank suggests that pollution should be stopped by setting up protection zones around groundwater abstraction points.

Absence of practices to deposit rainwater back into aquifers: In the past century, groundwater consumption has increased fivefold, resulting in freshwater scarcity. Rainwater is a basic source of potable water when it's available. Unfortunately, we have not been able to adopt practices that help to deposit rainwater from our rooftops to the surface. An increase of local deposits against the withdrawal will help to raise the groundwater level and bring back balance to avoid the ingress of salinity from the sea (Fig. 4).

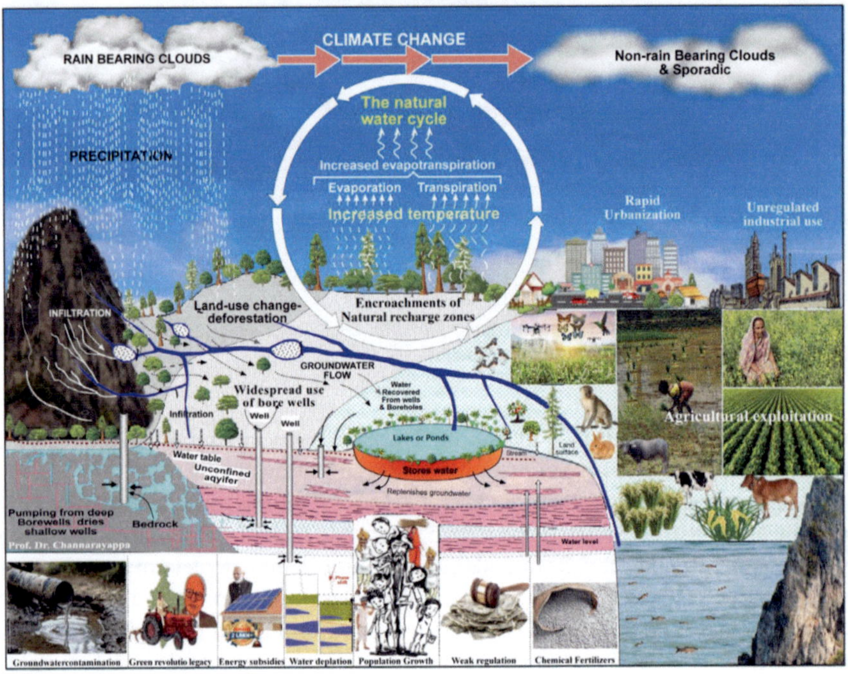

Fig. 4 The key drivers of groundwater depletion: notable causes, effects, and solutions

9 The Rate of Depletion of Groundwater Resources in Indian States

India is experiencing an accelerating groundwater depletion rate. The Central Groundwater Board (CGWB) reports that the groundwater depletion rate is alarming, with some regions exceeding the natural replenishment rates by as much as 65–100%. The rates of groundwater depletion vary significantly across Indian states, with some experiencing rapid depletion and others maintaining relatively stable levels (Fig. 5). Punjab, Haryana, and Rajasthan are known for high groundwater extraction, exceeding natural recharge rates. Bihar, Chhattisgarh, and Uttar Pradesh experience a moderate depletion rate. Arunachal Pradesh and Assam have relatively lower groundwater depletion rates.

The economic value of groundwater is increasing due to the worsening water quality and demand. Several methods have been used to price the groundwater. Unnecessary water use over the last decade has made the situation alarming, and the problem may further aggravate. To address the issue of the availability of quality water, feasible and facile methods include water storage, optimal use, and rainwater harvesting.

10 Conservation Techniques of Groundwater

Groundwater conservation involves protecting and managing underground water resources, ensuring their quality and quantity available for future use. Implementing the following methods can help conserve groundwater and water supply for future generations. There are several methods for groundwater conservation (Fig. 6), including collecting and storing rainwater for later use and reducing the demand for groundwater (rainwater harvesting). Water-efficient appliances and fixtures, fixing leaks, and reducing outdoor water use can all help conserve groundwater (Reducing water consumption). The right land use management can help reduce demand and recharge groundwater. For example, using drought-tolerant plants in landscaping can reduce outdoor water use. Artificial and natural recharging can help replenish aquifers and maintain groundwater levels. This can be achieved by diverting surface water to infiltration basins or injecting treated wastewater into the ground. Groundwater use can be controlled through permits and quotas to ensure sustainable resource use (regulating groundwater use). Proper disposal of waste, chemicals, plastics, etc. is essential to keep groundwater clean and safe (preventing pollution).

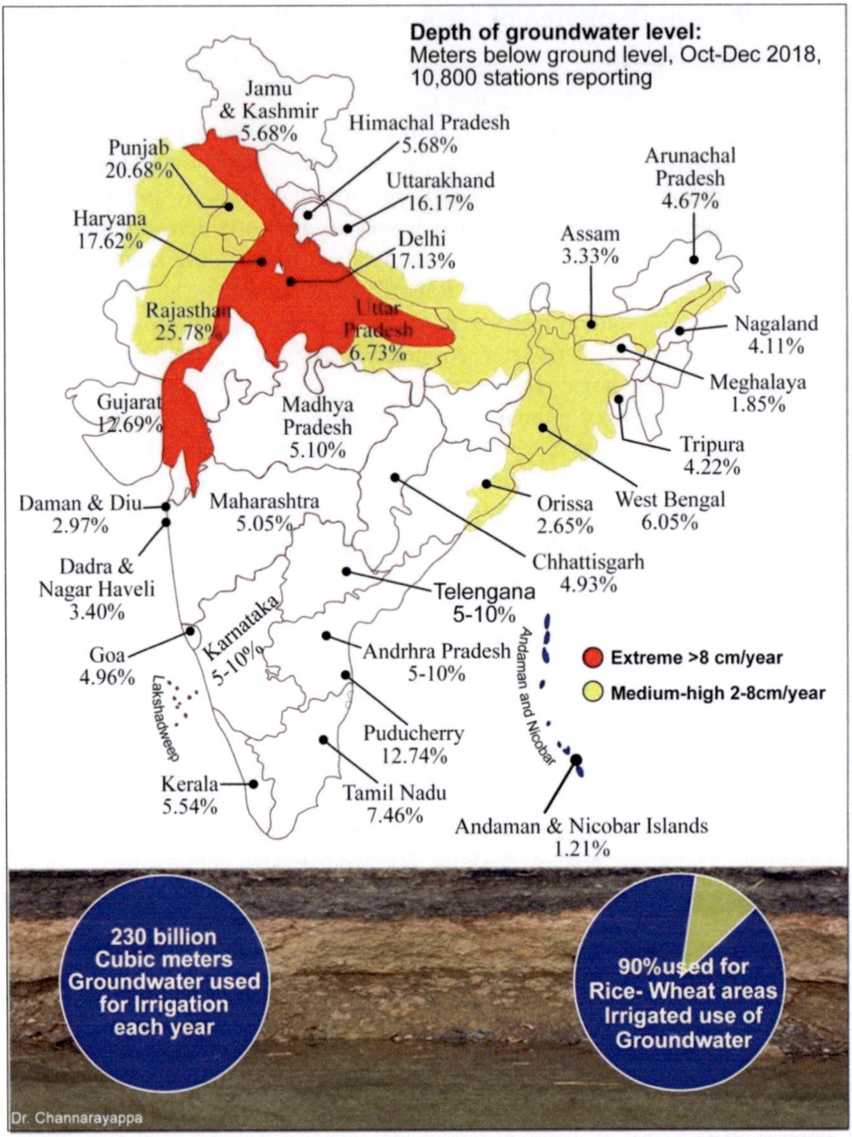

Sources: India Central Groundwater Board, WRI. Redrawn from Associated Press GRAPHIC NEWS

Fig. 5 The status of groundwater depletion in India (a case study). Groundwater depletion rates vary significantly across different states, with some experiencing severe over-extraction compared to natural recharge

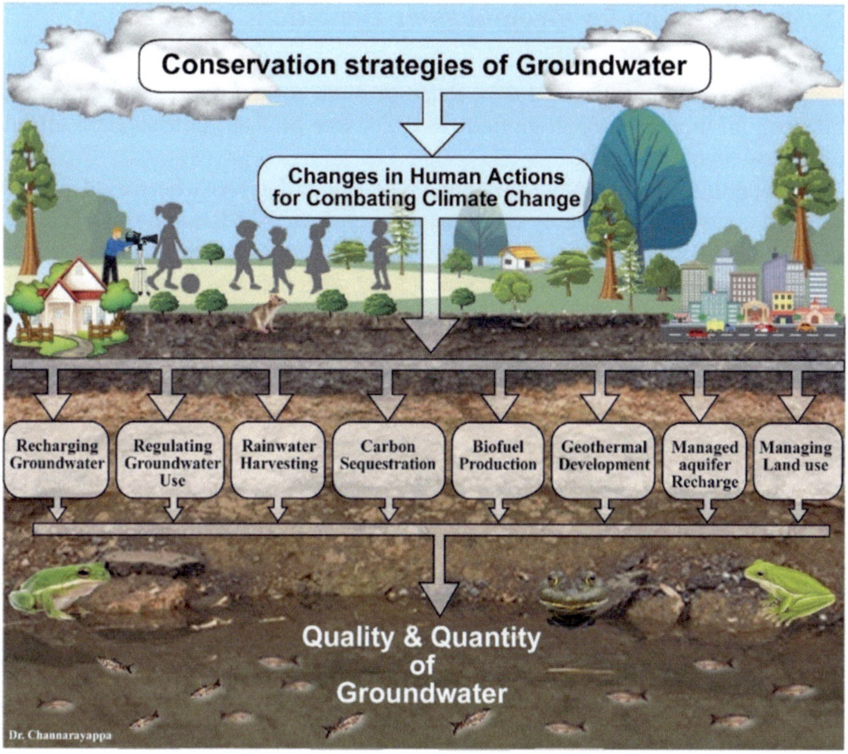

Fig. 6 Various conservation strategies followed in preservation of groundwater

11 Strategies to Rejuvenate Groundwater Depletion

Groundwater is a key factor in food production and security. The availability of freshwater is dependent on the economic growth of a nation. Climate alteration and population growth have seriously affected the global water resources, and are likely to pose more challenges in the future (Milly et al. 2005). The increase of irrigated agriculture in semi-arid areas with limited precipitation has greatly increased the dependence on groundwater withdrawal. This trend has been named the 'silent revolution of intensive groundwater use' (Lamas and Martínez-Santos 2005). A negative effect of groundwater depletion is a lowering of the water table (Fig. 6).

Water resource management personnel require positive attitudes and behavior from all stakeholders. The planning has to be community-driven, people-centric, and location-specific. The success of water-management strategies depends on good engineering, suitable agronomy, and socioeconomic considerations. The strategies focus on a project that needs additional water over the supply. The rise in groundwater or surface water supplies, conservation, reuse, demand management, weather alteration, and desalination are new strategies that need to be focused on.

12 Remediation for Groundwater Depletion

Groundwater depletion can be remediated through several approaches, including enhanced infiltration through groundwater recharge projects, conservation efforts, and management practices.

Check dams for stormwater management: Check dams have been used globally for flood control, water supply enhancement, and watershed restoration. They are small, temporary structures constructed within the concentrated-flow areas. Check dams are masonry structures for erosion and sediment control. They are appropriate where permanent channels are not yet vegetated or lining is infeasible. The rainwater stored in these check dams will improve groundwater levels and provide water for agriculture from open wells. Check dams are most effective when used in series, reducing erosion in concentrated-flow regimes (Chen et al. 2024).

Farm ponds: The farm ponds store the rainfall or the surface runoff from the surrounding areas. It solely depends on the rainfall present in the region. If the rainfall is not stored, most rainfall runs off the ground. The runoff causes water loss and washes away the fertile topsoil with it. Farm ponds improve agricultural water capture, storage, and irrigation, and use it for other purposes: recharging groundwater and providing economic and environmental benefits (Channarayappa and DP Biradar 2019).

Integrated demand and supply-side management (IDSM): The IDSM aims to coordinate energy demands and supplies within a multi-energy system (MES) to optimize efficiency and cost. The supply-side measures, like watershed management, aquifer recharging, and tank rehabilitation, these alone may not lead to long-term sustainable uses without demand-side management (DSM). Surface water withdrawing through farm ponds, check-dams, and efficient irrigation systems (e.g., more efficient drip and sprinklers), and growing less water-intensive crops need to be integrated on the demand side for improved management and reduced depletion. For example, check dams can be used where temporary channels are not vegetated. This practice is an emergency measure to limit flow in small open channels.

Use of Smart Water Systems: Traditionally, water waste is avoided using methods like drip irrigation and planting crops that use less water. Conservation efforts can be boosted to restore wetlands and green areas to increase water infiltration. However, smart irrigation systems utilize data-driven technology to optimize water usage in agriculture. It enhances water conservation efficiency by adopting intelligent technologies such as Topographic maps, GIS, and AI. It includes weather and soil moisture data to determine the irrigation needs of the landscapes and the products that maximize irrigation efficiency and reduce water waste while maintaining plant health. AI-based water management, along with the use of IoT systems, combines information obtained from several sources and optimizes water usage. These systems deliver the precise amount of water when and where the plants need it, maximizing crop yields and reducing costs. However, achieving high levels of precision requires combining the right tech stack with the right sensors, applications, platforms, and data (Mezouari et al. 2022).

Watershed management is a land area or a geohydrological unit that drains rainfall, snowmelt, and runoff into a common outlet, such as the outflow of a reservoir, a river, a creek, the mouth of a bay, or a large area covering thousands of square miles and multiple states or countries. Principally, the watersheds help replenish the groundwater. Watershed management aims to conserve the ecosystem and water resources while benefiting humanity. Watersheds are categorized by size, drainage, shape, and land use patterns.

We can identify the following parts or elements of watersheds (Fig. 2):

1. **Upper basin**: It constitutes the zone of birth of the river, which moves by a great slope.
2. **Middle basin**: It corresponds to the river valley area, and there is a balance between the solid materials carried by the current and deposited.
3. **Lower basin**: the lower zone of the river, where the material carried along the basin is deposited, due to the slower current speed. In general, plains are formed in this sector.

A natural watershed supports hydrologic and geomorphologic processes, native aquatic and riparian species, and healthy biological communities. Watersheds are critical for clean drinking water, outdoor recreation, a conducive environment, and a quality of life. The clean water is enormously sensitive to the conditions prevailing in the surrounding watershed. This is because pollutants can wash off from the land to the water sources and cause substantial harm (Fig. 2). Naturally, varying lake levels influence watersheds by moving water to and from groundwater. Other factors, such as forest fires, stormwater runoff patterns, location, and pollution, affect the quality. Protection of ecosystems recognizes their connectivity with each other and their surrounding watersheds. Unfortunately, human activities have greatly altered the water in the watershed areas.

13 Modern Water Conservation Techniques

Time-based assessment of water bodies using satellite imagery for pre- and post-rainy seasons can be recorded to elucidate the source of water, its quantity, quality, and the locations of water bodies. The maps obtained by these methods can be used as a preliminary reference in prioritizing the suitable sites for groundwater resources management and for rejuvenation.

13.1 Topographic Maps for Tracking Groundwater Sources

Topographic information is basic material for the science of geomorphology, and it includes values of various natural factors (Green et al. 2011).

Topographic maps are detailed, accurate graphical representations of features on the Earth's external surface area, culture, hydrography, relief, and vegetation. A topographic map of a large-scale detail and a quantitative representation of relief features usually consists of contour lines. A topographic survey is based on systematic observation and a map series. It depicts ground relief, forest cover, transportation routes, elevations, landfills, the contours, significant streams, water bodies, built-up areas, and the directions of stream flow.

The topography of a location significantly influences groundwater sources, flow patterns, and how it interacts with surface water bodies. Digital topographic data is recorded with the help of photogrammetry of stereo images taken from aircraft. Data collected from planetary orbits can result in spatial coverage that would take many years from airborne or terrestrial instruments. An alternative to topographic maps is contour lines to portray the shape and elevation of the land. Topographic maps provide the 3D elevations of the terrain on a 2D surface. Topographic maps depict geographic networks and a coordinate grid to determine relative and absolute positions of mapped features. The maps' names and locations of mountains, valleys, plains, lakes, rivers, and vegetation make them extremely useful to professional map users. Computer-assisted map production makes it easy to produce revised paper maps. Geographic information systems (GIS) make it possible to combine layers of digital data from several sources and to manipulate different layers related to each other.

13.2 Contour Maps with Contour Lines in Water Resources Engineering

Topographic maps with contour lines represent the shape of the Earth's surface, elevation, and terrain in a given region. These lines connect points of equal altitude, hills, valleys, and other landforms.

A contour map is a graphical depiction of a terrain area. Contour mapping is used in cartography and geography to represent the 3D Earth's surface on a two-dimensional map. Water table contours are drawn to join the area of equal heads, and the lines are called equipotential lines. The success of any groundwater study depends upon the availability and accuracy of measured data. Therefore, identifying and collecting data from integral parts of any groundwater exercise is essential. Contour plots are useful to depict functions of 2D input data with a one-dimensional output (Fig. 7). Contour maps have several important applications in water resources engineering: Delineation of the watershed, Flood mapping, Reservoir and dam planning, Groundwater modelling, Site selection, Erosion and sedimentation analysis, Geographic Information Systems, and Remote Sensing and Satellite Imagery.

Contour lines are continuous lines, connect points of equal elevation, close together, and are prepared by a contour marker (Nikalje et al. 2018). The contour lines are the function of two variables that share the same elevation. The topographic

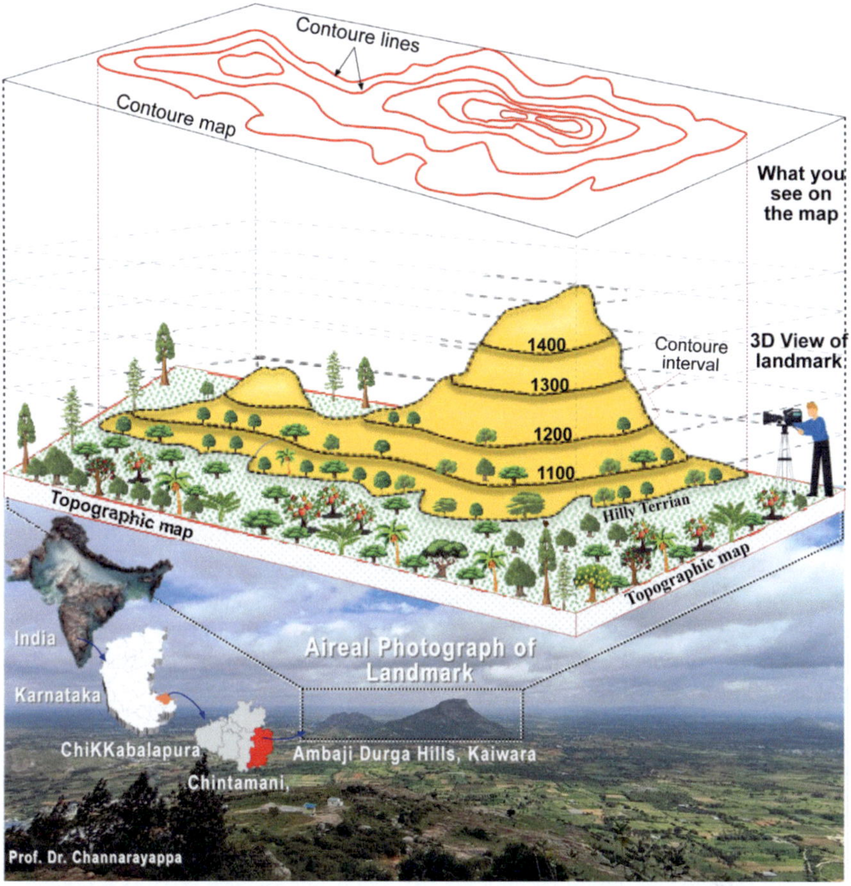

Fig. 7 The diagram represents hilly physical terrain along with its topographic map. The contour maps or contour lines depict the elevation of the terrain. Each curve in the figure above represents an increase in elevation of 100 ft

map displays valleys, hills, and the steepness of the slopes represented by contour lines. Contour lines depict the steepness of the terrain (Fig. 7). Contour line elevation changes rapidly in short distances if the terrain is steep. The difference in elevation between two successive contour lines is termed a contour interval. The different types of contour lines include Index contours, intermediate contours, and supplementary contours. Contour maps are an essential tool in various fields. They exhibit a way to visualize the topography of an area, making it easier to plan, study, and understand the terrain (Veldkamp et al. 2022).

Contour trenching for maximizing water conservation: Contour trenching is a commanding technique in water management and soil conservation areas, furnishing many environmental and agricultural benefits. Trenches are dug against a slope to capture and retain water flow in the direction of downhill and enhance infiltration rate into the soil below. This approach assists conservation and recharging of groundwater. However, the primary aim of contour trenching is to block the water flow across the landscape, allowing water to percolate into the soil rather than being lost as surface runoff (Poursaeid et al. 2022). Trenches are typically located in the upper reaches of a catchment area or recharge zones. This positioning allows them to intercept surface flow effectively. Trenches significantly contribute to groundwater recharge. For instance, in a 1-hectare area with clayey soil and a 25% slope, staggered trenches can increase water infiltration by up to 65%. The sustainability and effectiveness of trenches are enhanced by planting plants. Contour trenching is a powerful tool that offers significant environmental and agricultural benefits and supports resilience in climate change (Richards et al. 2023).

13.3 Village Tanks

A tank is a simple mud-based rainwater harvesting system traditionally followed by farmers, developed by indigenous wisdom. Most tanks are manmade and are called reservoirs. They are simple technological advances that accommodate the primary needs. The rainwater temporarily stored in these tanks for a season improves the groundwater, charges the open wells, and is used for domestic consumption, fisheries, and agricultural operations.

14 Artificial Intelligence in Groundwater Management

Recently established digital technology, Artificial Intelligence (AI), is a branch of computer science that deals with the simulation of intelligent behaviour displayed in a digital device. Groundwater restoration requires finding suitable locations for artificial recharge and reviving (Fig. 7). Since groundwater depository is a complex spatiotemporal phenomenon, it needs increased surface infiltration strategies to replenish aquifers. Several schemes exist for restoring groundwater supplies, but the traditional approaches alone may not be enough to discover suitable places for artificial recharge. In this regard, AI can augment supply insights, catchment area management, emergency response, treatment plant, and distribution networks, advanced service availability, and water justice (Poursaeid et al. 2022; Richards et al. 2023). AI systems are trained to create meaningful outputs from a defined dataset fed to the systems. These systems are first trained using descriptive data, i.e., supervised learning, to develop the ability to learn something. In the future, if similar kinds of data are fed to them, AI using ANNs will create meaningful data using their intelligence (Khare and Seth 2020). The technology used in AI and ICT-based nutrient and water management systems includes Sensors, Decision Support Systems, and Variable Rate Applicators. AI solutions can monitor and improve these functions while also increasing operational efficiency.

14.1 Groundwater Prediction Using AI Technology

Groundwater and surface water interplay with each other as the weather changes from summer to rainy seasons and vice versa. Groundwater resources safeguard against the depletion of freshwater resources due to climate fluctuations. Groundwater is critical for the survival of the global population, and they depend on it to meet their basic daily needs (UNESCO 2022). The rate of global groundwater depletion is more prominent in areas with denser populations, like China and India (Mukherjee 2016). As per the World Bank report, due to the world population growth, there will be a 40% shortfall between demand and available supply of water by 2030, which will lead to an additional 15% higher water withdrawal from already scarce groundwater resources (Sharma and Sharma 2024).

Groundwater in the region is highly dependent upon recharge from precipitation; hence, variability in precipitation patterns is important for sustainable groundwater management (Zaresefat et al. 2023). Climate change causes drastic effects on the frequency and intensity of extreme hydro-meteorological events and has a more complex impact on other variables such as annual precipitation and distribution. To create a preliminary model to predict future groundwater levels based on AI technology, the tree-based models (a type of ML algorithm used to make predictions suitable for classification and regression tasks) can correlate datasets like climate variables and changing demographics to predict future groundwater levels with the highest accuracy (Poursaeid et al. 2022).

14.2 Application of AI for Cities' Water Security

The existence of complex interrelations between forms of water and their uses has become a difficult task for ensuring urban water security. The Integrated Urban Water Management (IUWM) wants to intervene in the existing urban water use system, and recommends implementing nature-based solutions for wastewater treatment and disposal. These include rainwater harvesting, conserving natural lake systems, and using bio-swales instead of concrete-lined drains. The 'sponge city' concept, introduced in China in 2014, is an example of using the IUWM approach. Sponge cities can passively absorb, harvest, store, and purify rainwater and release it gradually when required, much like a sponge. Since water resource management poses challenges in actualising IUWM, the applications of AI technology may alleviate the problems. AI has the potential to understand complex interrelations and can build dynamic water models. Big data sets on weather patterns, water usage sectors, wastewater treatment networks, reservoir levels, groundwater availability, land use plans, and traffic data could be fed into AI models to identify patterns, inform decision-making, and suggest possible solutions.

14.3 Application of AI in the Water Sector

AI technologies have changed the technical landscape in many disciplines. AI in groundwater management is a transformative field described by innovation and challenges. It mainly emphasises augmenting resource management, improving water quality, and enhancing overall system efficiency. AI-driven innovations have been designed to reform groundwater management, provide precise predictive modelling, real-time monitoring, and data integration. AI innovations face several challenges, including interpretability issues, technical expertise requirements, and limited data for effective AI model performance. Advanced models can yield improved predictions of groundwater behaviour and offer guidance for water resource management (WRM). Digital twins enable prognostic maintenance of groundwater. These virtual simulations help analyse real-time data from multiple systems. The AI algorithms enable the repair of components before they fail. Because AI can detect drifts within large data, it can predict water demand to optimise supply throughout the day and minimise water waste and demand.

WRM uses diverse data, and fields like hydrology, meteorology, climatology, and geology can be integrated into a single platform using AI techniques. The ML algorithms analyse data in real-time and minimise losses. AI is taught to cogitate several weather parameters and respond to critical patterns. AI-enabled platforms are useful in water distribution networks and can identify anomalies in pressure flow data, an indication of leaks, and pinpoint their location. AI water monitoring systems can also inform the management about the status (Huntingford et al. 2019).

AI can analyse pollutants in water sources, ensuring public health. It leverages knowledge to remediate before they enter into crises. This enables data scientists and others to extract data, and the AI-enabled platforms allow utilities to monitor quality parameters. AI can also analyse sensor data to find anomalies and guide the taking of corrective actions to maintain safe and clean water supplies.

Overflows and floods are not unusual in the water cycle. In tapered regions, data on river levels, climate change, and historical flood patterns are used to predict and reduce the effects of runoff. AI-enabled water management is used to predict and mitigate the risks of floods, and develop various strategies to protect from floods. The industrial units are among the largest energy consumers. AI-driven optimisation algorithms assist in optimising treatment processes and minimising energy consumption, leading to significant cost savings. The digital twin lets individuals save energy and run tests using several variables.

Climate change makes it difficult to get enough water for a growing population. Building climate resilience involves taking proactive steps to anticipate, prepare for, and respond to the influences of climate vagaries. AI can enhance climate resilience by supporting adaptation and mitigation efforts. AI water management processes vast data to predict consumption patterns, detect leaks, and optimise distribution. AI-driven smart water management simulations exploit monetisation, and the processed data allows policymakers to evolve effective strategies for managing water sources and supplies (McIntosh 2025).

Digital twin tells sequences of events and refines the original data. These revelations allow planners to expect and prepare for strategies for unpredictable weather. It is possible to recognise probable susceptibilities so that pre-emptive arrangements can be made pre-emptive arrangements and groundwork for climate vagaries. Because vast conflicting data on weather, water tables, soil moisture, drought patterns, and other historical climate-related information that is not comprehensive. Policy makers can leverage combinations of AI-enabled technologies to predict and prepare water-related disasters, enabling early warnings and preventative measures.

Undoubtedly, the integration of AI technologies into water management enhances its efficiency in water resources management practices and decision-making. However, simultaneously raises questions about the ethical dimension of AI, and it extends beyond mere efficiency gains. The application of AI must be guided by principles of inclusivity, transparency, and responsibility to ensure that the rights of all personnel to access water are upheld. Misinterpretation and anomalies of water distribution may have profound effects on society. It requires the right attitude towards preserving and managing the water resource for all and forever.

15 Application of RS in Groundwater Management

The application of RS and GIS in the spatial analysis of a study area's groundwater is a more effective technique for evaluating and describing the spatial distribution.

15.1 RS and GIS Technologies Used for the Collection of Water Status in Large Areas

Remote sensing provides large amounts of data on the Earth's surface. RS and GIS automatically process a large amount of data on a computer. GIS is a tool that lets users create multi-layered interactive maps, which are used for visualising complex data and spatial analysis; it also includes surface, soil type, land cover, elevation, and weather conditions. The RS and GIS used for aerial photography and multispectral satellite imaging (Fig. 8). Application of RS and GIS technology in the spatial analysis and interpretation of the study area's groundwater is a tool for evaluating and describing the spatial distribution of the groundwater. GIS lets users to create multi-layered interactive maps, which includes surface topology, soil type, land cover, elevation, and weather conditions. The technologies and data sources used for geospatial analysis are: Aerial imagery, GPS, RS devices, monitoring tools, Public databases, Variable-rate technology, Agricultural mapping, Autonomous machinery, Water management, Assessing natural disaster damages, and Precision farming.

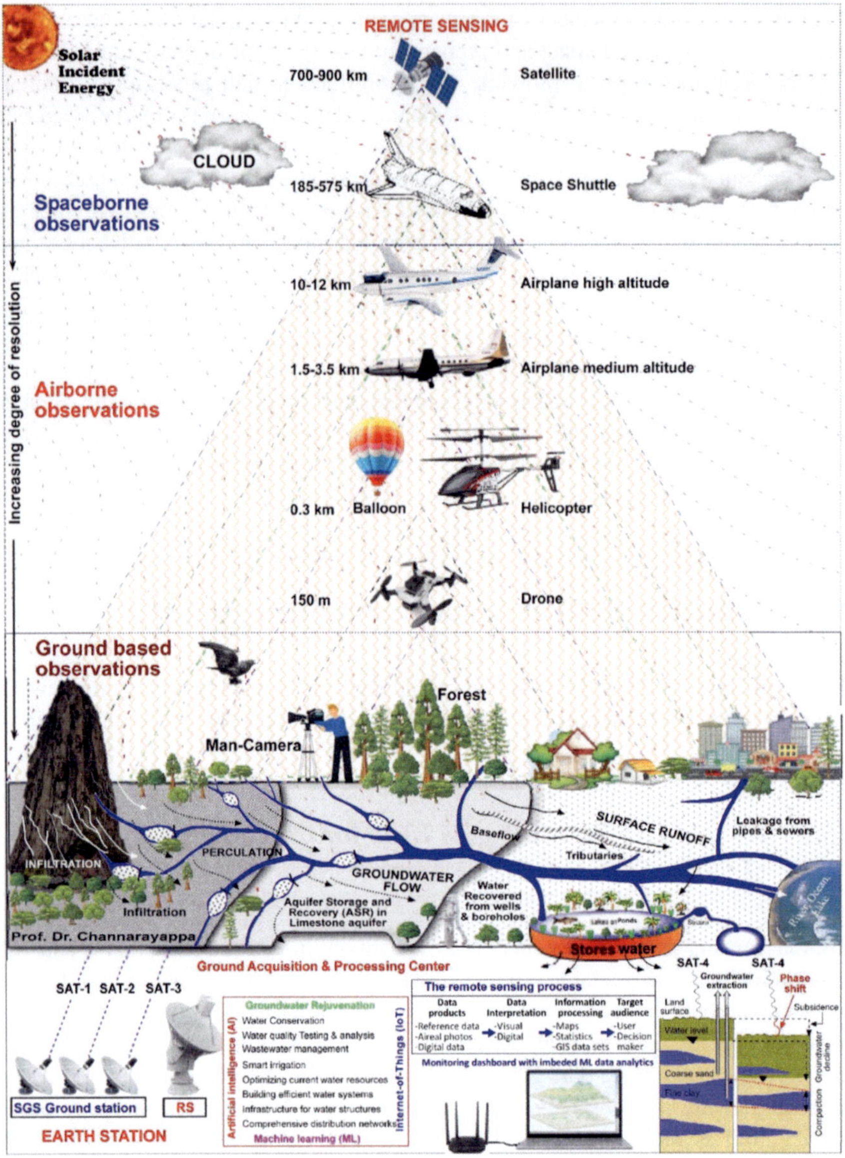

Fig. 8 Various platforms are used for data collection, storage, manipulation, and retrieval of spatially referenced data through remote sensing, GIS, and AI for natural resource management

15.2 Inverse Distance Weighted Method

Spatial analysis of geography refers to a set of methods used to analyse location-based data, to reveal the patterns, relationships, and distribution of geographic characteristics. Several spatial interpolation methods have been developed to support the transformation from point data to continuous surface maps (Chen et al. 2017). The simplest and frequently used method is inverse distance weighted (IDW), which is a simple and intuitive deterministic method for multivariate interpretation with a known homogeneously scattered set of points. The IDW is available in almost any GIS software and is applied frequently in various disciplines (Li and Heapc 2008). The main advantage of IDW is keeping the measured values at the sample location. Many studies have reported that IDW interpolation is inclined to produce "bull's eye" patterns in contour plotting. The IDW formula estimates a value of an unknown location (missing station) using a weighted average of known values (indexing stations) at surrounding locations. The weight of each known point is determined by its inverse distance from the unknown point Z (Fig. 9). In this example, IDW is used to estimate rainfall values at an unmeasured location (Z) based on known rainfall values at nearby four locations (A, B, C, D). However, IDW may not be suitable for regions with complex rainfall patterns that are influenced by elevation or other factors.

The sigma ($\sum$) notation means that whatever number of points is added will be interpolated. Here we are simply summing the average normal rainfall values at each indexing station (A, B, C, and D points), d_1, d_2, d_3, and d_4 distances between the indexing rain stations and the value missing station (Z) (Table 1).

15.3 Importance of GIS in Water Management

The water management sector depends more on GIS technologies than traditional methods. With GIS, various sensors, autonomous machines, and AI-powered analytics, organisations can predict crop yields, detect soil erosion, and assess environmental threats with remarkable accuracy. Implementation of GIS technology is a wise move for the future. GIS applications, particularly in water management, are becoming efficient in spatial data monitoring, simulations, interactive mapping, and visual presentation. This helps water management become productive, profitable, and sustainable (Preeja et al. 2011).

16 Conclusions

The information relevant to groundwater depletion and its consequences on agriculture and human well-being has been briefly discussed. Groundwater is being depleted at an alarming rate, it needs immediate attention to prevent and restore

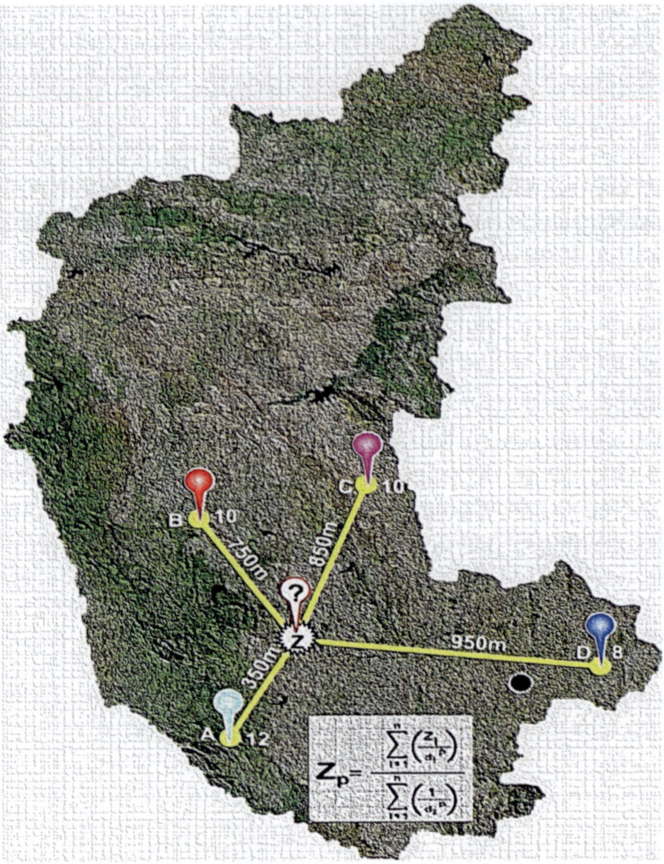

Fig. 9 Approximation of average rainfall received at an unknown location using the Inverse Distance Weighted spatial interpolation method. The four closest points or indexing stations (A, B, C, D) to the unknown point or value missing station (star shape, Z)

the damage it may cause soon. Various rejuvenation methods have been proposed for recharging the groundwater. Digital technology shared with satellite navigation systems, satellite-based remote sensors, and high-throughput software has ushered in a new era of surface water monitoring and finding appropriate solutions for mitigating groundwater depletion and recharging approaches. Among various modern techniques used in contour maps/lines, AI and GIS find effective solutions. These technologies provide high accuracy in saving time and cost, and their transformative potential in addressing the multifaceted complexities of water resources.

Table 1 Findings of average rainfall value for missing station (Z) from indexing stations (A, B, C, D), average normal rainfall values recorded as shown in the above diagram (Fig. 9)

	Distance (m)	Indexing station value	Weighted values		Numerator/Denominator	Missing station (Z) value
For a Power of 1, that cell value is equal to						
A	350	12	A/d = 0.0342857		$\sum$(A/d) + (B/d) + (C/d) + (D/d) = 0.0878044	(0.0878044)/(0.0064181) = **11.07** **13.68074**
B	750	10	B/d = 0.0133333			
C	850	10	C/d = 0.0117647			
D	950	8	D/d = 0.008421			
A	350	1	1/d = 0.002857		$\sum$(1/d) + (1/d) + (1/d) + (1/d) = 0.0064181	
B	750	1	1/d = 0.0013333			
C	850	1	1/d = 0.00117647			
D	950	1	1/d = 0.00105263			
For a Power of 2, that cell value is equal to:						
A	350	12	d^2 = 122,500	A/d^2 = 9.78598E-5	$\sum(A/d^2) + B/d^2) + (C/d^2) + (C/d^2)$ = 13.834257	(13.834257)/(1.242844) = **10.34** **11.1311**
B	750	10	d^2 = 562,500	B/d^2 = 1.77777E-5		
C	850	10	d^2 = 722,500	C/d^2 = 1.38408E-5		
D	950	8	d^2 = 902,500	D/d2 = 0.886427E-5		
A	350	1	d^2 = 122,500	$1/d^2$ = 8.16326E-6	$\sum(1/d^2) + (1/d^2) + (1/d^2) + (1/d^2)$ = 1.242844	
B	750	1	d^2 = 562,500	$1/d^2$ = 1.77777E-6		
C	850	1	d^2 = 722,500	$1/d^2$ = 1.38408E-6		
D	950	1	d^2 = 902,500	1/d2 = 1.10333E-6		

Conclusion: The missing station value is the power of 1 (**13.6807**) and the power of 2 (**11.1311**); both values are closely related.

References

Brown and Molly (2015) Satellite remote sensing in agriculture and food security assessment. Procedia Environ Sci 29:307

CGWB (2015) Aquifer mapping and ground water management. Chennai: central ground water board

Channarayappa, Biradar DP (2019) Soil basics, management and rhizosphere engineering for sustainable agriculture. Publisher: CRC PR INC. Taylor and Francis Group. ISBN-13: 978-1-138-48692-8 (Hardback). https://doi.org/10.1201/9781351044271

Chen T, Ren LL, Yuan F (2017) Comparison of spatial interpolation schemes for rainfall data and application in hydrological modelling. Water 9(5):342–342

Chen CF, Chen C-W, Jay FP, Lee AS-H, Hsu CW, Dong C-D (2024) Sediment organic matter predicts polycyclic aromatic hydrocarbon distribution in port sediments. Mar Pollut Bull 207(10):116869

Dehghani R, Poudeh HT, Izadi Z (2022) The effect of climate change on groundwater level and its prediction using modern meta-heuristic model. Groundw Sustain Dev 16:100702

Ebi KL, Vanos J, Baidwin JW, Bell JE, Hondula DM, Errett NA, Hayes K, Reid CE, Saha S, Spector J, Berry P (2021) Extreme weather and climate change: population health and health system implications. Annu Rev Public Health 42:293–315

Green TR, Taniguchi M, Gurdak JJ, Allen DM (2011) Beneath the surface of global change: impacts of climate change on groundwater. J Hydrol 405(3–4):532–560

Grogan DS, Xuidema S, Prusevich A, Wollheim WM, Gilidden S and Lammers RB (2022) Water Balance model (WBM) v.1.0.0: a scalable gridded global hydrologic model with water-tracking functionality. Geosci Model Dev 15(19):7287–323

Hartmann A (2022) The hydrology of groundwater systems—from recharge to discharge. In: Encyclopedia of inland waters. Elsevier, pp 324–330. https://doi.org/10.1016/b978-0-12-819166-8.00097-9

Huntingford C, Jeffers ES, Bonsall MB, Christensen HM, Lees T, Yang H (2019) Machine learning and artificial intelligence to aid climate change research and preparedness. Environ Res Lett 14(12):124007

Ibrahim A, Wayayok A, Shafri HZM, Toridi NM (2024) Remote sensing technologies for unlocking new groundwater insights: a comprehensive review. J Hydrol X 23:100175

Khare V, Seth A (2020) The case for AI in smallholder farmer water management. In: AI holds water: a case for AI-enabled water management in agriculture. White Paper, Wadhwani Institute for Artificial Intelligence, Mumbai, India

Lamas MR, Martínez-Santos P (2005) Intensive groundwater use: a silent revolution that cannot be ignored. Water Sci Technol 51(8):167–174

Li J, Heapc AD (2008) A review of spatial interpolation methods for environmental scientists, Canberra: Geoscience Australia

Lorenzo-Lacruz J, Garcia C, Moran-Tejeda E (2017) Groundwater level responses to precipitation variability in Mediterranean insular aquifers. J Hydrol 552(9):516–531

McIntosh R (2025) Water AI: 8 ways AI in water management creates a better future. Sand Technologies

El Mezouari A, El Fazziki A, Sadgal M (2022) Smart irrigation system. IFAC-Pap Online 55(10):3298–3303

Milly PC, Dunne KA, Vecchia AV (2005) Global pattern of trends in streamflow and water availability in a changing climate. Nature 438:347–350

Moeck C, Grech-Cumbo N, Podgorski J, Bretzler A, Gurdak JJ, Berg M, Schirmer MA (2020) Global-scale dataset of direct natural groundwater recharge rates: a review of variables, processes and relationships. Sci Total Environ 15(717):137042

Mukherjee D (2016) A review on artificial groundwater recharge in India. SSRG Int J Civ Eng (SSRG-IJCE) 3(1):57–62

Nikalje S, Shinde P, Chavhan S, Saruk D (2018) A water conservation technique: continuous contour trenches. VJER-Vishwakarma J Eng Res 2(4):186–194

Poursaeid M, Poursaeid AH, Shabanlou SA (2022) Comparative study of artificial intelligence models and a statistical method for groundwater level prediction. Water Resour Manag 36:499–1519

Preeja KR, Joseph S, Thomas J, Vijith H (2011) Identification of groundwater potential zones of a tropical river basin (Kerala, India) using remote sensing and GIS techniques. J Indian Soc Remote 39:83–94

Richards CE, Tzachorc A, Avin S, Fenner R (2023) Rewards, risks and responsible deployment of artificial intelligence in water systems. Nature Water 1:422–432

Rodell M, Famiglietti JS, Wiese DN, Reager JT, Beaudoing HK, Landerer FW, Lo M-H (2018) Emerging trends in global freshwater availability. Nature 557:651–659
Samadder RK, Kumar S, Gupta RP (2011) Paleochannels and their potential for artificial groundwater recharge in the western Ganga plains. J Hydrol 400:154–164
Sharma V, Sharma S (2024) Groundwater prediction using artificial intelligence: case study for Texas aquifers. J Emerg Investig 7:1–10
TWB (2010) Deep wells and prudence: towards pragmatic action for addressing groundwater overexploitation in India. The World Bank, Washington, DC
UNESCO (2022) World water assessment programme. The United Nations world water developmental report 2022: groundwater: making the invisible visible unesdoc. unesco.org/ark:/48223/pf00003807213.12
Veldkamp TIE, Wada Y, Aerts JCJH, Doll P, Gosling N, Liu J et al (2022) Water scarcity hotspots travel downstream due to human interventions in the 20th and 21st century. Nat Commun 8:15697. https://doi.org/10.1038/ncomms15697
Zaresefat M, Derakhshani R, Nikpeyman V, GhasemiNejad A, Raoof A (2023) Using artificial intelligence to identify suitable artificial groundwater recharge areas for the Iranshahr Basin. Water 15:1–12

Anammox Technology: Low-Carbon and Energy-Saving Solutions for Enhanced Water Pollution Control

Yuxiong He, Hanbing Ai, and Litong Shi

Abstract In recent years, partial denitrification-anammox (PDA) technology has emerged as an efficient and energy-saving solution for nitrogen removal in municipal wastewater treatment. However, challenges such as the retention of anammox bacteria (AnAOB), unstable nitrogen removal performance, high inoculation costs, and competition with heterotrophic bacteria hinder its widespread implementation. This chapter addresses these challenges by exploring the integration of PDA technology into traditional denitrifying biofilters (DNBFs). A partial denitrification filter was developed by inoculating conventional activated sludge to produce nitrite as a substrate for AnAOB growth. Two methods for inoculating anammox sludge were assessed, and the feasibility of in situ enrichment of AnAOB in single-stage and two-stage filter systems was examined. The results demonstrated the feasibility of applying anammox in DNBFs, although achieving self-enrichment of AnAOB remains difficult. In a single-stage system, NH_4^+-N removal was limited to 0.55 mg/L after 111 days, while a two-stage system achieved 6.45 mg/L after 164 days. Metagenomic analysis confirmed successful enrichment of Candidatus Brocadia in the secondary biofilter, with key AnAOB enzymes present. This study lays a foundation for advancing PDA technology in municipal wastewater treatment.

Keywords Partial denitrification · Anammox · Denitrification biofilter · Advanced nitrogen removal

Y. He (✉)
School of Environmental Science and Engineering, Sun Yat-Sen University, Guangzhou, China
e-mail: 15595668657@163.com

H. Ai
School of Geophysics and Geomatics, China University of Geosciences, Wuhan, Hubei, China

L. Shi
Key Laboratory of Agro-Forestry Environmental Processes and Ecological Regulation of Hainan Province, School of Environmental Science and Engineering, Hainan University, Haikou, China

A. Nath et al. (eds.), *Groundwater Depletion and Sustainability*, Advances in Geographic Information Science, https://doi.org/10.1007/978-3-032-09921-1_15

1 Introduction

Water pollution control is a critical aspect of environmental engineering aimed at mitigating the adverse effects of contaminants in water bodies (Chen et al. 2019). Traditional methods have relied heavily on chemical and physical processes, but biological technologies have emerged as sustainable and efficient alternatives. One such innovative approach is the use of biofilters for nitrogen removal, particularly focusing on the transformation of conventional denitrifying biofilters to PDA biofilters.

Traditional biological nitrogen removal technology includes nitrification and denitrification reactions (Saunders et al. 2016). The nitrification reaction occurs under aerobic conditions, where nitrifying bacteria use O_2 to oxidize ammonia nitrogen (NH_4^+-N) to nitrite nitrogen (NO_2^--N), which is then further oxidized to nitrate nitrogen (NO_3^--N). This process is completed by ammonia-oxidizing bacteria (AOB), ammonia-oxidizing archaea (AOA), and nitrite-oxidizing bacteria (NOB). Additionally, the nitrification process can also be completed independently by complete ammonia oxidizers (COMAMMOX) (Su et al. 2025). The denitrification reaction occurs under anoxic conditions, where denitrifying bacteria use reductive substances to reduce NO_3^--N to N_2, ultimately completing the nitrogen removal from wastewater.

Denitrification biofilters (DNBFs) are widely used in wastewater treatment plants (WWTPs) for efficient nitrate removal from secondary effluent (Zheng et al. 2018). These biofilters use microbial communities to convert nitrates into N2, reducing nitrogen loads in the treated water (Cui et al. 2020). DNBFs serve as secondary treatment technologies for advanced nitrogen removal, typically combined with mechanical filtration and nitrification filters. They are used for wastewater treatment before reuse. In practical applications, the need for added organic carbon to drive denitrification remains a challenge.

In 1977, Austrian chemist Broda hypothesized the existence of microorganisms capable of oxidizing ammonia with NO_2^- as an electron acceptor, which was confirmed nearly two decades later by Mulder et al. with the discovery of the anaerobic ammonium oxidation (Anammox) process (Mulder et al. 1995). This biological reaction, mediated by Planctomycetes bacteria, has since revolutionized our understanding of the nitrogen cycle and inspired innovative wastewater treatment technologies. Anammox bacteria, predominantly found in wastewater treatment systems and natural ecosystems, utilize NO_2^- to convert NH_4^+ into N_2 gas, offering a low-energy and sustainable alternative to conventional denitrification (Kartal et al. 2011).

Despite their potential, anammox bacteria face challenges in practical applications due to their long doubling times, sensitivity to environmental changes, and stringent operational requirements (Wang et al. 2025). By optimizing key factors such as substrate concentration, temperature, and dissolved oxygen levels, the efficiency and stability of Anammox systems can be improved, paving the way for broader applications in nitrogen removal and environmental protection.

Denitrification filters have long been regarded as an effective solution for removing nitrogen from wastewater, playing a vital role in improving water quality and protecting aquatic ecosystems. However, these traditional systems often rely on substantial external carbon sources, which increases operational costs and contributes to the environmental burden. In recent years, the emergence of the PDA process has provided a promising alternative. By integrating partial denitrification and anammox processes, PDA filters significantly reduce the need for carbon sources while enhancing nitrogen removal efficiency (Lu et al. 2021). This innovative approach has demonstrated the potential to transform denitrification filters into more sustainable and cost-effective solutions.

Nevertheless, the widespread application of PDA technology faces notable challenges. A key limitation lies in the scarcity of anammox sludge, which is essential for initiating the process. Anammox bacteria, the cornerstone of this technology, are known for their slow growth, requiring up to 11 days to double their population (Vigne et al. 2010). This, combined with their sensitivity to environmental conditions, makes the direct inoculation of sufficient anammox biomass impractical for large-scale wastewater treatment plants. As a result, researchers have shifted their focus towards enabling the self-enrichment of anammox bacteria within existing systems.

Interestingly, there have been encouraging examples of such self-enrichment in full-scale wastewater treatment plants. At Singapore's Changi Water Reclamation Plant, anammox bacteria were gradually enriched over time, contributing significantly to total nitrogen removal (Cao et al. 2017a). A similar achievement was reported in Xi'an, China, where anammox bacteria naturally developed within biofilms, contributing to approximately 15% of nitrogen removal (Li et al. 2019). These successes highlight the potential for anammox bacteria to thrive under specific operational conditions without the need for external inoculation.

Denitrification filters, with their anaerobic environments and large specific surface areas, appear to provide a suitable habitat for anammox bacteria (He et al. 2024). These systems create conditions conducive to biofilm development, which is essential for supporting the growth of slow-developing microorganisms like anammox bacteria. However, practical challenges remain. The inherent limitations of biofilm systems, such as restricted mass transfer, combined with operational requirements like periodic backwashing, pose significant obstacles to the successful enrichment of anammox bacteria in denitrification filters.

Given the growing need for sustainable wastewater treatment solutions, the exploration of anammox technology within denitrification filters holds great promise (Al-Hazmi et al. 2023; Sun et al. 2025). By building on existing knowledge and addressing these challenges, it is possible to develop innovative strategies that harness the unique capabilities of anammox bacteria. Such advancements could pave the way for a new generation of efficient and environmentally friendly wastewater treatment systems.

2 Materials and Methods

Lab-scale biofilters were employed to construct the DNBF system. Each biofilter had a height of 100 cm and a working volume of 0.7 L, with ceramsite (3.0–5.0 mm in diameter) used as the filter medium (Ma et al. 2020). The bed height for both biofilters was set at 35 cm. Initially, a single-stage DNBF was established for testing. However, due to unsatisfactory results, a two-stage DNBF system, consisting of a primary and a secondary biofilter connected in series, was reconstructed for further experiments (Fig. 1).

Synthetic wastewater was continuously supplied to the biofilters via a peristaltic pump, maintaining an influent flow rate of 0.84 L/h. This resulted in a hydraulic retention time (HRT) of 0.78 h for the single-stage system and 1.56 h for the two-stage system. As biofilm developed, filtration resistance increased, leading to potential clogging of the filter beds. To prevent this, a backwash system was incorporated into each biofilter (Wei et al. 2014). Backwashing was initiated when the pressure at the reactor's base exceeded 0.02 Mpa, with nitrogen gas pumped into the bottom of the biofilter to facilitate the process. Each biofilter was inoculated with an equal volume of returned sludge (MLSS: 2270 mg/L, MLVSS: 1180 mg/L) collected from the Baishamen WWTP (Haikou City, Hainan Province, China).

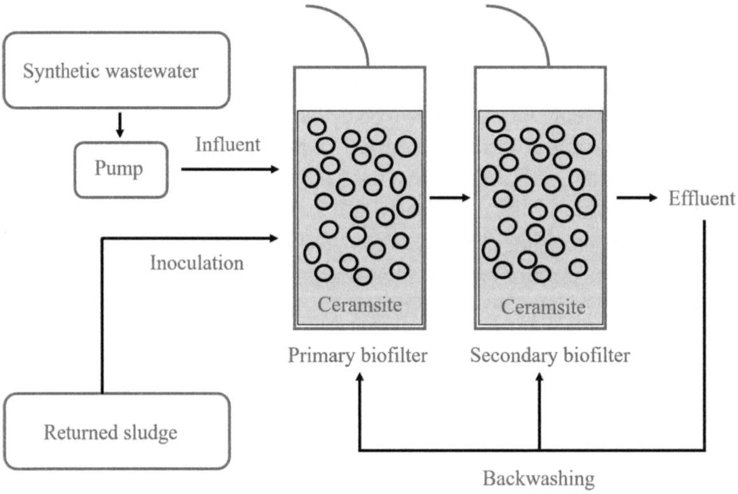

Fig. 1 Methodological flow diagram of the DNBF system

3 Results and Discussion

The research process adopted the method of directly inoculating the biofilm filter of the denitrification filter of a wastewater treatment plant to start the single-stage PDA filter in situ. This approach more closely simulates the actual working scenario of a wastewater treatment plant. The operation lasted for 111 days, with the influent carbon source controlled to ensure the effluent TN remained below 15 mg/L to meet discharge standards.

The first stage operated for 14 days without adding NH_4^+-N to the influent. From the first day, this reactor exhibited some nitrite accumulation, with an average nitrite accumulation of 3.73 mg/L during the first stage, indicating that the reactor inoculated with biofilm media from the wastewater treatment plant's denitrification filter had the capability for partial denitrification, providing substrates for anammox bacteria (Fig. 2a). In the second stage, NH_4^+-N was added to the influent to attempt in situ enrichment of anammox bacteria. Throughout the process, the average influent NH_4^+-N concentration was maintained at 2.6 mg/L, and the average effluent nitrite concentration was 5.17 mg/L. During the 111-day operation of the reactor, the in situ anammox activity was evaluated by the ammonia nitrogen removal ΔNH_4^+-N in the single-stage denitrification filter. However, the average ΔNH_4^+-N in the second stage was only 0.62 mg/L, still showing no significant increasing trend (Fig. 2b). Additionally, the average influent organic matter concentration throughout the process was 86.98 mg/L, and the average effluent organic matter concentration was 46.26 mg/L (Fig. 2c). Notably, the TN removal efficiency remained at 56% during the second stage, indicating poor nitrogen removal performance and no improvement during the anammox bacteria enrichment process (Fig. 2d), consistent with the previous study. During the entire 111-day operation, the single-stage denitrification filter was backwashed approximately once a week. It is speculated that the high influent organic matter concentration in the single-stage denitrification filter may be the direct cause limiting the self-enrichment of anammox bacteria, thus not showing a significant increase in ΔNH_4^+-N.

Based on the above studies, which were unable to achieve better ammonia nitrogen removal performance during the long-term operation of single-stage denitrification filters, it was concluded that the main influencing factor might be the abundant organic matter in the influent. This abundance of organic matter makes it difficult to establish an ecological niche suitable for anammox bacteria in single-stage denitrification filters. Therefore, in the next set of experiments, two sets of denitrification filters were connected in series to try to create a suitable space for the synergistic nitrogen removal of anammox bacteria and partial denitrifying bacteria in a two-stage denitrification filter system.

The primary denitrification filter was used to consume a large amount of carbon source in the influent and control the influent conditions to complete the partial denitrification function. This ensures that the secondary filter provides sufficient substrates (ammonium and nitrite) for the growth of anammox bacteria.

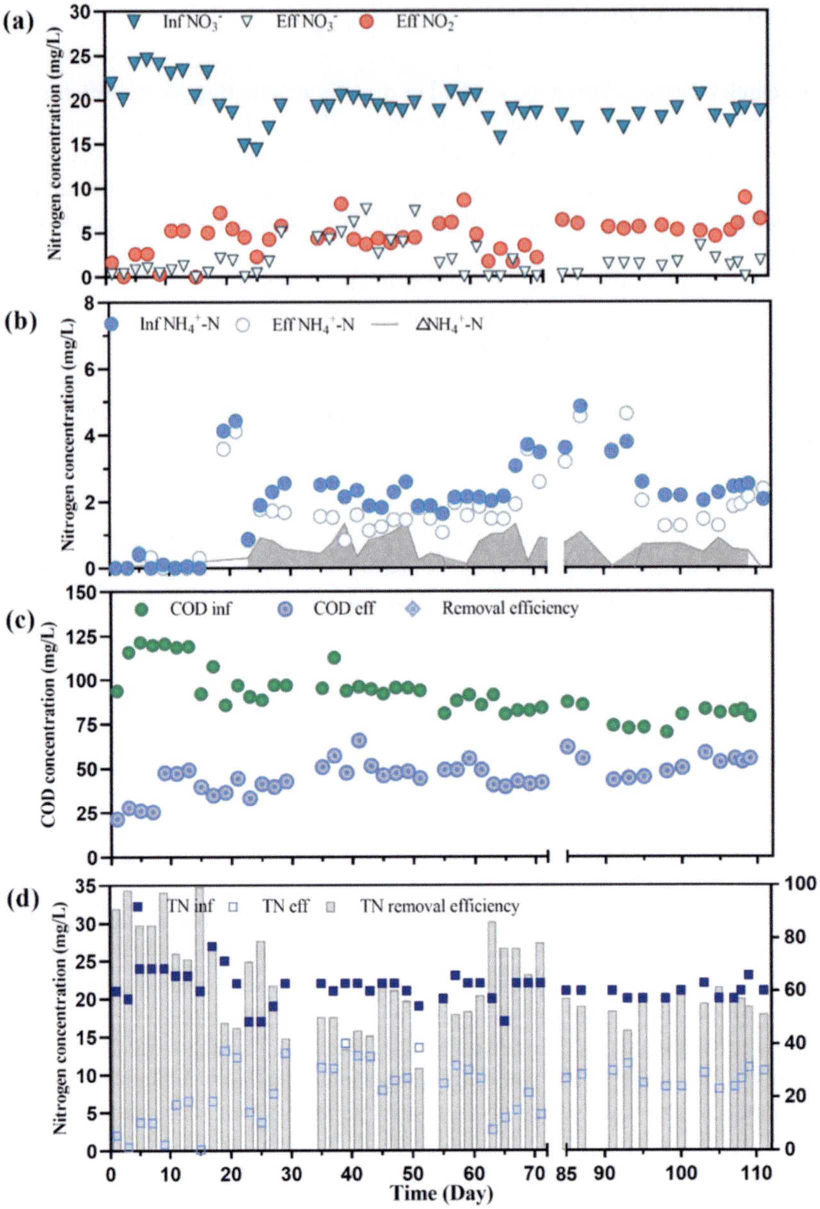

Fig. 2 Long-term nitrogen removal performance of a single-stage PDA biofilter started up in-situ by inoculating denitrification biofilm media

The two-stage denitrification filter system operated for 164 days, divided into five stages based on different influent ammonia concentrations (Table 1), with corresponding effluent conditions shown in Table 2. Notably, from day 30 to day 80, the two-stage denitrification filter system was idle at room temperature due to the winter break. During the reactor operation, the influent COD/NO_3^--N ratio was adjusted in real-time based on the effluent TN concentration. When there was a risk of effluent TN exceeding the Class A discharge standard of 15 mg/L specified in the "Discharge Standard of Pollutants for Municipal Wastewater Treatment Plant (GB 18,918–2002)," the COD/NO_3^--N ratio was increased to reduce the effluent TN concentration. Conversely, when there was no risk of effluent TN exceeding 15 mg/L, the COD/NO_3^--N ratio was decreased to maximize NO_2^- accumulation, providing essential substrates for the growth of anammox bacteria.

Under the control of an influent COD/NO_3^--N ratio of 4.21, the primary biofilter achieved a nitrite accumulation of 4.86 mg/L in the effluent on the fourth day. This is because a lower COD/NO_3^--N ratio favors nitrite accumulation. Studies have shown that using sodium acetate in the influent of the primary biofilter successfully achieved stable nitrite accumulation of 5.58 ± 1.42 mg/L at this stage. Subsequently, in the fourth stage, with an average influent COD/NO_3^--N ratio of 3.85 ± 0.51, a stable nitrite accumulation of 9.02 ± 2.15 mg/L was observed. However, the higher nitrite accumulation in the fourth stage made it difficult to remove nitrite in the subsequent secondary biofilter, leading to a risk of TN exceeding 15 mg/L. To address this issue, the addition of an external carbon source was increased in the fifth stage to reduce

Table 1 Influent conditions of two-stage denitrification biofilter system start up in-situ PDA biofilter

Phase	Time (days)	TN inf (mg/L)	NH_4^+-N inf (mg/L)	NLR (kg N/m^3/d)
III	1–30	26.96 ± 1.57	3.17 ± 0.71	0.39 ± 0.03
IV	80–98	30.68 ± 2.70	4.98 ± 1.81	0.44 ± 0.04
V	100–120	33.33 ± 1.33	10.71 ± 0.62	0.48 ± 0.03
VI	122–150	30.52 ± 2.11	7.27 ± 0.83	0.43 ± 0.03
VII	152–164	34.98 ± 1.19	11.01 ± 0.38	0.50 ± 0.02

Table 2 Effluent conditions of two-stage denitrification biofilter system start up in-situ PDA biofilter

Phase	Time (days)	TN eff (mg/L)	NH_4^+-N eff (mg/L)	NRR (kg N/m^3/d)
III	1–30	10.03 ± 2.70	0.93 ± 0.59	0.26 ± 0.06
IV	80–98	14.82 ± 4.34	1.43 ± 1.46	0.23 ± 0.04
V	100–120	11.51 ± 2.73	2.58 ± 1.81	0.32 ± 0.03
VI	122–150	6.82 ± 2.26	1.73 ± 1.45	0.34 ± 0.04
VII	152–164	7.25 ± 0.69	2.95 ± 0.83	0.39 ± 0.03

nitrite accumulation, resulting in an average influent COD/NO_3^--N ratio of 5.31 ± 0.35 and a reduction in nitrite accumulation to 3.03 ± 1.93 mg/L.

In the sixth stage, due to the enhanced ability of the secondary biofilter to remove nitrite, the amount of carbon source added was reduced to generate more nitrite in the primary biofilter. By maintaining an average influent COD/NO_3^--N ratio of 4.28 ± 0.50, stable nitrite accumulation of 5.33 ± 2.79 mg/L was achieved (Fig. 2a). In the seventh stage, further reduction of the carbon source resulted in a stable residual nitrite concentration of 8.61 ± 1.10 mg/L with an influent COD/NO_3^--N ratio of 4.23 ± 0.20.

Although lowering the COD/NO_3^--N ratio is generally considered favorable for nitrite accumulation, this study did not emphasize maximizing nitrite accumulation in the primary biofilter. High nitrite accumulation could pose challenges in meeting the effluent TN discharge standards. Considering the weak anammox capacity for removing additional nitrite in the reactor during the early third stage and the significant residual nitrite present in the effluent even in the seventh stage, we did not further reduce the COD/NO_3^--N ratio to increase nitrite accumulation. Overall, throughout the operation, the primary biofilter consistently provided the secondary biofilter with low-concentration influent containing nitrite and ammonium. This setup created a favorable substrate foundation for the enrichment of anammox bacteria in the secondary biofilter.

Based on the rapid achievement of partial denitrification functionality in the primary biofilter, significant anammox characteristics gradually emerged in the secondary biofilter over the 164-day operation period. TN removal was achieved through the combined efforts of the primary and secondary biofilters. With an influent TN concentration ranging from 27 to 35 mg/L, the effluent TN concentration consistently remained below 15 mg/L for an extended period, meeting China's Class A discharge standards. Ultimately, in stage VII, the average effluent TN was 7.28 mg/L.

In the primary biofilter, a substantial amount of COD was consumed for denitrification, leading to nitrate removal and ultimately benefiting TN removal. Additionally, the average ΔNH_4^+-N in the primary and secondary biofilters were 1.03 and 5.42 mg/L, respectively, resulting in a total ΔNH_4^+-N of 6.45 mg/L for the two-stage denitrification filter system in stage VII (Fig. 3b). The TN removal capacity of the secondary biofilter gradually increased, with the secondary biofilter eventually removing an average of 12.63 mg/L of TN, with the anammox process playing a dominant role, contributing 89% to TN removal (Fig. 3c). The excellent nitrogen removal capacity was attributed to the gradual enrichment of the anammox process in the secondary biofilter. The hydraulic retention time of the two-stage denitrification filter system was maintained at 1.65 h. Under an NLR condition of 0.44 ± 0.05 kg N/m^3/d, the NRR gradually increased from 0.26 kg N/m^3/d to 0.39 kg N/m^3/d, while the NRE increased from 65 to 79% (Fig. 3d).

The changes in Rs and Rp during the operation are shown in Fig. 0.4a, b. Rs and Rp represent the ratios of consumed NO_2^--N to consumed NH_4^+-N and produced NO_3^--N to consumed NH_4^+-N, respectively (Strous et al. 1998). Stable ratios close to theoretical values are also considered indicators of successful anammox startup.

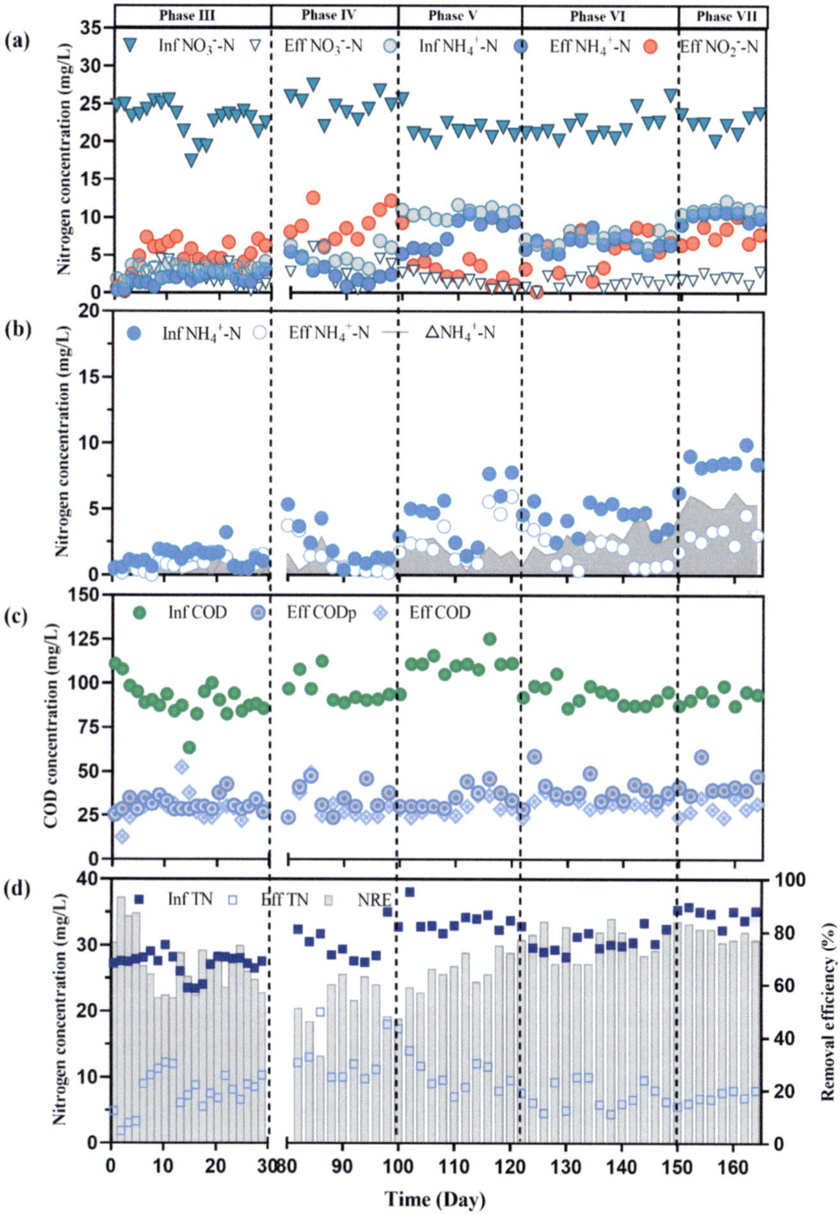

Fig. 3 Long-term nitrogen removal performance of two-stage PDA biofilter system started up in-situ by inoculating ordinary activated sludge

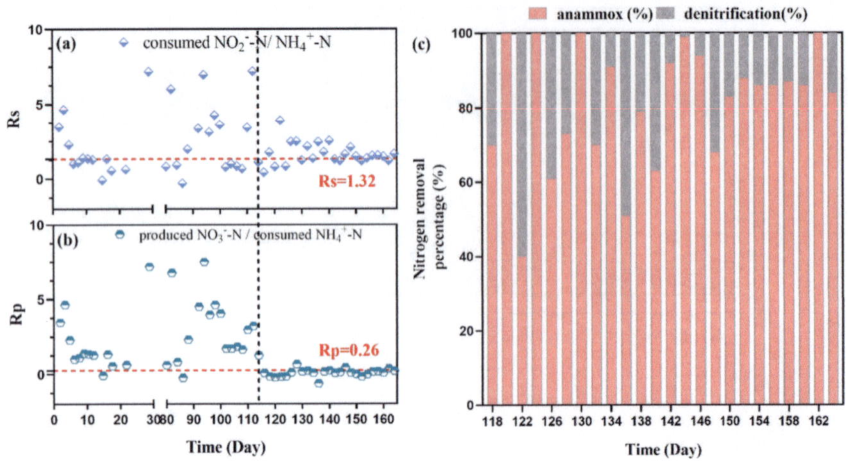

Fig. 4 Rs (**a**) and Rp (**b**) variations of the secondary biofilter during operation (**c**) Contribution ratios of anammox and denitrification pathways to TN removal in the secondary biofilter from day 118 to day 164

From day 118 onwards, the actual stoichiometric values gradually approached the theoretical values of the anammox reaction, indicating the initiation of the anammox process in the secondary biofilter. In this study, Rs was 1.43 ± 0.18 (Fig. 4a), higher than the theoretical value of 1.32, while Rp was 0.14 ± 0.16 (Fig. 4b), lower than the theoretical value of 0.26. This discrepancy was attributed to the denitrifying bacteria in the secondary biofilter. Due to the limited organic matter in the secondary biofilter influent, it could only serve as a supplement for denitrification. The differences between actual and theoretical stoichiometric coefficients could be attributed to the complex symbiotic relationships among denitrifying bacteria, anammox bacteria, and other coexisting bacteria.

After establishing the anammox process in the secondary biofilter from day 118 onwards, the contribution of anammox to nitrogen removal was analyzed. According to calculations, anammox accounted for $89.1 \pm 9.6\%$ of the nitrogen removal process after day 118, making it the primary contributor to TN removal in the secondary biofilter (Fig. 4c). The $\Delta COD/\Delta TN$ value reflected the system's ability to consume carbon sources. As anammox bacteria enriched in the secondary biofilter, the $\Delta COD/\Delta TN$ gradually decreased from 3.41 to 2.29 during long-term operation (Fig. 5), ultimately saving 32.8% of the carbon source addition. Notably, the $\Delta COD/\Delta TN$ in this study was lower than the reported 2.4–3.1 in partial denitrification-anammox systems (Cao et al. 2017b).

On the 113th day, an ex-situ partial denitrification biofilm batch test was conducted on the primary filter to verify the occurrence of partial denitrification. Within the first 30 min, all three tests showed a decrease in nitrate concentration and nitrite accumulation. The nitrite accumulation rates for the biofilm at the top, middle, and bottom of the primary biofilter were 59%, 60%, and 78%, respectively (Fig. 6a–c).

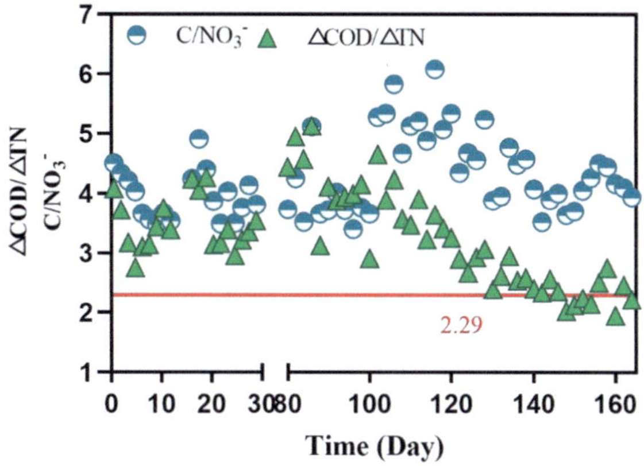

Fig. 5 Influence of COD/NO_3^--N and ΔCOD/ΔTN conditions in the two-stage biofilter system

The results indicated that the top, middle, and bottom sections of the primary biofilter all had the capability to accumulate nitrite through partial denitrification.

On the 161st day, an ex-situ anammox biofilm batch test was conducted on the secondary filter to verify the occurrence of the anammox reaction. The initial concentrations of ammonium nitrogen and nitrite nitrogen were set to approximately 5 mg/L, with trace elements and inorganic compounds added in proportion to the influent water quality. Under anoxic conditions, within 150 min, all three tests (top, middle, and bottom sections) showed simultaneous reductions in ammonium nitrogen and nitrite nitrogen, demonstrating the occurrence of the anammox reaction in the secondary biofilter (Fig. 6d–f). The NH_4^+-N removal rates for the anammox batch tests of the biofilm at the top, middle, and bottom sections were 0.45, 0.34, and 0.81 mg N/L/h, respectively. The NO_2^--N removal rates for the partial denitrification batch tests of the biofilm at the top, middle, and bottom sections were 0.55, 0.54, and 0.94 mg/L/h, respectively. These findings provide solid evidence for the establishment of anammox reactions in the secondary biofilter, continuously contributing to the removal of total nitrogen (TN) in the biofilter system.

To gain insights into the enrichment of functional bacteria from a microbial perspective, metagenomic analysis was conducted to identify bacterial communities at the genus level in the initial inoculated sludge and in the biofilm on the filter media in the top, middle, and bottom sections of the primary and secondary biofilters. Over the course of the biofilter system's operation from day 0 to day 161, Thauera was successfully enriched in the primary biofilter, with its abundance gradually increasing over time. The initial relative abundance of Thauera in the inoculated sludge was 1.44% (Fig. 7a), which contributed to the rapid accumulation of nitrite in the primary biofilter.

By day 161, metagenomic analysis revealed that the abundance of Thauera in the top, middle, and bottom of the primary biofilter increased to 14.89%, 5.98%, and 5.98%, respectively (Fig. 7b). The presence of sodium acetate in the influent was conducive to the enrichment of Thauera, and its successful enrichment in

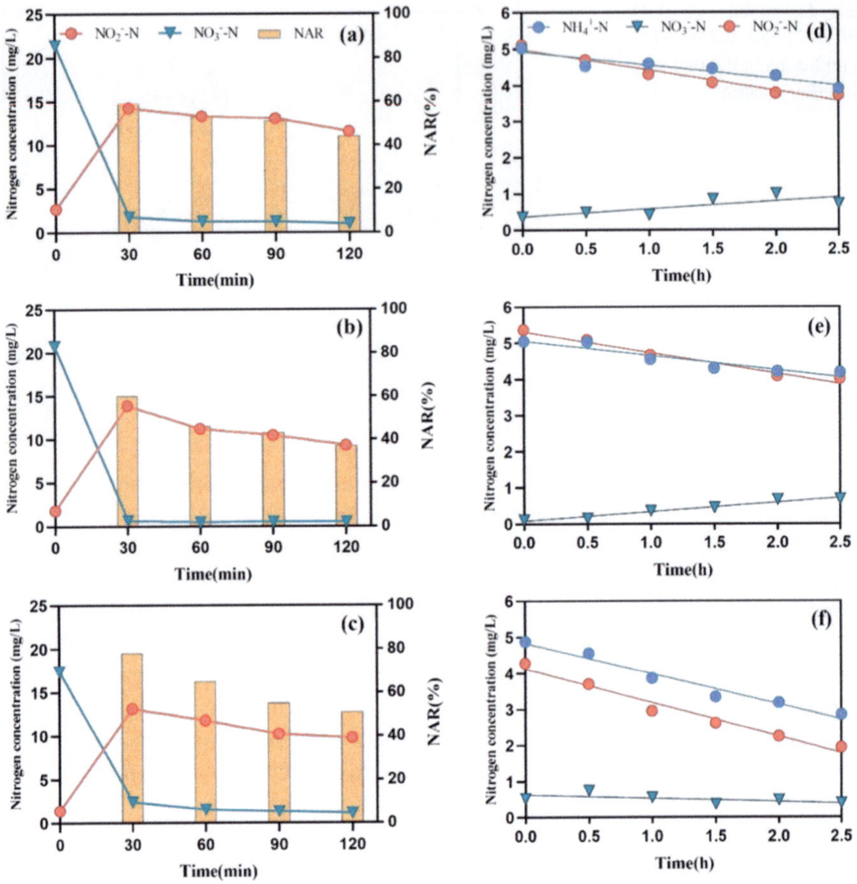

Fig. 6 Batch tests of partial denitrification (**a–c**) and anammox (**d–f**)

the primary biofilter facilitated the sustained occurrence of shortcut denitrification, thereby promoting a stable supply of nitrite to the secondary biofilter.

Simultaneously, Candidatus Brocadia was detected throughout the entire column of the primary biofilter, although at a very low level of 0.001%. This finding is consistent with the observed weak anammox performance in the primary biofilter. In the secondary biofilter, Candidatus Brocadia was identified as the sole anammox bacterial genus, playing a crucial role in the successful initiation of the anammox process. On day 161, the abundance of Candidatus Brocadia at the top of the secondary biofilter remained the highest at 23.81%, which is consistent with previous research findings. In the middle and bottom sections, the enrichment of Candidatus Brocadia increased from 0 to 11.16% and 14.09%, respectively (Fig. 7c).

It is noteworthy that the influent COD concentration in the secondary biofilter was relatively low, averaging 34.07 ± 5.08 mg/L. This resulted in limited availability of biodegradable organic matter for denitrifying bacteria to compete for nitrite with

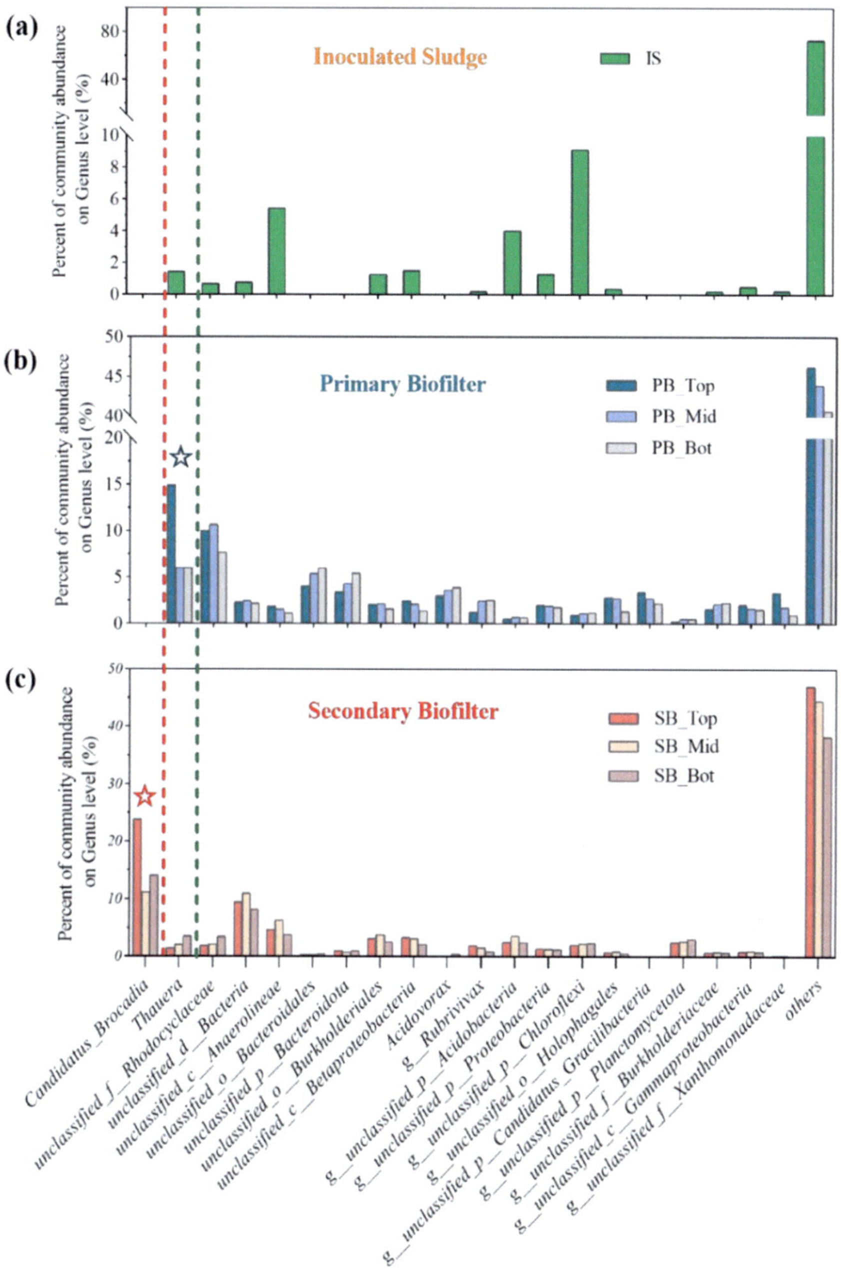

Fig. 7 Microbial genus-level composition of the two-stage biofilter system

anammox bacteria, as most of the organic matter in the wastewater was removed during pretreatment in the primary biofilter. Additionally, some degree of Thauera enrichment was observed in the secondary biofilter. Specifically, on day 161, the abundance of Thauera in the top, middle, and bottom sections of the secondary biofilter was recorded as 1.44%, 2.16%, and 3.57%, respectively. The presence of Thauera in the secondary biofilter likely promoted the growth of anammox bacteria to some extent, as the appropriate heterotrophic bacteria can stimulate anammox activity.

These results collectively underscore the successful establishment and enrichment of functional bacteria in the biofilter system, leading to enhanced nitrogen removal performance and providing critical insights for optimizing the design and operation of biofilters in wastewater treatment processes.

To gain deeper insights into the nitrogen metabolic pathways in the two-stage denitrification biofilter system, functional genes related to denitrification and anammox in the inoculated sludge (day 1) and in the two biofilters on day 161 were analyzed (Fig. 8). Genes associated with the denitrification process were detected in both biofilters, and these genes maintained a certain relative abundance, indicating that denitrification plays a significant role in nitrogen removal in this system. Interestingly, the relative abundance of genes associated with nitrate reductase (nar) involved in the denitrification process, namely narG/H/I and napA/B, was significantly higher than that of genes encoding nitrite reductase (nir) involved in the denitrification process, specifically nirK/S. This suggests that nitrate was primarily reduced to nitrite, implying that the shortcut denitrification process in the primary biofilter could provide substrates for the enrichment of anammox bacteria in the secondary biofilter.

These findings highlight the critical role of the denitrification process, especially the initial reduction of nitrate to nitrite, in facilitating the subsequent anammox process within the secondary biofilter. The consistent presence of denitrification-related genes across both biofilters further emphasizes the intricate interconnection of these biological processes in achieving effective nitrogen removal.

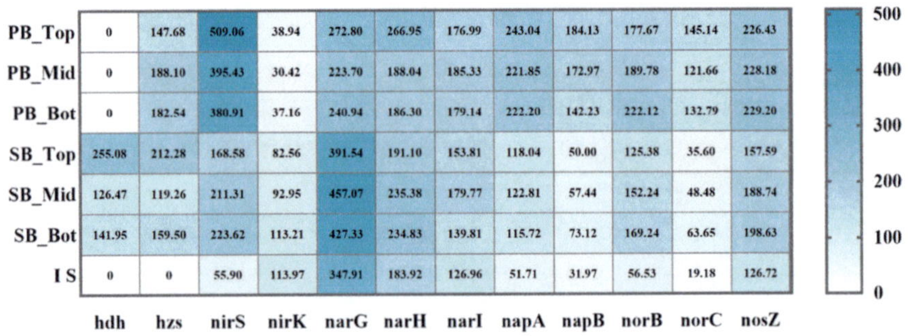

Fig. 8 Metagenomic analysis of the two-stage biofilter system

Specifically, in the primary biofilter, the relative abundances of nar in the top, middle, and bottom sections were 1144, 992, and 970, respectively, while the relative abundances of nir in these sections were 312, 426, and 418, respectively. In the secondary biofilter, the relative abundances of nar in the top, middle, and bottom sections were 904, 1052, and 990, respectively, while the relative abundances of nir were 312, 426, and 418, respectively (Fig. 8). In the inoculated sludge, the relative abundances of nar and nir were 742 and 170, respectively. Hdh (hydrazine dehydrogenase) and hzs (hydrazine synthase) are key genes involved in the anammox process. Notably, the relative abundances of these two genes were not detected in the inoculated sludge (Li et al. 2021). In the primary biofilter, the relative abundances of hzs in the top, middle, and bottom sections were 148, 188, and 173, respectively; hdh was not detected in the primary biofilter. In the secondary biofilter, the relative abundances of hzs in the top, middle, and bottom sections were 212, 119, and 160, respectively; meanwhile, the relative abundances of hdh in the top, middle, and bottom sections were 255, 126, and 142, respectively. This indicates that, compared to the inoculated sludge, the anammox process was enhanced in the secondary biofilter, and the anammox pathway plays a crucial role in nitrogen removal.

After 111 days of operation, we observed no significant increase in ΔNH_4^+-N in the single-stage denitrification biofilter. However, in the subsequent two-stage denitrification biofilter system experiment, successful in situ self-enrichment of anammox bacteria was achieved in the secondary biofilter after 164 days of operation. Interestingly, monitoring of both biofilter columns in the two-stage denitrification biofilter system showed the simultaneous presence of NH_4^+-N and NO_2^--N (Fig. 9). This finding indicates that both the primary and secondary biofilters have the potential to provide substrates for the in situ enrichment of anammox bacteria. However, compared to the secondary biofilter, the primary biofilter did not exhibit any significant anammox characteristics.

4 Conclusions

4.1 *Factors Contributing to the Failure of Anammox Bacteria Enrichment in the Primary Biofilter*

(1) Organic load and heterotrophic competition:

The abundant organic matter in the influent promoted the growth of heterotrophic bacteria, which outcompeted anammox bacteria for nitrite, an essential substrate for anammox bacteria growth. This competitive advantage of heterotrophic bacteria hindered the enrichment of anammox bacteria.

(2) Clogging and backwashing:

The clogging issues necessitated frequent backwashing, approximately once a week. During backwashing, significant amounts of biofilm, including any developing

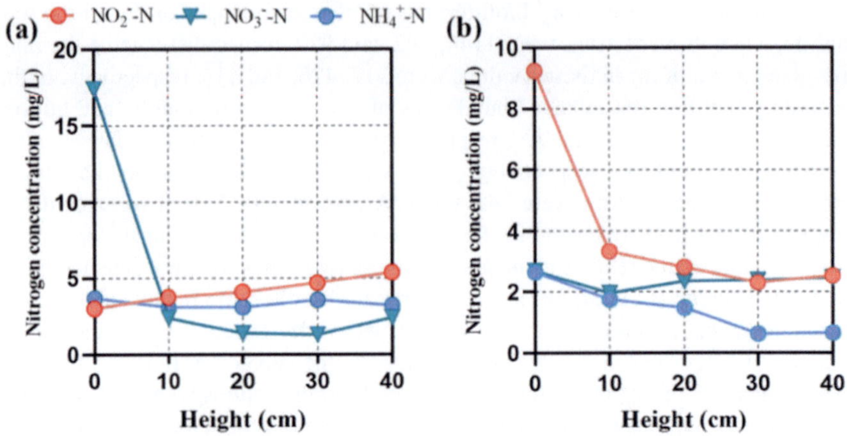

Fig. 9 Nitrogen concentration variations along the two-stage biofilter system **a** Primary biofilter **b** Secondary biofilter

anammox bacteria, were washed out of the primary biofilter. This frequent disturbance disrupted the retention and growth of anammox bacteria, further impeding their enrichment.

4.2 Factors Contributing to Successful Enrichment in the Secondary Biofilter

(1) Limited organic supply:

The limited availability of organic matter in the secondary biofilter restricted the growth of heterotrophic bacteria, reducing competition for nitrite. This allowed anammox bacteria to thrive without significant interference.

(2) Absence of clogging and backwashing:

The secondary biofilter did not experience clogging issues, eliminating the need for backwashing. The stable environment enabled continuous biofilm formation and the retention of enriched anammox bacteria.

(3) Optimized conditions for anammox bacteria enrichment

The two-stage biofilter system effectively managed influent organic matter, promoting rapid heterotrophic growth and biofilm thickening, thus avoiding potential mass transfer limitations. The key to successful anammox bacteria enrichment lies in creating optimal growth conditions and avoiding inhibitory factors.

(4) Environmental Conditions:

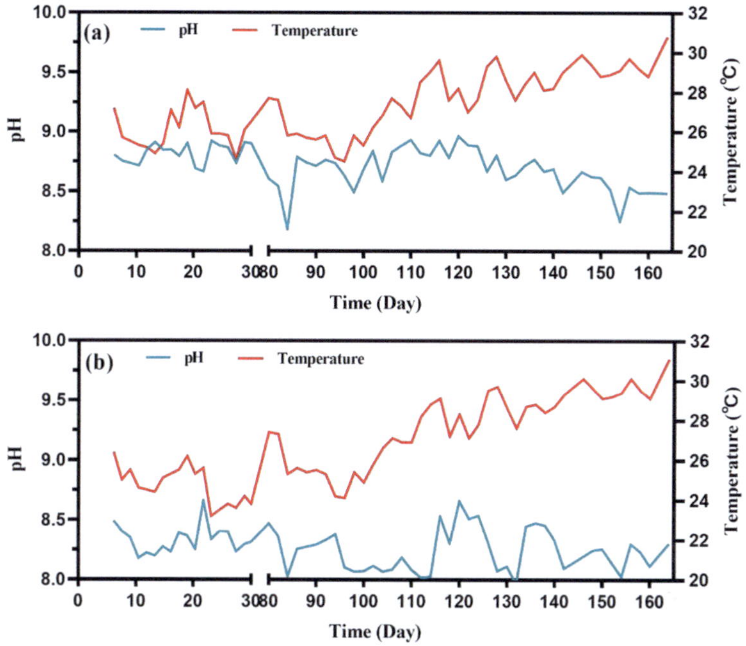

Fig. 10 Long-term pH and temperature monitoring of the two-stage biofilter system **a** Primary biofilter **b** Secondary biofilter

The influent ammonia and nitrite concentrations in the secondary biofilter were kept low, creating a favorable environment as low substrate concentrations do not inhibit anammox bacteria. The ceramic filter media in the biofilter formed a stable biofilm, retaining the enriched anammox bacteria and providing a stable growth platform. Anoxic biofilms can accommodate relatively high abundances of anammox bacteria. Close monitoring of reactor conditions revealed that the high temperatures and pH levels during the study period were conducive to the growth of anammox bacteria (Fig. 10). These factors collectively contributed to the successful enrichment of anammox bacteria in the secondary biofilter, highlighting the importance of optimizing conditions for efficient anammox process enrichment in wastewater treatment systems.

4.3 *Findings and Implications*

This study demonstrates the feasibility of transitioning from conventional denitrification biofilters (DNBFs) to partial denitrification-anammox (PDA) biofilters through in situ self-enrichment of anammox bacteria, eliminating the need for anammox sludge inoculation. Traditional DNBFs, commonly used in wastewater treatment plants (WWTPs), rely heavily on external carbon sources, which increases operational costs and carbon footprint. In contrast, PDA technology reduces the reliance on external carbon, improving nitrogen removal performance while addressing residual nitrate issues more efficiently.

The proposed strategy involves integrating a secondary biofilter in series with the conventional DNBF. By mixing partial mainstream wastewater (containing COD and NH_4^+-N) with secondary effluent (containing NO_3^--N) before it enters the primary denitrification biofilter, partial denitrification occurs, producing nitrite for the secondary biofilter where anammox bacteria are enriched under low-intensity substrate conditions. This study successfully enriched anammox bacteria in a two-stage DNBF system, maintaining the influent COD/ NO_3^--N ratio between 3.85 ± 0.51 and 5.31 ± 0.35, ensuring effluent total nitrogen remained below the discharge standard of 15 mg/L.

The approach offers notable benefits, such as reducing aeration energy consumption by efficiently removing ammonia through anammox, which lowers the energy required for conventional ammonia removal processes. Additionally, the strategy reduces external carbon source consumption compared to traditional denitrification, as the naturally biodegradable organic matter in the influent helps meet the carbon needs of the biofilters. Overall, the implementation of PDA biofilters in WWTPs offers a promising route to sustainable, cost-effective nitrogen removal.

4.4 Concept of Application in WWTPs

To implement this strategy in wastewater treatment plants (WWTPs), an additional set of denitrification biofilters is installed in series with the existing denitrification biofilter. The influent for the denitrification biofilter is a mixture of diverted mainstream wastewater and secondary effluent. By adjusting the proportions of these two streams and the final influent COD/ NO_3^--N ratio, it is possible to enrich anammox bacteria in the two-stage denitrification biofilters. This approach ultimately leads to reduced oxygen consumption in mainstream wastewater treatment and decreased external carbon source addition in the denitrification biofilters.

Transitioning from conventional denitrification to partial denitrification/anammox biofilters through in situ anammox bacteria enrichment presents a promising approach for achieving more sustainable and cost-effective nitrogen removal in wastewater treatment. This strategy not only enhances nitrogen removal efficiency but also reduces operational costs and the environmental impact associated with traditional denitrification processes.

Conflict of Interest Disclosure

The authors declare that there are no conflicts of interest regarding the publication of this paper. No financial relationships or personal interests, either directly or indirectly, influenced the design, conduct, analysis, or interpretation of this study. All authors have contributed equally to the research and writing of the manuscript, and no competing interests have been identified.

AI Disclosure

This manuscript was supported by artificial intelligence (AI) tools for certain aspects of the writing, editing, and data analysis processes. AI-based language models were used to assist with the refinement of the manuscript's text, ensuring clarity, conciseness, and grammatical accuracy. However, the interpretation of research findings, the formulation of conclusions, and the overall scientific content of this paper are the sole responsibility of the authors. The use of AI in this study did not involve the generation or interpretation of original data but was aimed at enhancing the presentation and flow of the research.

The authors confirm that the AI tools used in this study were not involved in the design, execution, or analysis of experiments or in the development of the research hypothesis. All experimental methodologies, data collection, and scientific discussions were carried out and approved by the research team.

References

Al-Hazmi HE, Maktabifard M, Grubba D, Majtacz J, Hassan GK, Lu X, Piechota G, Mannina G, Bott CB, Mąkinia J (2023) An advanced synergy of partial denitrification-anammox for optimizing nitrogen removal from wastewater: a review. Biores Technol 381:129168

Cao Y, Kwok BH, Van Loosdrecht MCM, Daigger GT, Png HY, Long WY, Chye CS, Ghani YABD (2017a) The occurrence of enhanced biological phosphorus removal in a 200,000 m3/day partial nitration and Anammox activated sludge process at the Changi water reclamation plant, Singapore. Water Sci Technol 75(3):741–751

Cao Y, van Loosdrecht MCM, Daigger GT (2017b) Mainstream partial nitritation–anammox in municipal wastewater treatment: status, bottlenecks, and further studies. Appl Microbiol Biotechnol 101(4):1365–1383

Chen R, Ji J, Chen Y, Takemura Y, Liu Y, Kubota K, Ma H, Li Y-Y (2019) Successful operation performance and syntrophic micro-granule in partial nitritation and anammox reactor treating low-strength ammonia wastewater. Water Res 155:288–299

Cui B, Yang Q, Liu X, Wu W, Liu Z, Gu P (2020) Achieving partial denitrification-anammox in biofilter for advanced wastewater treatment. Environ Int 138:105612

He Y, Feng D, Huang Y, Shi L, Zhao Y, Ma B (2024) Achieving transformation from denitrifying biofilter to partial denitrification/anammox biofilter through self-enrichment of anammox bacteria. J Water Process Eng 57:104686

Kartal B, Maalcke WJ, De Almeida NM, Cirpus I, Gloerich J, Geerts W, Op den Camp HJM, Harhangi HR, Janssen-Megens EM, Francoijs K-J (2011) Molecular mechanism of anaerobic ammonium oxidation. Nature 479(7371):127–130

Li J, Peng Y, Zhang L, Liu J, Wang X, Gao R, Pang L, Zhou Y (2019) Quantify the contribution of anammox for enhanced nitrogen removal through metagenomic analysis and mass balance in an anoxic moving bed biofilm reactor. Water Res 160:178–187

Li J, Peng Y, Gao R, Yang L, Deng L, Zhao Q, Liu Q, Li X, Zhang Q, Zhang L (2021) Highly enriched anammox within anoxic biofilms by reducing suspended sludge biomass in a real-sewage A2/O process. Water Res 194:116906

Lu W, Zhang Y, Wang Q, Wei Y, Bu Y, Ma B (2021) Achieving advanced nitrogen removal in a novel partial denitrification/anammox-nitrifying (PDA-N) biofilter process treating low C/N ratio municipal wastewater. Biores Technol 340:125661

Ma B, Xu X, Ge S, Li B, Wei Y, Zhu H, Nan X, Peng Y (2020) Reducing carbon source consumption through a novel denitratation/anammox biofilter to remove nitrate from synthetic secondary effluent. Biores Technol 309:123377

Mulder A, Van de Graaf AA, Robertson LA, Kuenen JG (1995) Anaerobic ammonium oxidation discovered in a denitrifying fluidized bed reactor. FEMS Microbiol Ecol 16(3):177–183

Saunders AM, Albertsen M, Vollertsen J, Nielsen PH (2016) The activated sludge ecosystem contains a core community of abundant organisms. ISME J 10(1):11–20

Strous M, Heijnen JJ, Kuenen JG, Jetten MSM (1998) The sequencing batch reactor as a powerful tool for the study of slowly growing anaerobic ammonium-oxidizing microorganisms. Appl Microbiol Biotechnol 50:589–596

Su R, Shi L, Wei Y, Ma B (2025) Comammox and AOA responses to ammonia loading rate in oligotrophic environments. Water Res 275:123191

Sun Y, Shad MF, Mansell B, Liu M, Hsia P, Coracero A, Tsai R, Danker B, Sun Y, Liao Z (2025) Leveraging primary effluent-and glycerol-driven partial denitrification-anammox within a pilot-scale tertiary step-feed moving bed biofilm reactor treating high-rate activated sludge systems effluent. Water Res 123505

Vigne E, Choubert J-M, Canler J-P, Heduit A, Sorensen K, Lessard P (2010) A biofiltration model for tertiary nitrification of municipal wastewaters. Water Res 44(15):4399–4410

Wang H, Dai H, Jiang D, Cao X, Wang R, Dai Z, Zhang W, Abbasi HN, Li B, Zhu G (2025) Screening, identification, and application of anaerobic ammonia oxidizing bacteria in activated sludge systems: a comprehensive review. J Environ Manag 375:124272

Wei N, Shi Y, Wu G, Hu H, Wu Y, Wen H (2014) Tertiary denitrification of the secondary effluent by denitrifying biofilters packed with different sizes of quartz sand. Water 6(5):1300–1311

Zheng X, Zhang S, Zhang J, Huang D, Zheng Z (2018) Advanced nitrogen removal from municipal wastewater treatment plant secondary effluent using a deep bed denitrification filter. Water Sci Technol 77(11):2723–2732

Assessing Groundwater Depletion in Marrakech, Morocco: A Remote Sensing Approach to Sustainable Water Management in Luxury Complexes

Hamza Ait Zamzami, Mohammed Elaanzouli, Ayoub Zahrani, Brahim Aachrine, Jamila Saidi, and Taieb Boumeaza

Abstract Water scarcity in Marrakech's semi-arid region is increasingly driven by the expansion of luxurious residential and recreational complexes, which demand substantial groundwater resources for irrigation and leisure amenities such as swimming pools and golf lakes. This study investigates the relationship between extensive vegetated areas, high irrigation demands, and groundwater depletion in ten prominent luxury complexes near Marrakech. Utilizing Earth observation techniques, including satellite imagery and water balance analysis, the research identifies critical areas of water overexploitation and assesses the impact of high-water consumption patterns. The findings highlight the urgent need for sustainable water management practices to mitigate groundwater depletion. Strategic solutions proposed include the implementation of precision irrigation systems, the adoption of drought-tolerant landscaping, and the integration of treated wastewater for irrigation and leisure purposes. Additionally, the study emphasizes the importance of robust policy frameworks to regulate groundwater extraction and promote efficient water use technologies. By adopting these measures, Marrakech can balance urban development and the preservation of vital water resources, ensuring long-term ecological and socio-economic sustainability.

Keywords Groundwater depletion · Marrakesh · Luxury complexes · Remote sensing · Sustainable water management

---

H. Ait Zamzami (✉) · M. Elaanzouli · A. Zahrani · J. Saidi · T. Boumeaza
Laboratory of Dynamics of Spaces and Societies (LADES), Department of Geography, Faculty of Arts and Humanities, HASSAN II University of Casablanca, Mohammedia, Morocco
e-mail: hamzaaitzamzami@gmail.com; hamza.aitzamzami-etu@etu.univh2c.ma

B. Aachrine
Expert in Irrigation, Water Management, Hydraulics, Environment, and Sustainable Development, Agronomic and Veterinary Institute (IAV Hassan II), Rabat, Morocco

A. Nath et al. (eds.), *Groundwater Depletion and Sustainability*, Advances in Geographic Information Science, https://doi.org/10.1007/978-3-032-09921-1_16

1 Introduction

The increasing demand for water in semi-arid regions, driven by urbanization and the development of luxurious residential and recreational complexes, poses a significant threat to groundwater sustainability (Bank 2017). In the vicinity of Marrakech, one of Morocco's most prominent urban centers, the proliferation of high-end buildings and complexes has led to acute water stress. This issue is exacerbated by the high irrigation demands required to maintain extensive green spaces and ornamental landscapes within these facilities (Berdai 2011; Mazirh 2023). Additionally, the widespread installation of swimming pools as leisure amenities further intensifies water consumption, as these pools necessitate substantial volumes of water for filling, regular maintenance, and compensating for evaporation losses. Such pools' aesthetic and recreational allure contributes to their prevalence, placing additional strain on the already limited water resources (Analy 2021a). Prolonged drought periods and declining annual rainfall further aggravate the region's water scarcity, underscoring the urgent need for sustainable water management practices.

This study focuses on ten renowned luxurious complexes near Marrakech, integrating Earth observation techniques with detailed assessments of irrigation practices and leisure amenities, including swimming pools. The research aims to analyze current water usage patterns within these complexes and identify key factors contributing to groundwater depletion. By utilizing satellite data and advanced analytical tools, the study seeks to pinpoint critical areas for intervention and develop strategies that promote efficient water allocation. Potential solutions explored include the implementation of sustainable irrigation technologies, optimization of water allocation, and the reuse of treated wastewater. Furthermore, the study emphasizes the necessity of strengthening regulatory frameworks to ensure the responsible and efficient use of water resources, thereby aligning luxury development projects with environmental sustainability goals.

Ultimately, this research highlights the pressing need for adaptive water management strategies to address the challenges posed by the overexploitation of groundwater resources in the region (Gagnol 2012). Ensuring the sustainability of these water resources is vital for maintaining ecological balance and supporting the continued development of luxurious complexes and their associated leisure activities (Engel-Di Mauro 2021), such as swimming pools, in semi-arid environments like Marrakech (Fig. 1).

2 Characteristics of the Study Region

The study region encompasses the urban and peri-urban areas surrounding Marrakech, one of Morocco's most significant cities, located at the foothills of the Atlas Mountains in the western part of the country (31.63°N, 7.99°W). Characterized by a semi-arid climate with hot, dry conditions and annual rainfall averaging between

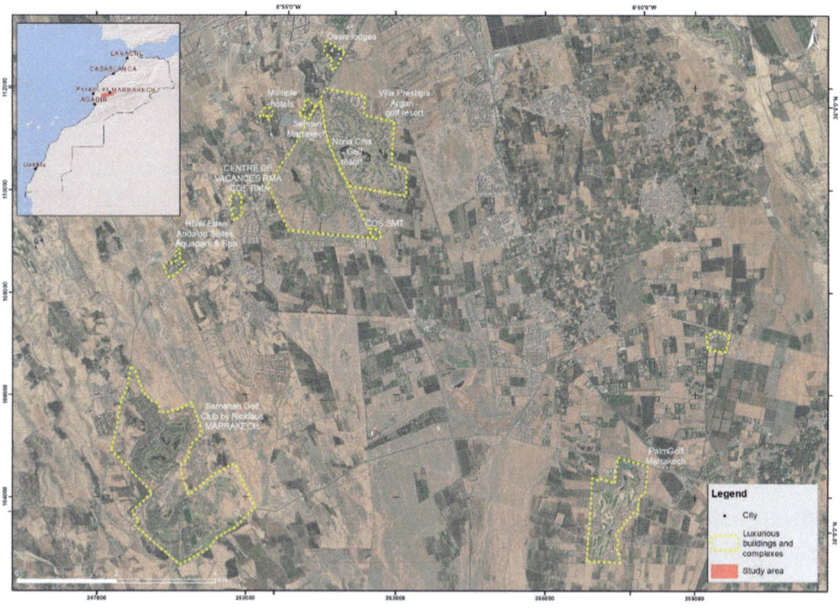

Fig. 1 Location of the luxurious complexes and buildings of Marrakesh, Morocco

300–400 mm, the area faces significant water scarcity exacerbated by prolonged droughts and declining precipitation patterns (Stockstill and Epilogue 2024).

The region's topography features a mix of mountainous terrains, fertile plains, and arid zones, influencing local weather patterns and water distribution. Rapid urbanization and population growth in Marrakech have led to the expansion of high-end residential and recreational complexes, including extensive green landscapes and leisure amenities such as swimming pools, golfs, and touristic investments (Ait Zamzami 2024; Gumuchian 2007). This development has intensified the demand for groundwater, resulting in over-extraction and declining water table levels. Additionally, traditional irrigation practices for maintaining ornamental gardens and pool facilities further strain the limited water resources, highlighting the critical need for sustainable water management strategies to ensure ecological balance and support ongoing luxurious developments in this semi-arid environment.

3 Material and Methods

3.1 Data Collection

To assess water overexploitation, we adopted an approach based on evaluating evapotranspiration using satellite imagery, specifically leveraging the WaPOR platform of the Food and Agriculture Organization of the United Nations (FAO). Rainfall data was obtained from CHIRPS (Climate Hazards Group InfraRed Precipitation with Station data) satellite imagery, supplemented with Sentinel-2 satellite images. Additionally, ground data on swimming pools, buildings, and surveys of luxurious complexes were incorporated. This method enables the assessment of groundwater withdrawals by integrating Earth observation data with ground-based information (Khalki 2023).

Vegetation: NDVI

Multispectral images from the Sentinel-2 satellites, operated by the European Space Agency (ESA), provide high-resolution data for analyzing vegetation, particularly by calculating the Normalized Difference Vegetation Index (NDVI). NDVI is calculated using the red (Band 4) and near-infrared (Band 8) bands, which are essential for assessing the health and density of vegetation in a given area. These images, available via the Copernicus Open Access Hub, require prior preparation, including atmospheric correction to eliminate unwanted effects and cropping to focus on the study area. Before calculating the NDVI, preparing the Sentinel-2 images by extracting the relevant spectral bands is important. The formula used to calculate the NDVI is:

$$\boldsymbol{NDVI} = \frac{(\boldsymbol{NIR} - \boldsymbol{Red})}{(\boldsymbol{NIR} + \boldsymbol{Red})}$$

where NIR represents the near-infrared band (band 8) and Red represents the red band (band 4). Once these bands have been extracted, the formula is applied to create a new NDVI image, where each pixel reflects the value of the index, allowing accurate spatial assessment of the vegetation. Interpretation of the NDVI results allows different vegetation classes to be distinguished (Fig. 2).

Values close to $+1$ indicate dense, healthy vegetation, while values close to 0 or negative indicate non-vegetated areas or water surfaces. To make it easier to analyze the NDVI image, we used a vegetation threshold to map the vegetated area in the tourist investment. This allows vegetated regions to be identified visually.

Evapotranspiration: WaPOR data

The evapotranspiration images provided by the WaPOR platform, developed by the FAO, are essential for estimating evapotranspiration in the study area using satellite data (FAO and WaPOR 2019). These images can be used to analyze water dynamics in soils, crops and vegetation, providing a clear picture of water use in agricultural areas (Allen 1998). WaPOR data is available online and requires specific preparation,

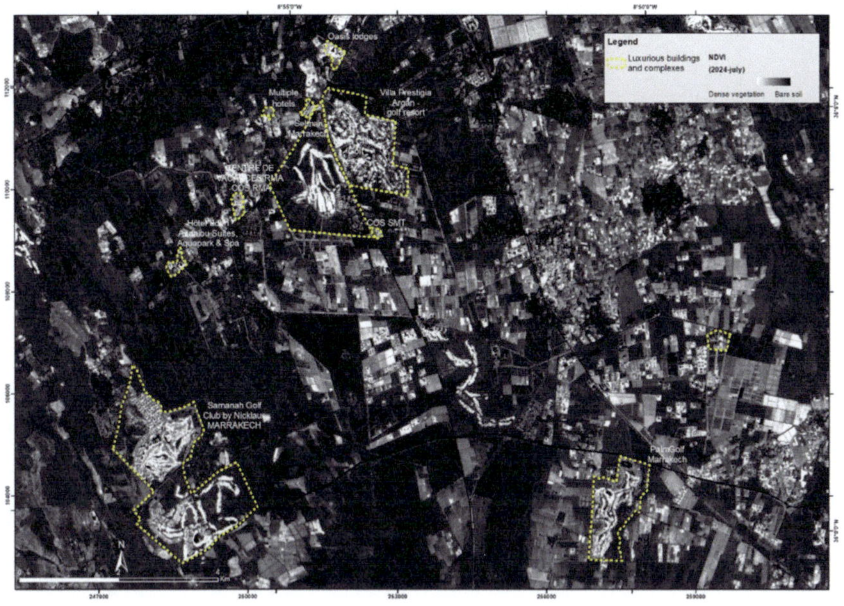

Fig. 2 NDVI raster of study area dated July 2024—Sentinel-2

including selecting the appropriate data layers and cropping the images to focus on the study area (Fig. 3).

$$\boldsymbol{AdjustedEvapotranspiration}(\boldsymbol{mm}) = [(\boldsymbol{RawDataValue} \times \mathbf{0.1}) \times \mathbf{2}]$$

The preparation of WaPOR data also includes the application of certain adjustments to ensure accurate results. The extracted evapotranspiration values must be multiplied by a transformation factor 0.1 to obtain consistent units. Then, a correlation factor of *2 must be applied to align the satellite data with station measurement in the field (Rabia 2020). This correlation factor is based on a survey of farmers, who measured their water consumption from meters installed on wells, thus ensuring a better match with field observations. These adjustments are important if we are to obtain a reliable estimate of evapotranspiration in the study area. (Courault 2005).

Precipitation data: CHIRPS imagery

The precipitation images provided by the CHIRPS database are invaluable tools for estimating rainfall at a regional scale. By combining satellite data with weather station observations, CHIRPS enables the analysis of precipitation patterns over extended periods, offering insights into climate variations within a given area. For this study, the annual precipitation data for 2024 was utilized. These data, available online, require careful preparation to ensure accurate and reliable results, including cropping the images to focus on the specific study area (Khettouch 2023).

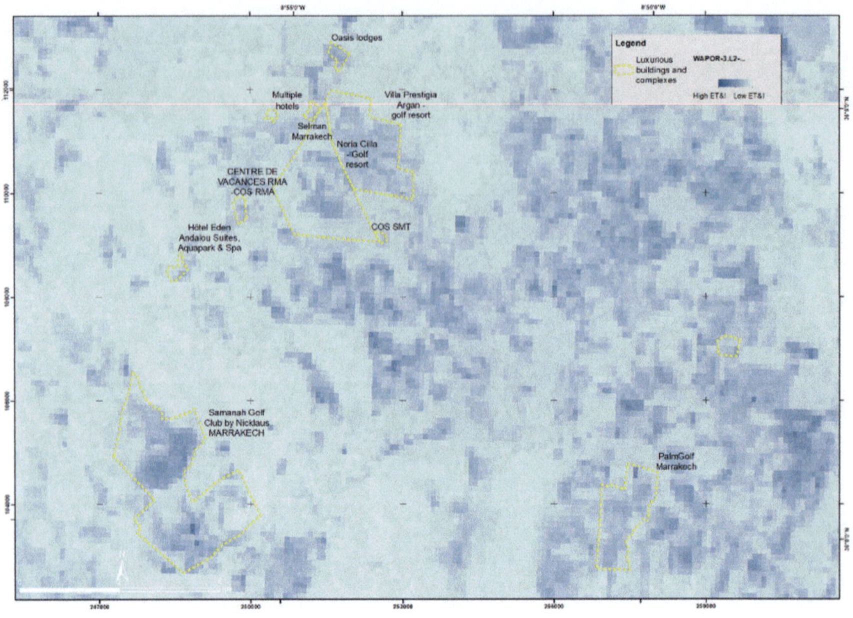

Fig. 3 Wapor raster data of 2024, presenting annual evapotranspiration

$$\boldsymbol{AdjustedPrecipitation(mm) = [RawDataValue \times 0.1]}$$

Preparing CHIRPS precipitation data involves several essential steps. Once the 2024 precipitation data has been downloaded, it is essential to crop it to the boundaries of the study region. Additionally, a transformation factor 0.1 must be applied to adjust the precipitation values to the correct units. This adjustment ensures consistency and accuracy in the analysis of precipitation data (Salles 2017).

Vector data for luxurious complexes and buildings.

For this analysis, we used multiple shapefiles representing 10 luxurious complexes in the study area and another list of water bodies such as lakes and swimming pools. The shapefile of luxurious complexes includes detailed information about the location of each facility, its type, capacity, and estimated water consumption. Meanwhile, the lakes and swimming pools shapefile provide details about their location, volumes, areas, and types of use. This combination of data is important for calculating and examining water depletion in these areas and assessing the potential impact on groundwater resources. By superimposing the shapefiles with Earth observation data (rasters), we aim to identify areas associated with high water overexploitation for leisure purposes, which could indicate direct reliance on groundwater.

4 Methodology

Methodology for Assessing Irrigation Water Requirements

In Marrakech, the annual reference evapotranspiration is estimated at approximately 1,345 mm/year using the Jensen-Haise method (Bouhlassa 2006), aligning closely with values obtained from the Penman–Monteith-FAO equation (Kaissi et al. 2024), which is recognized as the most reliable method for arid and semi-arid environments (Er-Raki 2010). This benchmark evaluates the adequacy of water conditions in the study area for maintaining healthy vegetation. Given the region's climatic trend towards increased drought over the past five years (Hadri 2024), an irrigation efficiency factor of 50% has been adopted (Kharrou 2011).

The following formula is used to determine the volume of irrigation water needed:

$$V = \frac{E * A}{\eta}$$

where:

- V: Volume of irrigation water needed (m^3)
- E: Evaporation (mm)
- A: Area to be irrigated (m^2)
- η: Irrigation efficiency = 50% = 0.5

This conservative estimate is informed by comprehensive consultations with experts from the Regional Office for Agricultural Development of Haouz (ORMVAH) and specialists from (CMGP) company in the agricultural sector, especially in irrigation, pumping, and crop protection, ensuring that the irrigation practices are tailored to the specific agricultural and green spaces-related in Marrakech.

Calculation of groundwater extraction and pumping

We adopted a water balance approach to estimate groundwater abstraction (Srinivasan 2020). After preparing the input data, we extracted area-specific information using shapefiles. The method focused exclusively on vegetated areas within the luxurious complexes, which were identified using NDVI images. We then extracted evapotranspiration and rainfall values and applied the irrigation parameters provided by ORMVAH (Office Régional de Mise en Valeur Agricole du Haouz). Additionally, we incorporated swimming pool water volumes, which were digitized using satellite and aerial imagery with GIS tools. To estimate the volume of water in swimming pools, we assumed a potential average depth of 1.5 m based on information provided by local experts familiar with typical swimming pool dimensions in the area. This assumption was used in conjunction with GIS analysis to estimate the total volume of water in swimming pools. Finally, we used GIS to map the differences observed within the study area (Fig. 4).

To summarize, groundwater withdrawal was calculated as follows:

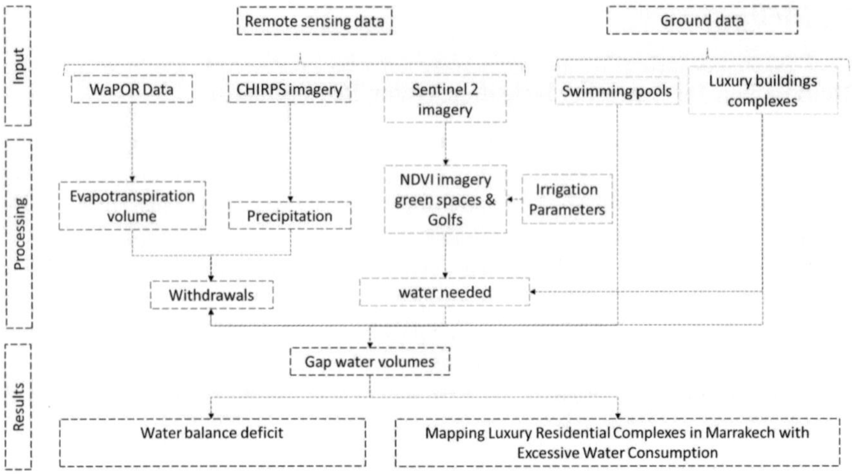

Fig. 4 Integrated remote sensing and ground data workflow for assessing excessive water consumption in luxury residential complexes

$$\textbf{\textit{Withdrawal}} = \left[\textbf{\textit{Irrigationwater}} + \textbf{\textit{Swimmingpoolsandlakes}}\right]$$

5 Results and Discussion

5.1 *Overview of Water Management and Consumption Patterns*

To evaluate the water consumption patterns and irrigation efficiency across luxurious complexes near Marrakech, Table 1 summarizes key metrics derived from the study. These include rainfall, evapotranspiration (ET&I), net irrigation requirements, vegetation coverage, irrigation water volumes (ET_a), potential water needs (ET_0), and water use for leisure amenities such as swimming pools and golf lakes. Additionally, the table provides the total pumped water and the difference between ET_0 and ET_a, offering insights into over-irrigation or deficits. This comprehensive dataset forms the basis for assessing the sustainability of water management practices and highlights areas requiring intervention to mitigate groundwater depletion.

5.2 Water Consumption and Groundwater Depletion: Spatial and Quantitative Analysis

Disparity Between Rainfall and Evapotranspiration

The semi-arid climate of Marrakech exacerbates water scarcity, with annual rainfall averaging only 130.5 mm across the study area. This is significantly lower than evapotranspiration values (ET&I), which range from 333.27 mm to 1002.87 mm. This pronounced disparity highlights the dependency on supplemental irrigation to sustain vegetation and leisure facilities. The highest ET&I values observed at Oasis Lodges (1002.87 mm) and Villa Prestigia Argan (778.64 mm) demonstrate the considerable water stress imposed by these complexes (Fig. 5).

Vegetation Coverage and Irrigation Demands

The vegetated areas across the complexes vary significantly (Mustapha EL Hannani 2017), from as little as 4,354 m^2 in Multiple Hotels complex to 471,525 m^2 in Villa Prestigia Argan. This variation correlates directly with irrigation water requirements (ET_a), from 1,765.72 m^3 to 611,228.43 m^3 annually. These findings reveal that larger complexes disproportionately demand groundwater resources to maintain their expansive landscapes.

Figure 10 illustrates the spatial distribution of vegetated areas, swimming pools, and leisure amenities across selected complexes, highlighting the scale of vegetation coverage and water-intensive features.

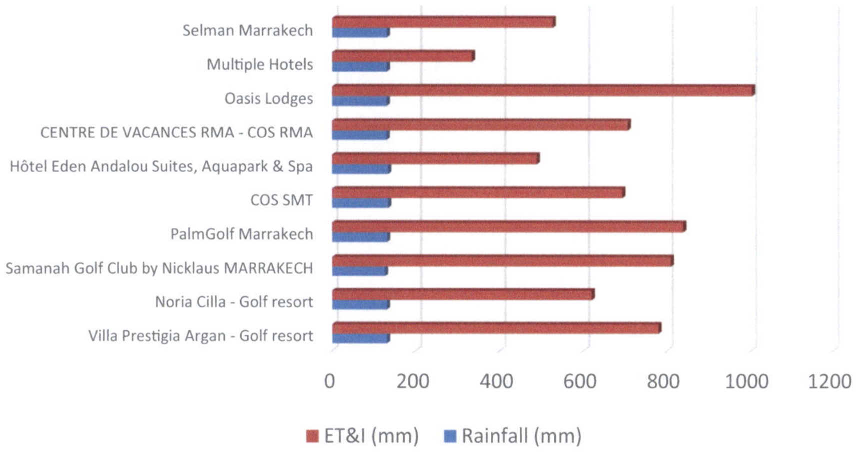

Fig. 5 Comparison of evapotranspiration (ET&I) and rainfall across luxury resorts and Golf Clubs in Marrakech

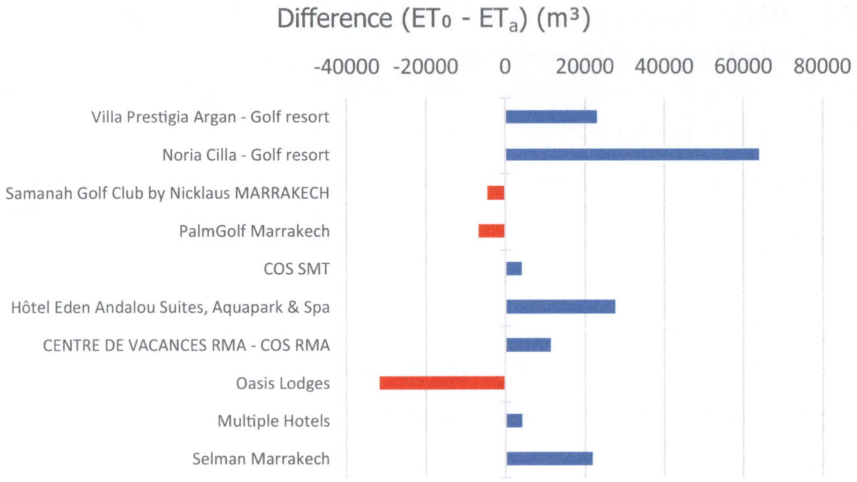

Fig. 6 Water balance analysis: difference between reference evapotranspiration (ET_0) and actual evapotranspiration (ET_a) by establishment

Water Use Efficiency: Surplus and Deficit Dynamics

The difference between potential water needs (ET_0) and actual irrigation volumes (ET_a) provides insights into water use efficiency. Positive differences, such as a surplus of 22,972.7 m^3 in Villa Prestigia Argan, indicate over-irrigation and inefficiencies. In contrast, deficits observed at Oasis Lodges (−31,675.4 m^3) and Palmgolf Marrakech (−6,715.4 m^3) highlight insufficient irrigation, potentially leading to vegetation stress (Fig. 6).

Contribution of Leisure Amenities to Water Demand

Swimming pools and golf lakes significantly contribute to total water demand. For instance, Samanah Golf Club allocates 212,723.5 m^3 annually to these features, while COS SMT uses only 750 m^3. The disparity highlights the varying impact of leisure amenities on complex water consumption (Fig. 7).

Groundwater Abstraction and Sustainability Challenges

The total groundwater abstraction required to meet irrigation and leisure demands underscores the over-reliance on this finite resource. Villa Prestigia Argan pumps the highest volume (623,541.8 m^3 annually), followed by Samanah Golf Club (505,153.9 m^3). Smaller establishments like Multiple Hotels exhibit lower abstraction rates due to limited vegetated areas and leisure amenities (Fig. 8).

Fig. 7 Pools and Golf Course Lake Volumes in m^3 by Location by establishment

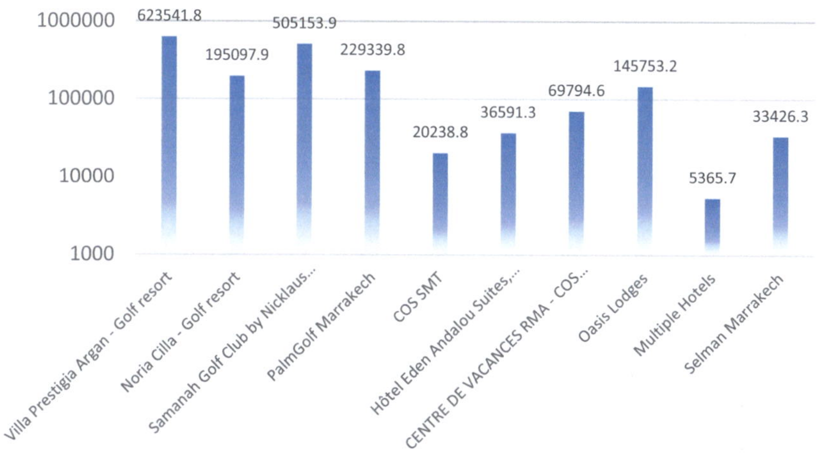

Fig. 8 Total volume of water pumped (m^3) by establishment

5.3 *Discussion of Results*

Groundwater Depletion and Rainfall-Evapotranspiration Disparity

The findings reveal a critical imbalance between rainfall and evapotranspiration (ET&I) in the Marrakech region, a disparity that directly contributes to the over-reliance on groundwater resources. With annual precipitation averaging around 130.5 mm and ET&I values reaching as high as 1002.87 mm, it is evident that natural precipitation is grossly insufficient to sustain the region's vegetation and leisure facilities. This imbalance forces luxury complexes to rely heavily on groundwater pumping to meet their irrigation needs, further exacerbating groundwater depletion in an already water-scarce environment (Analy 2021b).

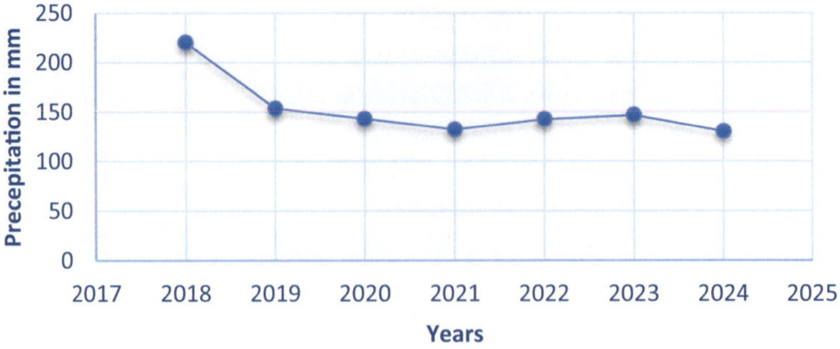

Fig. 9 Mean annual precipitation in the study area (2018–2024), (CHIRPS data)

Figure 9 highlights the declining trend in annual precipitation between 2018 and 2024, showcasing years of prolonged drought and increasingly limited surface water availability. This consistent reduction in rainfall has intensified the reliance on groundwater as the primary water source for irrigation and leisure amenities in luxury complexes. Notably, the sharp drop in precipitation from 2018 (over 200 mm) to 2019 (below 150 mm) reflects the beginning of persistent water stress in the region. Subsequent years, with annual precipitation stabilizing at low levels, underline the structural nature of water scarcity in Marrakech.

The insufficient precipitation fails to replenish aquifers at sustainable rates, particularly as evapotranspiration values remain exceedingly high (Liu et al. 2024). Complexes such as Oasis Lodges and Villa Prestigia Argan, with ET&I values of 1002.87 mm and 778.64 mm, respectively, exhibit significant water stress and consumption, further straining groundwater resources. Without intervention, this dependency will accelerate the depletion of aquifers (Hasan 2025), threatening their long-term viability as a sustainable water source.

To mitigate this growing reliance on groundwater, predictive models driven by AI and Earth observation systems can be employed to forecast water needs based on climatic patterns (Amanambu et al. 2020; Dhonthi 2024). These tools can help optimize irrigation schedules, identify periods of heightened vulnerability, and propose adaptive strategies such as reducing irrigation during peak evapotranspiration periods or transitioning to drought-resistant landscaping.

Vegetation Coverage, Irrigation Demands, and Groundwater Strain

Empirical data demonstrate a direct correlation between vegetated area extent and irrigation demands. For instance, Villa Prestigia Argan, covering 471,525 m^2, requires approximately 611,228.43 m^3 of water annually, compared to the Multiple Hotels' 4,354 m^2, which utilize only 1,765.72 m^3 per year. This disparity significantly pressures limited groundwater resources, especially in semi-arid regions like Marrakech, where aquifer recharge rates are naturally low.

Inefficient water use exacerbates groundwater depletion. Over-irrigation in complexes such as Villa Prestigia Argan and Noria Cilla results in the annual wastage of tens of thousands of cubic meters of water. Conversely, inadequate water allocation

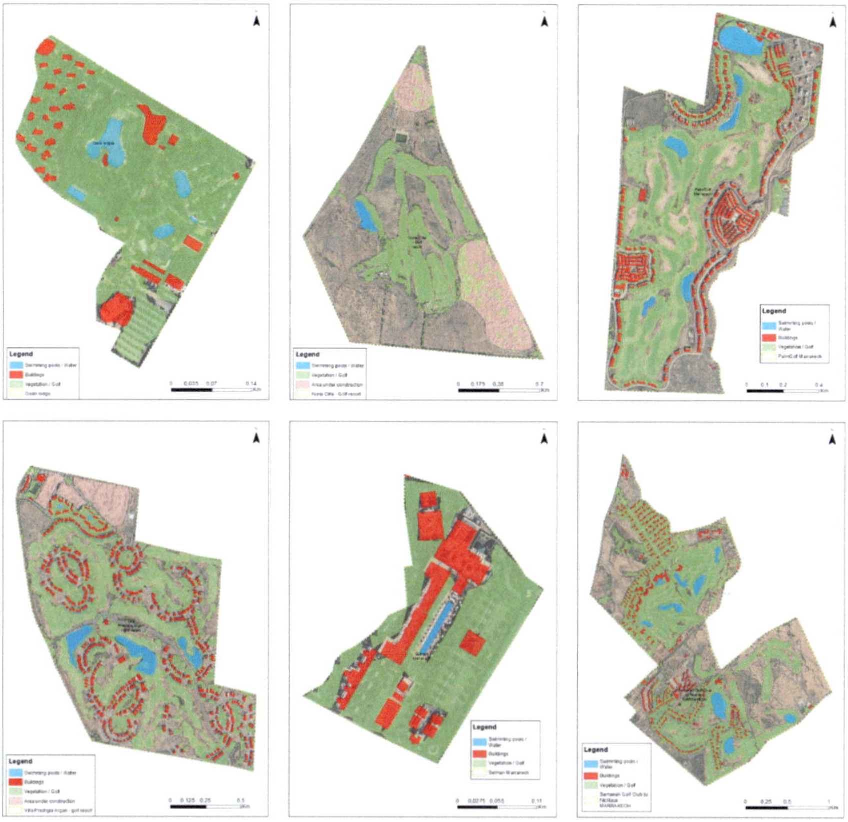

Fig. 10 Detailed land use analysis of luxury complexes: swimming pools, golf lakes, vegetation, and buildings

in facilities like Oasis Lodges and Palmgolf Marrakech leads to vegetation stress, necessitating additional and often unscheduled groundwater pumping to sustain plant health (Song et al. 2034).

Addressing these challenges involves adopting drought-tolerant landscaping to reduce irrigation demands by utilizing native or adaptive plant species requiring less water. Additionally, implementing earth observation techniques, including satellite imagery and remote sensing, can effectively monitor vegetation health and water usage. These technologies identify areas of overexploitation, enabling more efficient water redistribution and alleviating pressure on groundwater reserves.

Precision irrigation technologies are crucial for mitigating groundwater strain. These technologies facilitate accurate water allocation by analyzing soil moisture, weather patterns, and evapotranspiration rates, reducing over-extraction and under-supply. Even minor reductions in groundwater abstraction can significantly enhance aquifer sustainability in semi-arid regions.

Without these interventions, continued water abstraction to sustain extensive green spaces is likely to lead to aquifer depletion, triggering ecological and socio-economic challenges such as biodiversity loss, habitat degradation, reduced agricultural productivity, and increased water scarcity affecting local communities (Bentaleb et al.2024). Therefore, integrating earth observation methods with precision irrigation strategies is essential for the sustainable management of groundwater resources in semi-arid environments (Guemouria 2025).

Impact of Water-Intensive Leisure Amenities on Groundwater Depletion

Leisure amenities, such as swimming pools and golf lakes, significantly contribute to groundwater depletion in the region due to their high-water demands (Forrest and Williams 2010). For example, Samanah Golf Club utilizes 212,723.5 m^3 annually for these features, representing nearly 40% of its total water consumption. In contrast, smaller establishments like COS SMT and Selman Marrakech use only 750 m^3 and 2,250 m^3 annually, respectively.

Figure 10 presents the spatial layout of these water-intensive features, showing the location and extent of pools, lakes, and landscaped areas within the complexes.

Elevated evaporation rates in semi-arid climates like Marrakech exacerbate the overuse of groundwater by these facilities. Earth observation techniques, including satellite imagery and water balance analysis, can pinpoint areas where water-intensive leisure features lead to excessive water extraction (Michailovsky 2023; Abida 2024). By examining spatial water usage patterns, managers can evaluate evaporation losses and develop strategies for more efficient resource allocation.

Implementing measures such as reducing the surface area of water features, incorporating shaded structures to lower evaporation rates, and substituting fresh groundwater with treated wastewater for pools and lakes can substantially alleviate pressure on aquifers (Bihadassen 2020). These strategies, guided by earth observation data, offer practical solutions to harmonize the maintenance of luxury amenities with sustainable water resource management.

Groundwater Abstraction: A Sustainability Crisis

The total volume of groundwater abstraction, particularly by large complexes such as Villa Prestigia Argan (623,541.8 m^3 annually) and Samanah Golf Club (505,153.9 m^3), raises significant concerns about the long-term sustainability of groundwater resources in the Marrakech region. Smaller establishments, such as Multiple Hotels (5,365.7 m^3), demonstrate the potential for reducing groundwater reliance through scaled-down vegetation and leisure facilities (Berahmani 2012).

The unsustainable abstraction rates threaten to lower the water table further, increasing the costs and energy required for pumping and potentially leading to aquifer collapse. AI-driven groundwater monitoring systems could provide real-time data on abstraction rates, aquifer levels, and recharge dynamics to address this. This would enable proactive regulatory measures and encourage developers to adopt sustainable practices.

6 Conclusion

The escalating water demand in Marrakech's semi-arid environment, driven by the expansion of luxurious residential and recreational complexes, poses a significant threat to groundwater sustainability. This study identified a direct correlation between extensive vegetated areas, high irrigation demands, and substantial groundwater depletion. Large-scale complexes like Villa Prestigia Argan and Samanah Golf Club exhibit disproportionate water consumption, intensifying groundwater strain amid limited aquifer recharge and prolonged drought periods.

Earth observation techniques, including satellite imagery and water balance analysis, enabled precise monitoring of vegetation health and water usage patterns. These methods identified areas of overexploitation, facilitating informed decision-making and efficient water redistribution. The findings highlight the critical need for strategic interventions to optimize water use and preserve groundwater resources.

To mitigate groundwater depletion, the following strategic solutions are recommended:

1. **Precision Irrigation Systems**: Implement automated drip irrigation and soil moisture sensors to enhance irrigation efficiency and reduce water wastage.
2. **Drought-Tolerant Landscaping**: Adopt xeriscaping and native or adaptive plant species to lower irrigation demands and increase landscape resilience.
3. **Treated Wastewater Utilization**: Integrate treated wastewater for irrigation and maintaining leisure amenities, decreasing reliance on fresh groundwater.
4. **Earth Observation Techniques**: Continue employing satellite imagery and water balance analysis to monitor and manage water usage effectively.
5. **Robust Policy Frameworks**: Develop and enforce regulations to control groundwater extraction, mandate efficient water usage, and promote sustainable water management technologies.
6. **Redesign Leisure Facilities**: Incorporate water-efficient designs in swimming pools and golf lakes to minimize evaporation losses and conserve water resources.
7. **Integrated Water Management Plans**: Formulate comprehensive strategies that combine precision irrigation, drought-tolerant landscaping, treated wastewater use, and continuous monitoring to ensure holistic groundwater sustainability.
8. **Stakeholder Collaboration and Education**: Engage policymakers, developers, technology providers, and local communities in collaborative efforts and educate stakeholders on sustainable water practices.

Addressing groundwater depletion in Marrakech requires immediate and coordinated action to balance urban development and luxury amenities with preserving vital water resources. Implementing these strategic solutions will ensure the sustainable management of groundwater, maintain ecological balance, and support the socio-economic well-being of the region.

References

Ahalli T, Ouakhir H, Nouyati N, Khattach D, Ennaji N, Goumih M, Maziane A, Aynaou A, Talbi FZ (2025) Impact of precipitation intensity and drought patterns on groundwater storage fluctuations within the Ghis-Nekor aquifer (Morocco). Ecol Eng Environ Technol 26:341

Ait Zamzami H, Elaanzouli M, Zahrani A, Saidi J, Boumeaza T, Musiyam M (2024) Tourism investment and land use change in the Northern Piedmont of the Western High Atlas (1989–2022), Morocco. Forum Geogr 38(1):83–93

Allen RG, Pereira LS, Raes D, Smith M (1998) Crop evapotranspiration: guidelines for computing crop water requirements, FAO

Amanambu AC, Obarein OA, Mossa J, Li L, Ayeni SS, Balogun O, Oyebamiji A, Ochege FU (2020) Groundwater system and climate change: present status and future considerations. J Hydrol 589

Analy M, Laftouhi NE (2021) Groundwater depletion in an urban environment under semiarid climate and persistent drought–City of Marrakesh (Morocco). Water (MDPI) (1). https://doi.org/10.3390/w13223253

Analy M, Laftouhi NE (2021) Groundwater depletion in an urban environment under semiarid climate and persistent drought–city of Marrakesh (Morocco). Water 13(122):3253

Bank W (2017) Managing urban water scarcity in Morocco: summary report, World Bank

Bentaleb MEM, El Adnani MJ, Bentaleb FZ (2024) Economic and environmental implications of solar photovoltaic pumping systems in agri-water management: a case study of Marrakesh Orchards. Int J Account Financ Audit Manag Econ (IJAFAME) 5(110):88–109

Berahmani A, Faysse N, Errahj M, Gafsi M (2012) Chasing water: diverging farmers' strategies to cope with the groundwater crisis in the coastal Chaouia region in Morocco. Irrig Drain 61

Berdai H, Karrou M, Chati MT, Boutfirass M, Bekaoui A (2011) Irrigation water management in Morocco: a review. Mel CGIAR

Bihadassen B, Mimouni R, Bourouache M, El Boulani A, Laktib A, Hassi M, Hamadi F, Aitalla A (2020) Golf courses irrigation with UV-treated wastewater in Agadir city: impact on the quality of local groundwater. J Water Environ Sci 4(11):554–564

Bouhlassa S, Paré S (2006) Évapotranspiration de référence dans la région aride de Tafilalet au sud-est du Maroc. Afr J Environ Assess Manag 11:1–16

Courault D, Seguin B, Olioso A (2005) Review on estimation of evapotranspiration from remote sensing data: From empirical to numerical modeling approaches. Irrig Drain Syst 19:223–249

Dhonthi A, Fonseca Aponte FH, Nemer Z, Hadj Ali C, Tebbouche MY, Van Der Kwast H (2024) Advancing water productivity monitoring: WaPLUGIN for the analysis and validation of FAO WaPOR data in QGIS. Int Arch Photogramm Remote Sens Spat Inf Sci 48:35–41

Engel-Di Mauro S (2021) Socialist states and the environment: lessons for eco-socialist futures: environmental impacts in context, Pluto Press

Er-Raki S, Chehbouni A, Khabba S, Simonneaux V, Jarlan L, Ouldbba A, Rodriguez JC, Allen R (2010) Assessment of reference evapotranspiration methods in semi-arid regions: can weather forecast data be used as alternate of ground meteorological parameters? J Arid Environ 1587–1596

FAO and WaPOR (2019) A satellite-based, open-access database for mapping water productivity in agriculture, Food and agriculture organization of the united nations

Forrest N, Williams E (2010) Life cycle environmental implications of residential swimming pools. Environ Sci Technol 44(114):5601–5607

Gagnol L, Soubeyran O (2012) S'adapter à l'adaptation. La condition sahélienne à l'épreuve de l'injonction au changement climatique. Géographie et cultures

Guemouria A, Tanji R, El Harraki W, Elhassnaoui I (2025) Management of public irrigation systems in Morocco: challenges and innovations for resilient agriculture, Tanding technical committee on water (CW), World federation of engineering organizations (WFEO)

Gumuchian BPH (2007) La ressource territoriale, Paris: Anthropos

Hadri A, Ndiaye AS, Khadir L, Jaffar O, Zamzami HA, Mahdi El Khalki E, Amazirh A, Bouchaou L, Chehbouni A Spatio-temporal analysis of meteorological drought return periods in a Mediterranean arid region, the center of Morocco. J Water Clim Chang
Kaissi O, Er-raki S, Bouras E, Belaqziz S, Chehbouni A Assessment and comparison of crop growth models for estimating wheat production in a semi-arid region of Morocco. chez EGU General Assembly 2024, Vienna, Austria
El Khalki EM, Tramblay Y, Saidi ME, Marchane A, Chehbouni A (2023) Hydrological assessment of different satellite precipitation products in semi-arid basins in Morocco. Front Water (1). https://doi.org/10.3389/frwa.2023.1243251, 2023.
Kharrou MH, Er-Raki S, Chehbouni A, Duchemin B, Simonneaux V, Le Page M, Ouzine L, Jarlan L (2011) Water use efficiency and yield of winter wheat under different irrigation regimes in a semi-arid region. Agric Sci 2(13):273–282
Khettouch A, Hssaisoune M, Hermans T, Aouijil A, Bouchaou L (2023) Ground validation of satellite-based precipitation estimates over poorly gauged catchment: the case of the Drâa basin in Central-East Morocco. MediterrEan Geosci 10(13)
Liu Y, Wu G, Ma B, Wu T, Wang X, Wu Q (2024) Revealing climatic and groundwater impacts on the spatiotemporal variations in vegetation coverage in marine sedimentary basins of the North China Plain, China. Sci Rep 14
Mazirh K, El Goumi S, Ibnoussina M, Witam O, Nocairi M, Kasimi R, Er-Raki S (2023) Using remote sensing to address green heritage of the city of Marrakech, Morocco. Can J Remote Sens 49(11)
Michailovsky CI, Coerver B, Mul M, Jewitt G (2024) Investigating sources of variability in closing the terrestrial water balance with remote sensing. Hydrol Earth Syst Sci 27:4335–4354
El Hannani M, Taïbi AN, Brabra N, Giffon S (2017) Les enjeux du végétal dans une ville du Sud. The importance of plant life in a city of the "south"—the case of Marrakesh and the end of "Garden-City" model: Le cas de la ville de Marrakech et la fin d'un modèle de cité-jardin, Projets de paysage 16(116)
Rabia OE (2020) Étude de modélisation hydrogéologique de la nappe Bahira. Agence de bassin hydraulique Oum Er Rabia
Salles D, Le Treut H (2017) Comment la région Nouvelle Aquitaine anticipe le changement climatique? Sci Eaux Territ 22(11):14–17
Song D, Pei X, Mao L, Wang J, Tian Y, An X (2024) An study on the spatial–temporal variation of groundwater depth and its impact on vegetation coverage in Ejina Oasis. Forest 15(111):2034
Srinivasan R (2020) Water security issues of India. Curr Sci 119(19):1395–1396
Stockstill A (2024) Epilogue: the genius loci of Marrakesh. In: Marrakesh and the mountains: landscape, urban planning, and identity in the medieval Maghrib, Penn State University Press, pp 127–132
Trabelsi R, Abida H (2024) Wetland water balance estimation in an arid region using remote sensing technology and hydrological modelling. Hydrol Sci J pp 905–922

Groundwater Recharge Assessment with Integrated Approaches

C. Thivya, S. Venkatramanan, K. Radhakrishnan, K. Sivakumar, Karung Phaisonreng Kom, and P. Karthick

Abstract Groundwater recharge is a crucial process for the conservation of water resources, especially in regions with water scarcity. This review aims to evaluate the different methods used to assess groundwater recharge, highlighting the benefits of integrated approaches. Traditional methods such as the water balance method, hydrograph analysis and tracer techniques are discussed alongside modern techniques such as remote sensing, geophysical methods and numerical modeling. By combining these methods, integrated approaches provide a more robust and comprehensive assessment of groundwater recharge. The paper presents case studies from different geographical areas to demonstrate the application and effectiveness of these integrated methods. Technological advances such as GIS, machine learning and IoT sensor networks are explored for their potential to improve groundwater recharge assessment. Environmental and policy implications are discussed, highlighting the importance of sustainable water management and informed policy making. The report identifies current research gaps and suggests future directions to improve the assessment of groundwater recharge. Through this comprehensive analysis, the paper aims to contribute to the development of more accurate and sustainable groundwater management practices.

C. Thivya (✉) · S. Venkatramanan · K. Radhakrishnan · K. P. Kom · P. Karthick
PG and Research Department of Geology, National College (Autonomous), Tiruchirappalli, Tamil Nadu, India
e-mail: thivyac1988@gmail.com; thivya@nct.ac.in

S. Venkatramanan
e-mail: venkatramanan@nct.ac.in

K. Radhakrishnan
e-mail: radhakrishnan@nct.ac.in

K. P. Kom
e-mail: kom@nct.ac.in

P. Karthick
e-mail: karthickp@nct.ac.in

K. Sivakumar
Department of Geology, Faculty of Science, Alagappa University, Karaikudi, Tamil Nadu, India
e-mail: siva.karthi90@yahoo.com

A. Nath et al. (eds.), *Groundwater Depletion and Sustainability*, Advances in Geographic Information Science, https://doi.org/10.1007/978-3-032-09921-1_17

Keywords Groundwater recharge · Integrated approaches · Water balance method · Remote sensing · GIS · Sustainable water management · Hydrology

1 Introduction

Groundwater supplies around 2.5 billion people worldwide with drinking water (WHO 2021). Groundwater is used extensively for irrigation, especially in regions with insufficient rainfall and surface water supplies. In many countries, it accounts for the majority of water used in agriculture (FAO 2021). Groundwater is one of the most important natural resources for domestic, agricultural, industrial and other development activities (Gautam et al. 2023). Groundwater is a complex natural resource that occurs in underground formations. It cannot be observed directly from the earth's surface and is difficult to measure due to its temporal and spatial fluctuations (Ghorbani Nejad et al. 2017). Groundwater levels are declining at an alarming rate over time due to uncontrolled extraction of groundwater for various purposes. This problem is particularly prevalent in areas with fresh groundwater. Hence, there is an urgent need to develop effective surface water management strategies to improve groundwater recharge (Chaudhary and Kumar 2018).

Groundwater recharge refers to the process by which groundwater is replenished by the infiltration of surface water into underground aquifers (Todd 2007). Understanding groundwater recharge dynamics is essential for effective water resources management, especially in the context of increasing population pressure, climate change and groundwater overexploitation (Sharma et al. 2010). The study of groundwater recharge involves assessing the rates at which aquifers are recharged, the variability of recharge over time and space, and the factors that influence the recharge process. These factors include precipitation patterns, land use changes, soil types, geologic and topographic drainage, and climate variability (Scanlon et al. 2006). Enrichment is influenced by natural processes such as precipitation and infiltration as well as human activities such as irrigation and urbanization (Cai et al. 2016).

The long-term sustainability of freshwater resources today depends on a range of groundwater recharge strategies. Without compromising groundwater quality, these techniques help to maintain constant water supplies (Richard-Ferroudji et al. 2018). The assessment of potential recharge zones and appropriate recharge methods is necessary to address the problems of overexploited and contaminated aquifers. According to recent studies (Wu et al. 2020), the Central Valley in California, the North China Plain, the Guarani aquifer in South America, the Canning Basin in northwest Australia and the aquifers beneath the headwaters of the Indus and Ganges rivers in northwest India are just a few examples of groundwater-dependent areas where climate change is adversely affecting domestic, industrial and agricultural water demands. Therefore, initiatives to improve recharge and changes in groundwater management in certain areas can contribute to the recovery of aquifers by slowing down depletion rates. Adapting to climate change and increasing water demand requires the preservation of natural recharge areas, the introduction of artificial recharge and the implementation of sustainable land use policies. Green infrastructure and sound policy frameworks are essential for sustainable groundwater management (Pourghasemi et al. 2020).

Given the critical role of groundwater for global water supply, it is essential to use comprehensive recharge assessment methods. The holistic combination of hydrological modeling, remote sensing, field monitoring and data assimilation has proven to be an effective strategy for estimating recharge rates and ensuring sustainable groundwater management (Voss et al. 2013). The demand for water resources is increasing daily due to the effects of urbanization and climate change. Climate change brings with it fluctuations in precipitation, humidity and temperature, which reduce groundwater recharge and groundwater levels. At the same time, urbanization leads to higher demand for freshwater to support economic development and ensure food security. In India, groundwater extraction plays an important role in food security and contributes 9% of GDP. However, many aquifers are overexploited and contaminated, as reported by the Central Groundwater Board of India (Central Ground Water Board 2023). To overcome these challenges and improve the quality and quantity of water resources, it is crucial to identify potential recharge zones and apply appropriate recharge techniques based on terrain conditions. Sustainable groundwater management urgently requires the implementation of effective recharge systems, and various techniques are available to address this issue. In this context, the present chapter provides a comprehensive overview of the different approaches used worldwide to assess groundwater recharge. These techniques can be broadly categorized into quantitative and qualitative methods.

2 Quantitative Methods

Empirical methods are used by researchers around the world to quantitatively estimate groundwater recharge (Meinzer 1923; Theis 1938). These techniques are based on observed relationships between groundwater recharge and factors such as precipitation, runoff, evapotranspiration, land use/land cover, soil properties and aquifer properties (Fig. 1). The following empirical techniques are commonly used to assess groundwater recharge at specific spatial and temporal scales.

2.1 Water Balance Method

Groundwater recharge can be estimated using the traditional water balance method. This empirical approach to estimating groundwater recharge was first proposed by Thies in 1935. The recharge component is calculated by balancing the total water input (precipitation) against the outputs (evapotranspiration, surface runoff and groundwater discharge) using the principle of conservation of mass. Therefore, the water balance can be expressed as follows (Thapa et al. 2017):

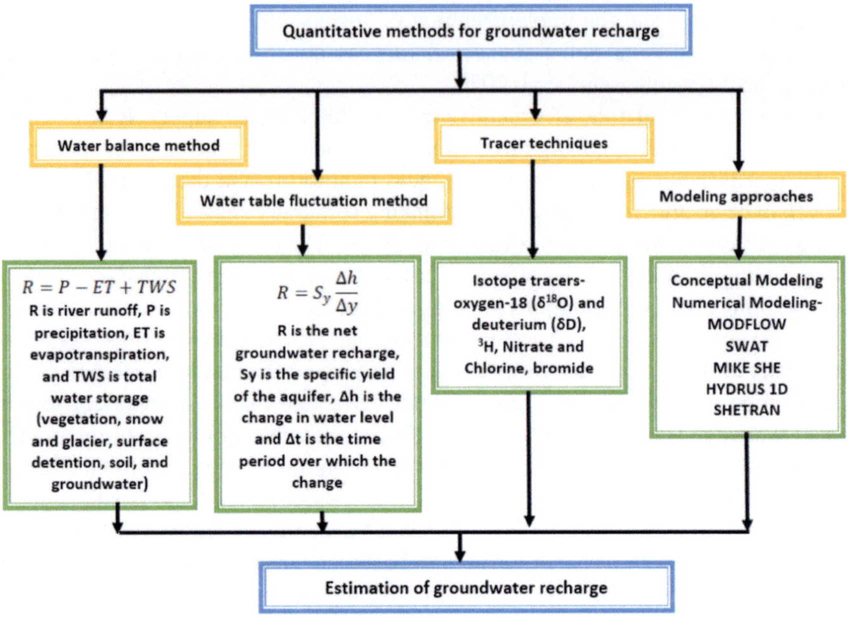

Fig. 1 Flowchart of various quantitative methods used for estimating groundwater recharge

$$R = P - ET + TWS$$

where, R is river runoff, P is precipitation, ET is evapotranspiration, and TWS is total water storage (vegetation, snow and glacier, surface detention, soil, and groundwater).

Although this approach is simple, it requires precise and comprehensive data on hydrological parameters. In agricultural regions, the water balance approach is often used to assess recharge from irrigation. The infiltration of water into the subsurface as recharge in semi-arid areas has been determined using precipitation data, infiltration rates and runoff coefficients (Atalla et al. 2024; Li et al. 2023).

2.2 *Water Table Fluctuation Method*

This estimation method has been used for centuries to assess groundwater recharge. Due to its simplicity, it is widely accepted and used in many parts of the world to estimate recharge. It is based on the assumption that a rise in the water table in groundwater wells is the result of recharge occurring throughout the aquifer (Meinzer 1923). The following equation for estimating groundwater recharge was derived on the basis of this principle:

$$R = S_y \frac{\Delta h}{\Delta y}$$

where R is the net groundwater recharge, Sy is the specific yield of the aquifer at a given point, Δh is the change in water level and Δt is the time period over which the change is measured. The Sy value for different aquifer media can be taken here from Freeze and Cherry (Freeze and Cherry 1979). Chaturvedi (Chaturvedi 1973) derived the following empirical relationship based on groundwater fluctuations and annual precipitation for estimating recharge in the Indian context:

$$R = 2.0(P - 15)^{04}$$

where R is the net recharge due to precipitation during the year (in inches), and P is the annual precipitation (in inches).

2.3 Hydrograph Analysis

Baseflow Separation

The contributions of groundwater are reflected in the base flow, i.e. the steady flow of rivers and streams during dry periods (Gnann et al. 2023). It is possible to derive groundwater recharge rates by applying statistical or graphical methods to separate the base flow.

Recession Curve Analysis

This method is used to examine the downward sloping section of a hydrograph during dry periods. Recharge can be estimated from the rate of decline or recession, which provides information on the characteristics of groundwater discharge (Becke et al. 2024).

2.4 Tracer Techniques

Isotope Tracers

Groundwater recharge is often investigated using stable isotopes such as oxygen-18 ($\delta^{18}O$) and deuterium (δD) (Masoud et al. 2025; Qu et al. 2024). The distinction between older groundwater and younger infiltration is facilitated by seasonal changes in isotope characteristics. The monitoring of recharge processes in arid and semi-arid areas is facilitated by the use of this technique.

Hamidi et al. (Hamidi et al. 2023) investigated the stable isotope composition ($\delta^{18}O$ and $\delta^{2}H$) of surface and groundwater for the Upper Kabul and Logar rivers in the city of Kabul. The contribution of river water to groundwater, including the uncertainty analysis of its estimation, was investigated using the hydrograph separation technique. Based on the stable isotopes, the study revealed that precipitation is the main source of groundwater recharge in the Central Kabul sub-basin. In the Logar sub-basin, groundwater recharge was controlled by a combination of precipitation, irrigation return flow and river recharge.

Thivya et al. (Thivya et al. 2016) conducted a study in the hard rock aquifers of Madurai district in Tamilnadu to identify the recharge processes using stable isotopes. In the study, the interpretation of $\delta^{18}O$ and δD indicates enrichment by meteoric water in charnockite, quartzite, granite and some samples of fissile hornblende-biotite gneiss. It is also inferred that the enrichment is due to evaporated water in alluvium and fissile hornblende-biotite gneiss.

Chemical Tracers

Accumulation rates are estimated using chemical makers such as nitrate or chloride. To calculate the rate of infiltration into the groundwater system, for example, the chloride mass balance technique is used to investigate the accumulation of chloride in soil profiles (Bekele et al. 2024; Li et al. 2025). Where there is little surface drainage, this technique works well.

Groundwater recharge was estimated from historical tracers (^{3}H and ^{36}Cl) resulting from human activities or historical events in the past by locating the concentration peak of the tracers, assuming a flask flux for the soil water. Numerical simulation is an additional technique for estimating groundwater recharge. Hydrus-1D has been used extensively and consecutively to estimate soil hydraulic parameters using the global optimization approach, soil water and solvent flux, and groundwater recharge (Gogolev 2002; Bethune 2008). The model may not work at field scale because surface conditions including evapotranspiration, soil evaporation and irrigation are too complex to obtain accurate data (Lu 2011). This problem could be solved by setting the upper boundary of the model at a depth below the root zone and evaluating the water content and water potential of the soil over time.

Applied Tracers

These are used as pulses on the ground surface or in the subsurface and are recognized as a valuable technique for estimating groundwater recharge as historical tracers. The primary tracers used were fluorescein dyes, chlorine and bromide, all of which have been successful and efficient in determining current groundwater recharge (Flint et al. 2002; Dassi 2010; Wang et al. 2008).

2.5 *Modeling Approaches*

Conceptual Models

Simplified representations of groundwater systems derived from assumptions and observed data are referred to as conceptual models (Secci et al. 2024). These models help researchers to understand recharge mechanisms and formulate research hypotheses.

Numerical Models (E.G., MODFLOW)

MODFLOW and other numerical models simulate groundwater flow and recharge using mathematical equations (Abraham et al. 2021; Rahman et al. 2025). The geometry of the aquifer, hydraulic properties and temporal variations in recharge are just some of the complicated elements included in these models. They are widely used in water resource management and provide accurate, quantitative insights into groundwater dynamics. Aslam et al. (2025) integrated Certainty Factors (CF), Evidential Belief Function (EBF), Frequency Ratio (FR) and Convolutional Neural Network (CNN) with Logistic Model Tree (LMT) and Random Forest (RF) to map groundwater recharge zones using machine learning and bivariate statistical models. Their results show that CNN-RF produced the most accurate maps of groundwater recharge potential with an AUC of 0.923, demonstrating the value of integrating data mining methods with bivariate statistical models. Rajasekhar et al. (2021) identified suitable zones for artificial groundwater recharge in a semi-arid area in southern India by combining geospatial and decision techniques such as fuzzy logic (FL), frequency ratio (FR) and analytic hierarchy processes (AHP). By categorizing the recharge zones into five groups, their study showed that the integrated approach gave the best results (82%) when validated with Receiver Operating Characteristic (ROC) and Area under the Curve (AUC) techniques and provided important information for sustainable groundwater management.

Using MODFLOW, a transient-state groundwater flow model was created in the Cachar area of Assam to evaluate groundwater recharge. The study estimated a total inflow of 28.47 million cubic meters (mcm) and an outflow of 27.45 mcm, highlighting the possible agricultural zones (Singh et al. 2022). The impacts of pumping and recharge on the groundwater system were assessed using MODFLOW (Sahoo and Jha 2017). According to their model, pumping has a major effect on the constrained aquifer system, highlighting the necessity of sustainable groundwater management techniques. In the Hailiutu River Catchment in Northwest China, a modified SWAT (Soil and Water Assessment Tool) model was utilized to model groundwater recharge in a semi-arid area. The alteration made it possible to simulate the interactions between deep and shallow aquifers, giving the groundwater cycle a more realistic depiction (Shao et al. 2019). Groundwater recharge was estimated using MIKE SHE (European Hydrological System) model. Based on the rainfall and soil characteristics, the modeling results demonstrated that daily groundwater recharge varied from 0 mm/day in January to 29.85 mm/day in August (OKE, Muritala. 2020). Deep percolation and net groundwater use under different farming patterns in India's semi-arid Indo-Gangetic Plain were simulated using the HYDRUS 1D model. During the wet season, the model projected deep percolation rates of 159.1 mm for soybeans and 293.8 mm for cotton, suggesting substantial groundwater recharge potential (Tonkul et al. 2019). Catchment hydrology has been modeled using SHETRAN in in geologically complicated terrains. A watershed with an extent of 586 km^2 with complicated geology has had its hydrology simulated, allowing for a thorough evaluation of groundwater recharge and flow dynamics (Vázquez et al. 2022).

3 Qualitative Methods

All over the world, quantitative methods are used to determine the groundwater recharge potential at the earth's surface instead of estimating the amount of recharge. These methods help to locate zones where significant recharge is expected and areas where the recharge rate is comparatively lower than before. The identification of zones with recharge potential is crucial for determining suitable locations for the implementation of recharge structures to support controlled groundwater recharge. In this context, advanced remote sensing and GIS technologies are widely used as effective tools to delineate potential recharge zones in any region (Vellaikannu et al. 2021; Arumugam et al. 2023) (Fig. 2). GIS enables the integration of topography, land use, soil and precipitation data for the spatial analysis of enrichment zones. Techniques such as multi-criteria decision analysis (MCDA) prioritize the enhancement of areas. Remote sensing data integrated with GIS helps in high-resolution mapping and predictive modeling (Shaikh and Birajdar 2024a). These tools support sustainable water management by identifying critical recharge zones and assessing the impact of land use. Landsat, Sentinel and MODIS are examples of satellite-based remote sensing systems that provide geographical and temporal data on vegetation, land cover and soil moisture. Possible enrichment zones are identified and quantified by integrating these data sets with hydrological models. In addition, satellite images facilitate large-scale evaluations and are particularly useful in areas with few ground-based monitoring systems (Nguyen et al. 2020). Meng et al. (2024) mapped and identified groundwater recharge zones in Pakistan's Mohmand district using remote sensing and GIS-based multi-criteria decision analysis (MCDA).

Using Analytical Hierarchy Process (AHP) and Multi-Influencing Factor (MIF) approach, the study evaluated thematic layers such as geomorphology, slope, lithology, rainfall, lineament density, drainage density, land use/land cover and hydrological soil group. Artificial recharge zones were obtained by overlaying all the thematic maps in the terms of weighted overlay analysis methods using remote sensing and GIS Techniques (Karthick and Lakshumanan 2018). The results confirmed the importance of spatial models in the assessment of groundwater

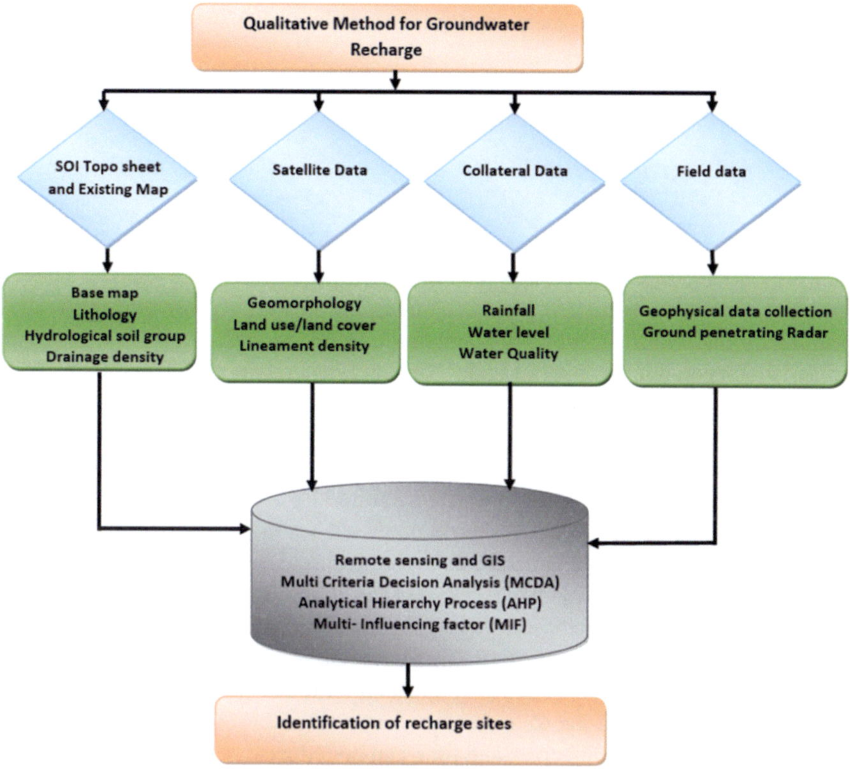

Fig. 2 Flow chart explaining the identification of recharge sites through qualitative method

resources by showing differences in groundwater recharge capacity. MIF recommends 1.20% and AHP identifies 1.81% of the region as extremely high recharge zones.

3.1 Geospatial Techniques

Detailed representations of surface and sub-surface features, including vegetation patterns, water bodies, drainage density, lineament density, slope and land use & land cover, are extracted from the high-resolution satellite data of Sentinel-2A, LISS- IV and Cartosat-DEM which are available in NRSC Bhuvan website. These thematic layers are crucial in defining enrichment sites and assessing their potential. Combination of remote sensing and GIS techniques are often used for integrating of all the thematic layers for accurate mapping and identifying the suitable of recharge zones. All these techniques allow a qualitative assessment of groundwater recharge. However, to assess groundwater recharge quantitatively, they need to be combined with appropriate conventional methods, which are described in the quantitative techniques.

3.2 Geophysical Methods

Geophysical methods are not used directly to estimate groundwater recharge, but they provide important information on the storage properties of the aquifer, which is crucial for the assessment of recharge. Among the various geophysical exploration techniques, the electrical resistivity method is used by hydrogeologists to investigate groundwater potential in many regions worldwide (Prabakaran et al. 2020; Sekar et al. 2021; Sivakumar et al. 2022).

Electrical resistivity: In this technique, resistivity is measured by passing an electric current into the soil. The presence of water means that enrichment sites often have a lower resistivity. This method is well suited for locating recharge zones and aquifer boundaries. Shafiqullah Wahab et al. (Wahab et al. 2021) used the 2D inversion results of the pole-pole arrangement for electrical resistivity to identify three layers of electrical resistivity in particular: (1) a shallow low resistivity layer ($<50\ \Omega$ m), which is thought to be topsoil; (2) a groundwater aquifer ($50\ \Omega$ m to $170\ \Omega$ m); and (3) a highly resistive bedrock ($> 200\ \Omega$ m). These results can be used to locate well drillings in the study area and are important for future groundwater exploration. By knowing the depth of the water-saturated zone (aquifer) and the lateral extent of the shallow aquifer, drilling costs and time can be reduced.

3.3 Ground-Penetrating Radar (GPR)

High frequency electromagnetic waves are used by ground penetrating radars (GPR) to visualize underground structures (Tassiopoulou and Koukiou 2024). It is particularly useful for mapping stratigraphic layers, water table changes and shallow recharge zones (Jiang et al. 2024).

4 Role of Machine Learning and AI

ML and AI are central to analyzing large data sets and improving charge predictions. Algorithms such as random forest, support vector machines and neural networks classify recharge zones and simulate hydrological processes (Mosavi et al. 2021). AI-based tools optimize artificial recharge structures, detect anomalies and integrate different data sets for comprehensive modeling. Challenges include data quality and model interpretability, which are addressed by hybrid approaches that combine ML and physical models (Pourghasemi et al. 2020). Recent advances have revolutionized the assessment of groundwater recharge. AI and machine learning analyze complex datasets and predict groundwater recharge under different scenarios (Mosavi et al. 2021; Jaafarzadeh et al. 2021). High-resolution tools such as LiDAR and hyperspectral imaging facilitate precise mapping of recharge areas (Dong and Chen 2017). IoT-enabled smart sensors monitor soil moisture and aquifer conditions in real time, complementing hydro-informatics platforms that promote data visualization and collaboration (Singh et al. 2020; Jonoski 2022).

Integrating technology with traditional measures such as infiltration tanks improves recharge potential and addresses challenges such as climate change and over-abstraction of water (Bouwer 2002). Mapping brings together topographic, soil, land use, climate and hydrological data. Techniques such as weighted overlay analysis and isotopic studies confirm the enrichment zones. AI models improve prediction accuracy and support resource management and urban planning (Mosavi et al. 2021). Despite challenges such as data scarcity and complexity, advances in remote sensing and machine learning offer promising solutions. Long-term monitoring and stakeholder collaboration are crucial for effective management. Technologies such as GIS, remote sensing and IoT improve precision, while MAR techniques mitigate shortcomings in enrichment (Shaikh and Birajdar 2024b). Integration of ML with traditional methods improves predictive capabilities and planning. Zaresefat et al. (2023) used artificial intelligence (AI) approaches to find suitable sites for artificial groundwater recharge (AGR) in the Iranshahr Basin. By evaluating nine critical parameters, including precipitation, geology, land use and soil permeability, a 97 accurate map of AGR suitability was created by combining machine learning algorithms with geospatial data analysis. The results illustrate the efficiency of AI in optimally selecting groundwater recharge sites, providing a sustainable solution to reduce aquifer depletion in arid regions.

4.1 IoT and Real-Time Monitoring

IoT-enabled sensor networks are revolutionizing the assessment of groundwater recharge by providing real-time data on groundwater levels, soil moisture and precipitation. These networks can be integrated with GIS and remote sensing for detailed spatial models, facilitating dynamic management and early detection of anomalies (Zulkifli et al. 2022). Challenges such as high costs and the complexity of data management are mitigated by advances in sensor technology and connectivity.

5 Managed Aquifer Recharge (MAR)

The ultimate goal of quantitative and qualitative groundwater recharge estimation is to support managed groundwater recharge that raises groundwater levels and promotes sustainable development in regions of water scarcity. In some

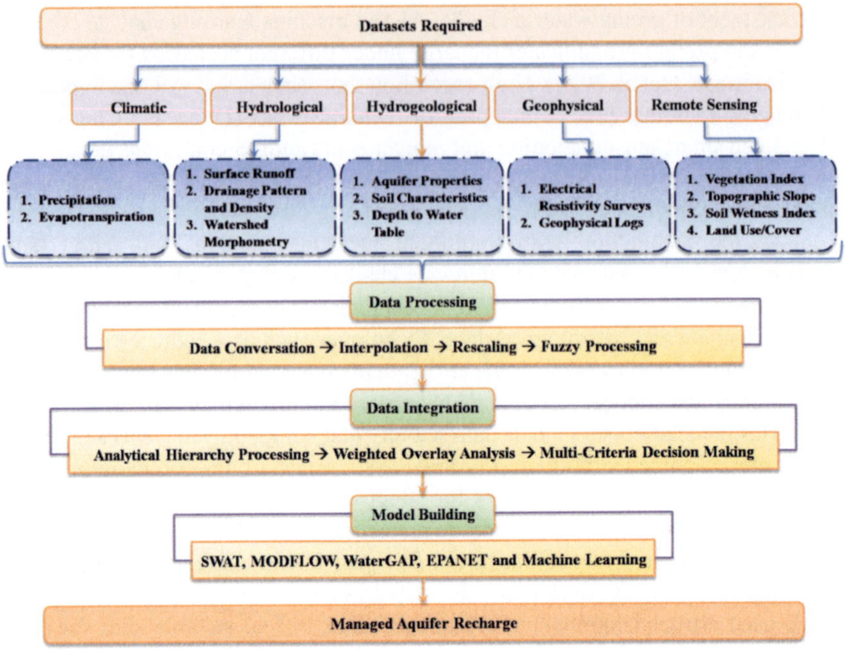

Fig. 3 Shows the flow chart for the integrated approach to estimating groundwater recharge

studies, aquifer recharge through appropriate artificial recharge structures has been proposed to improve groundwater quality in areas where pollution has been detected (Karthikeyan et al. 2016) (Fig. 3).

Sheik et al. (2024) investigated the monitoring and design of controlled aquifer recharge (MAR) systems for sustainable groundwater management using machine learning (ML) approaches. The study demonstrated the effectiveness of different ML models, including supervised, semi-supervised and ensemble learning techniques, in predicting MAR parameters and improving water quality control. The results suggest that the integration of ML with MAR improves groundwater recharge strategies and provides better decision support for sustainable water resources management.

6 Policy and Management Recommendations

Although these scientific interventions promote the controlled recharge of aquifers for the improvement of groundwater and sustainable development, efforts at the policy and management level are required to implement them effectively for the benefit of society at the grassroots level. Based on this fact, suggestions, recommendations, roles of various organizations and future directions for further studies are outlined below.

7 Strategies for Sustainable Groundwater Use

The sustainable use of groundwater is an essential prerequisite for securing the water supply for future generations. The National Water Act of 1998, integrated surface and groundwater governance and encouraged stakeholder interaction, established the ideas of Integrated Water Resource Management (IWRM), which has been successfully applied in nations like South Africa. According to the Global Water Partnership (GWP)(GWP 2000) this strategy makes sure that water allocations balance conflicting sectoral demands while taking ecological sustainability into account. Demand-side management techniques have been effective in India, where groundwater extraction has been greatly decreased without sacrificing agricultural productivity using micro-irrigation technologies like drip and sprinkler systems. These techniques improved water-use efficiency in water-limited areas like Gujarat (Shah et al. 2007). By requiring permits for extraction and setting sustainable yield limitations, regulatory measures—such as India's Model Groundwater Bill (Ministry of Water Resources 2011)—have established frameworks to control groundwater pollution and prevent over-abstraction. In Andhra Pradesh, India, community-based methods have demonstrated significant promise in participatory groundwater management projects. Local farmers, assisted by NGOs and government organizations, participated in water budgeting and aquifer mapping, resulting in more efficient and equitable water use (Kemper et al. 2005). Australia and India have effectively implemented artificial recharge and aquifer restoration techniques. For instance, Dillon et al. (2019) describe managed aquifer recharge (MAR) projects that have restored depleted aquifers and enhanced water quality by allowing harvested rainwater and treated storm water to infiltrate, particularly in arid and semi-arid zones.

8 Role of Government and International Organizations

In order to ensure the sustainable use of groundwater, Governments and International organizations must develop and execute coordinated, cooperative policies. One important strategy is creating all-encompassing regulations that match groundwater management with more general plans for water resources. Transparency, stakeholder engagement, and adaptive governance are highlighted in the OECD Principles for Water Governance (OECD 2015), which offer a framework for this kind of integration. Furthermore, these organizations make financial and infrastructure investments, assisting groundwater recharge projects and updating water systems to minimize losses and improve efficiency, as the World Bank (2018) noted in its proposals for bolstering global water security. Reliable mechanisms for gathering and monitoring data are also essential; programs such as those run by the International Groundwater Resources Assessment Centre (IGRAC 2017) assist nations in setting up networks for real-time groundwater quality and level monitoring. Another crucial element is capacity building. Organisations like UNESCO's International Hydrological Programme (IHP) concentrate on research, education, and training to support sustainable groundwater management, especially in areas dealing with complicated issues like transboundary aquifers (UNESCO-IHP. 2020). Lastly, the management of shared groundwater resources requires international cooperation. For the sustainable regulation of transboundary aquifers, successful cross-border cooperation is demonstrated by agreements such as the Guarani Aquifer Agreement in South America (Foster et al. 2009).

9 Conclusion

Emerging technology and interdisciplinary approaches are essential for future groundwater management research to meet the increasing problems of groundwater contamination and depletion. The developments in GIS and remote sensing have already shown promise in tracking groundwater levels and pinpointing hotspots for recharge and depletion. The artificial intelligence and machine learning models presents fresh possibilities for forecasting aquifer behavior and maximizing resource utilization. With the integration of IoT-enabled sensor networks, real-time data gathering for intelligent groundwater monitoring is being improved, and isotopic tracking techniques offer important insights into contamination pathways and recharge sources. Furthermore, especially in environments with limited groundwater resources, decentralized water treatment technologies are showing promise as local groundwater cleaning and reuse solutions. Future research should concentrate on comprehending how climate change affects groundwater systems, investigating the social and economic factors that affect water usage patterns, and creating collaborative approaches to transboundary aquifer management. In order to guarantee that conservation initiatives are in line with sustainable water usage, research should also deepen our understanding of ecosystems that depend on groundwater. These paths will be essential to developing robust and comprehensive groundwater governance.

Acknowledgements (If any) Authors like to express their sincere gratitude to Department of Geology, National college for providing an intellectually stimulating environment and the necessary academic support that greatly contributed to the development of this book chapter. The encouragement and resources offered by the institution have been instrumental in the completion of this work.

AI Discloser No AI has been used.

Disclose Conflict of Interest No conflict of interest.

References

Abraham J, Mathew A, Thomas A (2021) Simulation of groundwater flow and recharge using MODFLOW: A case study in Kerala. India. J Hydrol 592:125815

Arumugam M, Kulandaisamy P, Karthikeyan S, Thangaraj K, Senapathi V, Chung SY, Manimuthu S (2023) An assessment of geospatial analysis combined with AHP techniques to identify groundwater potential zones in the Pudukkottai District, Tamil Nadu. India. Water 15(6):1101. https://doi.org/10.3390/w15061101

Aslam MA, Pradhan B, Jebur MN (2025) Integration of machine learning models with bivariate statistical approaches for groundwater recharge potential mapping. Water Resour Manage 39(4):1123–1140

Atalla MA, Shebl A, Đurin B, et al (2024) Assessment of groundwater potential zones in Kuwait's semi-arid region: A hybrid approach of multi-criteria decision making, Google Earth Engine, and geospatial techniques. Scientific Reports, 14, 29938. https://doi.org/10.1038/s41598-024-76989-4

Becke AL, Solorzano-Rivas SC, Werner AD (2024) The watertable fluctuation method of recharge estimation: A review. Adv Water Resour 189:104635. https://doi.org/10.1016/j.advwatres.2024.104635

Bekele D, Alemayehu T, Kebede S (2024) Estimation of annual and seasonal groundwater recharge by using Wetspass-M and chloride mass-balance methods: Upper Bilate Catchment, Rift Valley Basin. Ethiopia. Int J Environ Sci 5(3):123–135. https://doi.org/10.1155/2024/2724015

Bethune MG (2008) Understanding and predicting deep percolation under surface irrigation. Water Resour Res 44(12):W12430. https://doi.org/10.1029/2007WR006380

Bouwer H (2002) Artificial recharge of groundwater: Hydrogeology and engineering. Hydrogeol J 10:121–142

Cai X, Rosegrant MW, Cline SA (2016). World water and food to 2025: Dealing with scarcity. Int Food Policy Res Inst

Central Ground Water Board (2023) Central Ground Water Year Book. https://cgwb.gov.in/cgwbpnm/public/uploads/documents/17014272111704550895file.pdf

Chaturvedi RS (1973) A note on the investigation of groundwater resources of Uttar Pradesh. Annual Report, U.P. Irrigation Research Institute, India

Chaudhary BS, Kumar S (2018) Identification of groundwater potential zones using remote sensing and GIS of KJ Watershed, India. J Geol Soc India 91(6):717–721

Daniele Secci, Ali Kerem Saysel, İzel Uygur, Onur Cem Yoloğlu, Andrea Zanini, Nadim K. Copty, Modeling for sustainable groundwater management: Interdependence and potential complementarity of process-based, data-driven and system dynamics approaches, Science of The Total Environment, Volume 951, 2024, 175491, ISSN 0048–9697, https://doi.org/10.1016/j.scitotenv.2024.175491.

Dassi L (2010) Assessment of groundwater recharge in a semi-arid groundwater basin using chloride mass balance, tritium, and carbon-14 methods. Environ Earth Sci 60(4):703–713. https://doi.org/10.1007/s12665-009-0204-8

Dillon P, Pavelic P, Page D, Beringen H, Ward J (2019) Managed aquifer recharge: Overview and recent developments. Water 11(5):1001. https://doi.org/10.3390/w11051001

Dong P, Chen Q (2017) LiDAR remote sensing and applications. CRC Press

FAO (2021) The state of the world's land and water resources for food and agriculture: Systems at breaking point. Food Agric Organ U N

Flint AL, Flint LE, Hevesi JA, Blainey JM (2002) Regional ground-water recharge in the Basin and Range Province, northwestern Nevada. U.S. Geological Survey Water-Resources Investigations Report, 02–4145. https://doi.org/10.3133/wri024145

Foster S, Garduño H, Tuinhof A (2009) The Guarani Aquifer Initiative—Sustainable management of a transboundary aquifer. Hydrogeol J 17(5):1213–1228

Freeze RA, Cherry (1979) A stochastic conceptual model of groundwater flow in a leaky aquifer system. Water Resour Res, 15(4), 1019–1033. https://doi.org/10.1029/WR015i004p01019

Gautam VK, Pande CB, Kothari M, Singh PK, Agrawal A (2023) Exploration of groundwater potential zones mapping for hard rock region in the Jakham river basin using geospatial techniques and aquifer parameters. Adv Space Res 71(6):2892–2908. https://doi.org/10.1016/j.asr.2022.11.009

Ghorbani S, Falah F, Daneshfar M, Haghizadeh A, Rahmati O (2017) Delineation of groundwater potential zones using remote sensing and GIS-based data-driven models. Geocarto Int 32(2):167–187. https://doi.org/10.1080/10106049.2015.1115763

Gnann S, Reinecke R, Stein L, Wada Y, Thiery W, Müller Schmied H, Wagener T (2023) Functional relationships reveal differences in the water cycle representation of global water models. Nature Water 1:1079–1090. https://doi.org/10.1038/s44221-023-00160-y

Gogolev A (2002) Assessing groundwater recharge with two unsaturated zone modeling technologies. Environ Geol 42(2–3):232–241. https://doi.org/10.1007/s00254-002-0545-3

GWP (2000) Integrated water resources management. Glob Water PartnShip Tech Comm

Hamidi DM, Gröcke DR, Joshi SK, Greenwell C (2023) Investigating groundwater recharge using hydrogen and oxygen stable isotopes in Kabul City, a semi-arid region. J Hydrol 617:129087. https://doi.org/10.1016/j.jhydrol.2023.129087

IGRAC (2017) Global Groundwater Monitoring Network. International Groundwater Resources Assessment Centre

Jaafarzadeh MS, Tahmasebipour N, Haghizadeh A, Pourghasemi HR, Rouhani H (2021) Groundwater recharge potential zonation using an ensemble of machine learning and bivariate statistical models. Sci Rep 11:5587. https://doi.org/10.1038/s41598-021-85127-w

Jiang Y, Wang X, Li Z (2024) Discriminating underground cavities by numerical simulation of ground-penetrating radar. J Appl Remote Sens 19(1):014521. https://doi.org/10.1117/1.JRS.19.014521

Jonoski A (2022) Michael Abbott's Hydroinformatics: Poiesis of new relationships with water. IWA Publishing

Karthick P, Lakshumanan C (2018) Identification of groundwater recharge sites and suitable recharge structures for Thuraiyur Taluk using Geospatial Technology. Indian J Geo Mar Sci 47(10):2117–2125

Karthikeyan B, Jagadeshan G, Kalpana L, Elango L (2016) Fluoride in weathered rock aquifers of southern India: Managed aquifer recharge for mitigation. Environ Sci Pollut Res 23:8302–8316

Kemper KE, Foster S, Garduño H, Tuinhof A (2005) Participatory aquifer management. World Bank Discussion Papers

Li J, Pang Z, Liu Y, Hu S, Jiang W, Tian L, Tian J (2023) Changes in groundwater dynamics and geochemical evolution induced by drainage reorganization: Evidence from 81Kr and 36Cl dating of geothermal water in the Weihe Basin of China. Earth Planet Sci Lett 623:118425. https://doi.org/10.1016/j.epsl.2023.118425

Li X, Zhang Y, Wang J (2025) Use and misuse of the chloride-mass balance method in estimating groundwater recharge. Hydrogeol J 33(1):45–58. https://doi.org/10.1016/j.hydro.2025.01.003

Lu Z (2011) Estimating groundwater recharge using HYDRUS-1D. Eng Geol Hydrogeol 48(3):245–253. https://doi.org/10.1016/j.enggeo.2011.06.002

Masoud M, Al-Farrah N, Sen Z (2025) Groundwater recharge evaluation using stable isotopes and the NETPATH model in Al-Madinah Al-Munawarah Province. Saudi Arabia. Water 17(2):211. https://doi.org/10.3390/w17020211

Meinzer OE (1923) The occurrence of ground water in the United States, with a discussion of principles (Water-Supply Paper 489). U.S. Geological Survey

Meng F, Khan SJ, Ullah I (2024) Identification and mapping of groundwater recharge zones using multi-influencing factor and analytical hierarchy process. Sci Rep 14:70324. https://doi.org/10.1038/s41598-024-70324-7

Ministry of Water Resources (2011) Model bill for the conservation, protection, regulation and management of groundwater. Government of India

Mosavi A, Sajedi F, Choubin B, Goodarzi M, Dineva AA, Rafiei E (2021) Ensemble boosting and bagging based machine learning models for groundwater potential prediction. Water Resour Manage 35:23–37

Mrinal Kumar Singh, Susmita Ghosh (2022) Groundwater flow modeling for cachar, India using MODFLOW: a case study. ISH Journal of Hydraulic Engineering, 28, 2

Nguyen A, Pham H, Tran D (2020) Application of remote sensing and GIS to assess groundwater potential zones: A case study in Vietnam. J Hydrol: Regnal Stud 29:100678. https://doi.org/10.1016/j.ejrh.2020.100678

OECD (2015) OECD principles on water governance. OECD Publishing

OKE, Muritala (2020) Application of MIKE SHE software for estimation of groundwater recharge in Ogun and Oshun basins, southwestern Nigeria. Journal of Water and Land Development. 86–93. https://doi.org/10.24425/jwld.2020.133049

Pourghasemi HS, Sadhasivam N, Yousefi S, Tavangar S, Nazarlou HG, Santosh M (2020) Using machine learning algorithms to map the groundwater recharge potential zones. J Environ Manage 265:110525. https://doi.org/10.1016/j.jenvman.2020.110525

Prabakaran S, Rajendran S, Elango L (2020) Delineation of groundwater potential zones and seawater intrusion areas using earth resistivity imaging in coastal regions. J Environ Geol 78(5):1234–1245

Qu W, Zhang Y, Li X (2024) Stable isotopes reveal the water conversion and transition dynamics in the Baoxing River Basin. Environ Earth Sci 83(5):125. https://doi.org/10.1016/j.envsci.2024.125

Rahman MA, Lee S, Kim Y (2025) Numerical modeling of groundwater recharge in response to climate change using MODFLOW. Environ Earth Sci 84(3):65

Rajasekhar M, Ajaykumar K, Sudarsana Raju G, Bhagat V (2021) Identification of artificial groundwater recharge zones in semi-arid region of Southern India using geospatial and integrated decision-making approaches. Environ Chall 5:100278. https://doi.org/10.1016/j.envc.2021.100278

Richard-Ferroudji A, Raghunath TP, Venkatasubramanian G (2018) Managed aquifer recharge in India: Consensual policy but controversial implementation. Water Alternatives 11(3):749–769

Sahoo S, Jha MK (2017) Numerical groundwater-flow modeling to evaluate potential effects of pumping and recharge: implications for sustainable groundwater management in the Mahanadi delta region. India. Hydrogeol J 25:2489–2511

Scanlon BR, Healy RW, Cook PG (2006) Choosing appropriate techniques for quantifying groundwater recharge. Hydrol Process 20(3):777–787. https://doi.org/10.1002/hyp.6131

Sekar M, Kumar P, Stanly S (2021) Application of electrical resistivity method in groundwater exploration: A case study from Tamil Nadu. India. Int J Geophys 16(3):567–578

Shah T, Burke J, Villholth KG (2007) Groundwater: A global assessment of scale and significance. IWMI Research Report

Shaikh M, Birajdar F (2024a) Advancements in remote sensing and GIS for sustainable groundwater monitoring: Applications, challenges, and future directions. Int J Res Eng, Sci Manag 7(1):45–52. https://doi.org/10.5281/zenodo.1234567

Shaikh M, Birajdar F (2024b) Artificial intelligence in groundwater management: Innovations, challenges, and future prospects. Int J Sci Res Arch 11(01):502–512

Shao G, Zhang D, Guan Y, Xie Y, Huang F (2019) Application of SWAT model with a modified groundwater module to the semi-arid hailiutu river catchment. Northwest China Sustain 11:2031. https://doi.org/10.3390/su11072031

Sharma R, Tiwari S, Sharma P (2010) Groundwater recharge in arid and semi-arid regions: A case study of the Thar Desert. Springer, India

Sheik AG, Kumar A, Sharanya AG, Amabati SR, Bux F, Kumari S (2024) Machine learning-based monitoring and design of managed aquifer rechargers for sustainable groundwater management: Scope and challenges. Environ Sci Pollut Res. https://doi.org/10.1007/s11356-024-35529-3

Singh R, Mishra V, Narasimhan B, Ghosh S, Sharma A, Dutta S, Mujumdar PP (2020) March). Hydrological modeling in India. In Proceedings of the Indian National Science Academy 86:479–494

Sivakumar V, Vijayaprabhu M, Gopinath S (2022) Identification of groundwater potential zones using electrical resistivity imaging in a coastal aquifer system. Hydrogeol J 30(2):789–803

Tassiopoulou S, Koukiou G (2024) Fusing ground-penetrating radar images for improving image characteristics fidelity. Appl Sci 14(15):6808. https://doi.org/10.3390/app14156808

Thapa BR, Ishidaira H, Pandey VP, Shakya NM (2017) A multi-model approach for analyzing water balance dynamics in Kathmandu Valley. Nepal. Journal of Hydrology: Regional Studies 9:149–162. https://doi.org/10.1016/j.ejrh.2016.12.080

Theis CV (1938) The significance and nature of the cone of depression in ground-water bodies. Econ Geol 33(8):889–902. https://doi.org/10.2113/gsecongeo.33.8.889

Thivya C, Chidambaram S, Rao MS, Gopalakrishnan M, Thilagavathi R, Prasanna MV, Nepolian M (2016) Identification of recharge processes in groundwater in hard rock aquifers of Madurai District using stable isotopes. Environ Process 3(2):463–477. https://doi.org/10.1007/s40710-016-0137-3

Todd DK (2007) Groundwater hydrology (3rd ed.). Wiley

Tonkul S, Baba A, Şimşek C, Durukan S, Demirkesen AC, Tayfur G (2019) Groundwater recharge estimation using HYDRUS 1D model in Alaşehir sub-basin of Gediz Basin in Turkey. Environ Monit Assess. 191(10) 610

UNESCO-IHP (2020) Transboundary aquifers: Challenges and new directions. UNESCO

Vázquez R, Brito J, Hampel H, Birkinshaw S (2022) Assessing the performance of SHETRAN simulating a geologically complex catchment. Water 14:3334. https://doi.org/10.3390/w14203334

Vellaikannu S, Muralitharan S, Palanivel K (2021) Assessment of groundwater potential zones using AHP and GIS techniques. Int J Environ Sci 5(3):123–135

Voss CI et al (2013) Estimating groundwater recharge in arid and semi-arid areas using remote sensing and hydrological modeling. J Hydrol 486:191–207. https://doi.org/10.1016/j.jhydrol.2013.01.014

Wahab S, Saibi H, Mizunaga H (2021) Groundwater aquifer detection using the electrical resistivity method at Ito Campus, Kyushu University (Fukuoka, Japan). Geoscience Letters 8(1):15. https://doi.org/10.1186/s40562-021-00188-6

Wang X, Zhang Y, Li Z (2008) Estimating groundwater recharge in Hebei Plain, China under varying land use practices using tritium and bromide tracers. J Hydrol 356(1–2):209–222. https://doi.org/10.1016/j.jhydrol.2008.04.009

WHO (2021) Drinking water. World Health Organization. https://www.who.int/news-room/fact-sheets/detail/drinking-water.

World Bank (2018) Water security for resilient economies. World Bank Publications

Wu WY, Lo MH, Wada Y et al (2020) Divergent effects of climate change on future groundwater availability in key mid-latitude aquifers. Nat Commun 11:3710. https://doi.org/10.1038/s41467-020-17581-y

Zaresefat M, Derakhshani R (2023) Revolutionizing groundwater management with hybrid AI models: A practical review. Water 15(9):1750. https://doi.org/10.3390/w15091750

Zulkifli CZ, Garfan S, Talal M, Alamoodi AH, Alamleh A, Ahmaro IY, Chiang HH (2022) IoT-based water monitoring systems: A systematic review. Water 14(22):362

MIX
Papier aus verantwortungsvollen Quellen
Paper from responsible sources
FSC® C105338

If you have any concerns about our products,
you can contact us on
ProductSafety@springernature.com

In case Publisher is established outside the EU,
the EU authorized representative is:
Springer Nature Customer Service Center GmbH
Europaplatz 3, 69115 Heidelberg, Germany

Printed by Libri Plureos GmbH
in Hamburg, Germany